Indium Phosphide: Crystal Growth and Characterization

SEMICONDUCTORS AND SEMIMETALS
Volume 31

Semiconductors and Semimetals

A Treatise

Edited by R. K. Willardson
ENIMONT AMERICA INC.
PHOENIX, ARIZONA

Albert C. Beer
BATTELLE COLUMBUS LABORATORIES
COLUMBUS, OHIO

Indium Phosphide: Crystal Growth and Characterization

SEMICONDUCTORS
AND SEMIMETALS
Volume 31

Volume Editors

R. K. WILLARDSON
ENIMONT AMERICA INC.
PHOENIX, ARIZONA

ALBERT C. BEER
BATTELLE COLUMBUS LABORATORIES
COLUMBUS, OHIO

ACADEMIC PRESS, INC.
Harcourt Brace Jovanovich, Publishers

*Boston   San Diego   New York
London   Sydney   Tokyo   Toronto*

This book is printed on acid-free paper. ∞

COPYRIGHT © 1990 BY ACADEMIC PRESS, INC.
ALL RIGHTS RESERVED.
NO PART OF THIS PUBLICATION MAY BE REPRODUCED OR
TRANSMITTED IN ANY FORM OR BY ANY MEANS, ELECTRONIC
OR MECHANICAL, INCLUDING PHOTOCOPY, RECORDING, OR
ANY INFORMATION STORAGE AND RETRIEVAL SYSTEM, WITHOUT
PERMISSION IN WRITING FROM THE PUBLISHER.

ACADEMIC PRESS, INC.
1250 Sixth Avenue, San Diego, CA 92101

United Kingdom Edition published by
ACADEMIC PRESS LIMITED
24-28 Oval Road, London NW1 7DX

The Library of Congress has cataloged this serial title as follows:

Semiconductors and semimetals.—Vol. 1—New York: Academic Press, 1966-

  v.: ill.; 24 cm.

  Irregular.
  Each vol. has also a distinctive title.
  Edited by R. K. Willardson and Albert C. Beer.
  ISSN 0080-8784 = Semiconductors and semimetals

  1. Semiconductors—Collected works. 2. Semimetals—Collected works.
I. Willardson, Robert K. II. Beer, Albert C.
QC610.9.S48   621.3815'2—dc19   85-642319
                                          AACR 2   MARC-S
Library of Congress   [8709]

ISBN 0-12-752131-3 (v. 31)

Printed in the United States of America

90 91 92 93   9 8 7 6 5 4 3 2 1

Contents

LIST OF CONTRIBUTORS ix
PREFACE xi

Chapter 1 Growth of Dislocation-free InP   1
J. P. Farges

List of Acronyms	1
I. Introduction	2
II. Dislocations in InP	4
III. Different Alternatives to Reduce the Dislocation Density in InP . . .	11
IV. Conclusion	31
References	32

Chapter 2 High Purity InP Grown by Hydride Vapor Phase Epitaxy   37
M. J. McCollum and G. E. Stillman

List of Acronyms	37
I. Introduction	38
II. Growth	40
III. Characterization of High Purity Indium Phosphide	53
IV. Conclusions and Summary	66
Acknowledgments	66
References	67

Chapter 3 Direct Synthesis and Growth of Indium Phosphide by the Liquid Phosphorous Encapsulated Czochralski Method   71
Tomoki Inada and Tsuguo Fukuda

I. Introduction	71
II. Direct Synthesis System	73
III. Direct Synthesis and Czochralski Growth	82
IV. Composition of Liquids and Crystals	86
V. Conclusion	90
Acknowledgments	91
References	91

Chapter 4 InP Crystal Growth, Substrate Preparation and Evaluation 93

O. Oda, K. Katagiri, K. Shinohara, S. Katsura, Y. Takahashi, K. Kainosho, K. Kohiro, and R. Hirano

List of Symbols	94
I. Introduction	94
II. Polycrystal Synthesis	97
III. Single Crystal Growth	111
IV. Evaluation	138
V. Substrate Preparation	148
VI. Evaluation	159
VII. Summary	167
Acknowledgments	168
References	169

Chapter 5 InP Substrates: Production and Quality Control 175

Koji Tada, Masami Tatsumi, Mikio Morioka, Takashi Araki, and Tomohiro Kawase

List of Acronyms	175
I. Introduction	176
II. Applications and Requirements	176
III. Material Features	178
IV. Production of InP Crystal	180
V. New Developments in InP Crystal Growth	193
VI. Evaluations	204
VII. Quality Control of Crystal Growth	221
VIII. Wafer Processing	222
IX. Summary	238
References	238

Chapter 6 LP-MOCVD Growth, Characterization, and Application of InP Material 243

Manijeh Razeghi

List of Acronyms	244
I. Introduction	245
II. Energy Band Structure of InP	248
III. Growth Technology	252
IV. Low Pressure Metalorganic Chemical Vapor Deposition (LP-MOCVD)	264
V. InP-InP Systems	272
VI. Growth and Characterization of InP Using TMIn	295
VII. Incorporation of Dopants	301

VIII. Microwave Applications	. . . .	307
IX. LP-MOCVD Growth of InP on Alternative Substrates.	. . . .	313
X. Optoelectronic Applications .	. . . .	339
XI. Conclusion	. . . .	351
Acknowledgments .	. . . .	351
References	. . . .	351

Chapter 7 Stoichiometric Defects in InP   357

T. A. Kennedy and P. J. Lin-Chung

List of Symbols	357
I. Introduction	358
II. Theory of Stoichiometric Defects	359
III. Experiments That Reveal the Atomic Structure of Stoichiometric Defects .	368
IV. Areas That Manifest Stoichiometric Defects	376
V. Conclusion	383
Acknowledgments	383
References	383

INDEX	391
CONTENTS OF PREVIOUS VOLUMES	395

List of Contributors

Numbers in parentheses indicate the pages on which the authors' contributions begin.

TAKASHI ARAKI (175), *Semiconductors Materials Research and Development, Basic High Technology Laboratories, Sumitomo Electric Industries, Ltd., 1–3, Shimaya 1-chome, Konohana-ku, Osaka, 554 Japan*

J. P. FARGES (1), *Laboratoires d'Electronique et de Physique Appliquée, 3 avenue Descartes, 94451 Limeil-Brevannes Cedex, France*

TSUGUO FUKUDA (71), *Institute for Materials Research, Tohoku University 2-1-1 Katahira, Sendai, Miyagi, 980 Japan*

R. HIRANO (93), *1st Department, Electronics Laboratory, Electronic Materials and Components Research Laboratories, Nippon Mining Company, Ltd., 3–17–35, Niizo-Minami, Toda-Shi, Saitama-Ken, 335 Japan*

TOMOKI INADA (71), *4th Department, Metal Research Laboratory, Hitachi Cable, Ltd., 5–1–1 Hitaka-cho, Hitachi, Ibaraki, 319–14 Japan*

K. KAINOSHO (93), *1st Department, Electronics Laboratory, Electronic Materials and Components Research Laboratories, Nippon Mining Company, Ltd., 3–17–35, Niizo-Minami, Toda-Shi, Saitama-Ken, 335 Japan*

K. KATAGIRI (93), *1st Department, Electronics Laboratory, Electronic Materials and Components Research Laboratories, Nippon Mining Company, Ltd., 3–17–35, Niizo-Minami, Toda-Shi, Saitama-Ken, 335 Japan*

S. KATSURA (93), *1st Department, Electronics Laboratory, Electronic Materials and Components Research Laboratories, Nippon Mining Company, Ltd., 3–17–35, Niizo-Minami, Toda-Shi, Saitama-Ken, 335 Japan*

TOMOHIRO KAWASE (175), *Semiconductors Materials Research and Development, Basic High Technology Laboratories, Sumitomo Electric Industries, Ltd., 1–3, Shimaya 1-chome, Konohana-ku, Osaka, 554 Japan*

T. A. KENNEDY (357), *Electronics Science and Technology Division, Naval Research Laboratory, Washington, DC 20375*

K. KOHIRO (93), *1st Department, Electronics Laboratory, Electronic Materials and Components Research Laboratories, Nippon Mining Company, Ltd., 3–17–35, Niizo-Minami, Toda-Shi, Saitama-Ken, 335 Japan*

P. J. LIN-CHUNG (357), *Electronics Science and Technology Division, Naval Research Laboratory, Washington, DC 20375*

M. J. McCollum (37), *NIST 724.02, 325 South Broadway, Boulder, Colorado 80303*

Mikio Morioka (175), *Semiconductors Materials Research and Development, Basic High Technology Laboratories, Sumitomo Electric Industries, Ltd., 1–3, Shimaya 1-chome, Konohana-ku, Osaka, 554 Japan*

O. Oda (93), *1st Department, Electronics Laboratory, Electronic Materials and Components Research Laboratories, Nippon Mining Company, Ltd., 3–17–35, Niizo-Minami, Toda-Shi, Saitama-Ken, 335 Japan*

Manijeh Razeghi (243), *Thomson- C.S.F., Laboratoire Central de Recherches, Domaine de Corbeville -BP 10, 91401 Orsay, France*

K. Shinohara (93), *1st Department, Electronics Laboratory, Electronic Materials and Components Research Laboratories, Nippon Mining Company, Ltd., 3–17–35, Niizo-Minami, Toda-Shi, Saitama-Ken, 335 Japan*

G. E. Stillman (37), *Center for Compound Semiconductor Microelectronics, Materials Research Laboratory and Coordinated Science Laboratory, University of Illinois at Urbana–Champaign, Urbana, Illinois 61801*

Koji Tada (175), *Semiconductors Materials Research and Development, Basic High Technology Laboratories, Sumitomo Electric Industries, Ltd., 1–3, Shimaya 1-chome, Konohana-ku, Osaka, 554 Japan*

Y. Takahashi (93), *1st Department, Electronics Laboratory, Electronic Materials and Components Research Laboratories, Nippon Mining Company, Ltd., 3–17–35, Niizo-Minami, Toda-Shi, Saitama-Ken, 335 Japan*

Masami Tatsumi (175), *Semiconductors Materials Research and Development, Basic High Technology Laboratories, Sumitomo Electric Industries, Ltd., 1–3, Shimaya 1-chome, Konohana-ku, Osaka, 554 Japan*

Preface

In the past decade, advances in bulk crystal growth of InP and epitaxial growth techniques have made InP-based materials very useful for ultra-high speed devices and as a substrate for high-performance optoelectronic circuits. Advantages of InP, when compared with GaAs, include higher thermal conductivity; superior surface passivation capabilities, which enhance circuit reliability; and improved thermal properties that can render the materials suitable for higher current densities.

Advances in epitaxial growth techniques have allowed scientists to create complex, high-performance layered structures. These structures include strained-layer superlattices, npn transistors with two AlInAs/GaInAs resonant tunneling double barriers, modulation-doped FETs, and heterostructure bipolar transistors based on GaInP/InP materials, as well as high electron mobility transistors (HEMTs), and radiation resistant solar cells.

In this volume, scientists from France, Japan, and the United States summarize the advances in direct synthesis, in large crystal growth with low dislocation densities, and in epitaxial layer growth. Especially significant is a complete description of substrate preparation and evaluation. Methods for the reduction of dislocations and measurement of stoichiometric defects are treated in detail.

The state of the art in liquid encapsulated Czochralski (LEC) growth of dislocation-free InP is discussed in Chapter 1, which includes a treatment of dislocation generation and propagation. The importance of low thermal gradients, as well as heavy doping to suppress dislocations, is explained.

In Chapter 2, details are given regarding epitaxial InP grown by the hydride vapor phase process. Characterization of these layers by photoluminescence, magneto-photoluminescence, and CC-DLTS is described. Of special interest is a summary of the 77 K electron mobilities in layers grown by various techniques.

A new LEC process, designed to lower the cost of producing InP single crystals, is introduced in Chapter 3. The idea is to synthesize and grow the single crystal in one step, as is done with GaAs, and to use liquid phosphorus as the encapsulant. The liquid interaction parameter is shown to be significant in the synthesis reaction. Industrial applications of this process could make low-cost InP available.

In Chapters 4 and 5, commercial production of InP single crystals and substrates is described. Synthesis, crystal growth by high-pressure LEC techniques, substrate processing and cleaning, as well as quality control and characterization procedures, are discussed in detail. Together these two chapters are of special importance. Each has a different approach, and thus great insight into actual production conditions and challenges is revealed.

Chapter 6 concentrates on the use of low pressure MOCVD to prepare epitaxial layers of InP, GaInAs and GaInP heterostructures on InP substrates. The preparation and characteristics of ultra-high purity InP are described. Of special note is the application of these layers to devices, including the growth and utilization of strained layer heterostructures.

The final chapter is devoted to a discussion of defects in InP. Magnetic resonance experiments that reveal growth-related, intrinsic, metastable, and stoichiometric defects, as well as defect complexes, are introduced. Antisite defects are characterized.

The importance of InP for device applications is highlighted by the international technical conferences dedicated specifically to InP. The second international conference, titled "InP and Related Materials," was held in Denver, Colorado, on April 23-25, 1990.

The editors are indebted to the many contributors and their employers who made this treatise possible. They wish to express their appreciation to Enimont America Inc. and Battelle Columbus Laboratories for providing the facilities and the environment necessary for such an endeavor. Special thanks are also due the editors' wives for their patience and understanding.

R. K. WILLARDSON
ALBERT C. BEER

CHAPTER 1

Growth Of Dislocation-free InP

J. P. Farges

LABORATOIRES D'ELECTRONIQUE ET DE PHYSIQUE APPLIQUÉE
PHILIPS RESEARCH ORGANIZATION

	LIST OF ACRONYMS	1
I.	INTRODUCTION	2
II.	DISLOCATIONS IN InP	4
	1. Dislocation Generation and Propagation	6
	2. Dislocation Evaluation	10
III.	DIFFERENT ALTERNATIVES TO REDUCE THE DISLOCATION DENSITY IN InP	11
	3. Low Thermal Gradient Bulk-Growth Techniques	11
	4. High Pressure Liquid Encapsulated Czochralski (HPLEC) Techniques	14
	5. Doping to Suppress the Dislocation	24
	6. State of the Art in LEC-Grown Dislocation-free InP	29
IV.	CONCLUSION	31
	REFERENCES	32

List of Acronyms

APD	avalanche photodiode
CRSS	critical resolved shear stress
CZ	Crochralski
EDG	electrodynamic gradient
EPD	etch pit density
HB	horizontal Bridgman
HEMT	high electron mobility transistor
HGF	horizontal gradient freeze
HPLEC	high pressure liquid encapsulated Czochralski
JFET	junction field effect transistor
LEC	liquid encapsulated Czochralski
LED	light emitting diode
MXD	multiple x-ray diffraction
PBN	pyrolytic boron nitride
TEGFET	two-dimensional electron gas field effect transistor
TXRT	transmission x-ray topograph
VGF	vertical gradient freeze
VMF	vertical magnetic field

I. Introduction

In the past decade, opto-electronic devices based on Indium Phosphide substrates have gained an increasing interest, and there has been a growing demand for InP bulk material with a high purity and a low dislocation density. This wave of interest starts from the development of low-loss optical fibers, with optimum characteristics in the 1.1 to 1.6 μm wavelength region. Figure 1 illustrates two important parameters for state-of-the-art optical

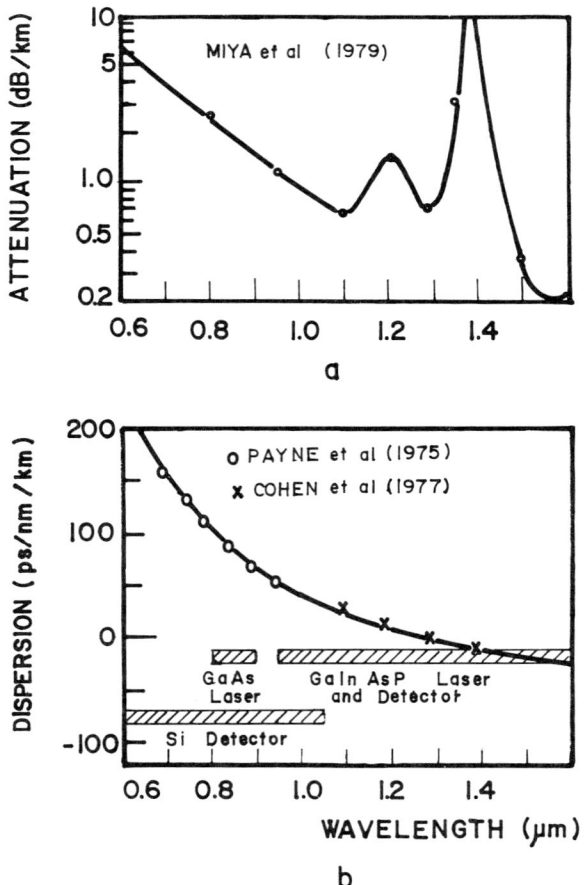

FIG. 1. Transmission characteristics of state-of-the-art optical fibers: (a) attenuation versus wavelength. (Reprinted with permission from IEF, Miya, T., Terunuma, Y., Hosaka, T., and Miyashita, T. (1979). *Elect. Lett.* **15**.), and (b) dispersion versus wavelength. (Reprinted with permission from IEE, Payne, D. N., and Gambling, W. A. (1975). *Elect. Lett.* **11**.)

fibers: attenuation versus wavelength shows a minimum at 0.85, 1.27 and 1.55 μm, according to Miya et al. (1979); and dispersion versus wavelength equals zero for 1.27 μm, according to Payne et al. (1975) and Cohen et al. (1977).

Devices that are working in the 1.1 to 1.6 μm wavelength region are well suited to benefit from these improvements. Figure 2 shows the energy gap variation versus the lattice parameter for different III-V compounds.

By alloying binary compounds, such as GaAs or InAs for example, a continuous range of energy gap and lattice constants can be obtained. It appears that GaInAs ternary compounds or GaInAsP quaternary compounds are well suited for latticematched epitaxial growth on InP substrates, in the wavelength range from 0.9 to 1.6 μm.

The feasibility of numerous devices has already been demonstrated with detectors: PIN, APD; and emitters: LEDS, heterojunction lasers, and quantum-well lasers.

Furthermore, GaInAs or GaInAsP compounds are also suitable for making high-frequency transistors, for example HEMT/TEGFET or JFET, which permit a monolithic integrated device including the photodetector and the FET on the same substrate (Miura et al., 1987).

In devices based on GaAs subtrates, it is clear that dislocations affect their performances and particularly on the device's lifetime. Dark lines or dark spots observed in double-heterostructure GaInAsP/InP devices have been shown to be related to substrate defects (Ueda et al., 1982).

The effect of dislocation density of the GaAs substrate on the threshold voltage V_{th} of FET's transistors fabricated by direct ion implantation has

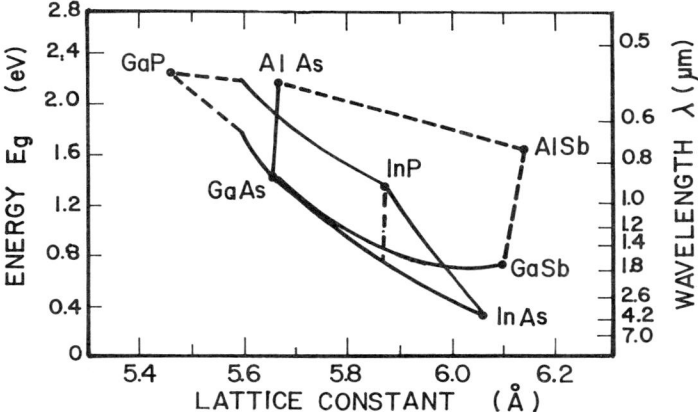

FIG. 2. Energy gap versus lattice constant for different III-V compounds. (Reprinted with permission from the Optical Society of America, Cohen, L. C., and Lin, C. (1977). *Applied Optics* **16**, 3136.)

been examined (Suchet et al., 1987); and although there have not yet been any equivalent studies on InP-based devices, we can reasonably think that such effects are quite possible.

As a consequence, there is an increasing demand for low dislocation density InP substrates, and the main goal of the InP crystal grower is to be able to provide such material.

II. Dislocations in InP

Dislocation generation and propagation in semiconductors have been extensively studied via the thermoelasticity theory. Numerous models for the heat transfer in a high-pressure LEC puller have already been presented. Several factors complicating this study include:

(1) Some thermal and mechanical properties of InP, which are essential to crystal growth modeling are not well known at temperatures near the melting point. In recent studies, Jordan (1985 a; b) gives some semiempirical formulae to obtain these parameters.

(2) The heat transfer conditions throughout the boric oxide layer are not well known. Models usually assume that the encapsulant transmission is one (transparent medium) or zero (opaque medium). In a more realistic study, Martin et al. (1987) suppose the boric oxide to be a semitransparent medium, i.e., transparent for $\lambda < 3.5$ μm and opaque for $\lambda > 3.5$ μm. Their assumption

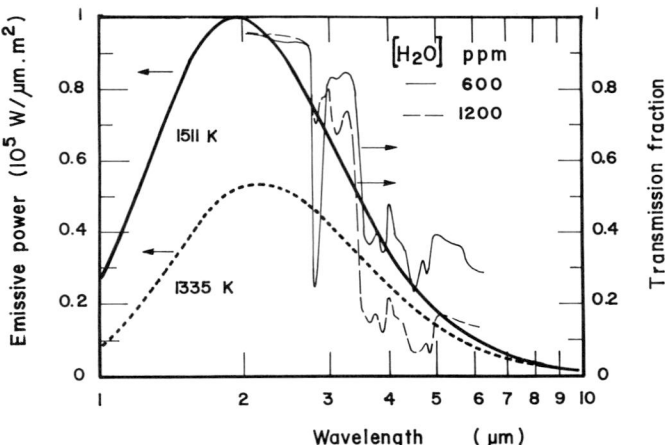

FIG. 3. Transmission spectrum of solid boric oxide at room temperature, depending on the water content, compared with black body spectral emissive power distribution at 1511 K (GaAs) and 1335 K (InP).

is based upon the I.R. transmission curves for solid B_2O_3 at room temperature (Fig. 3), but it is not clear that these curves are still valid for molten B_2O_3, the temperature of which ranges between the melting point of the compound T_m and $(T_m - 400 \text{ K})$.

(3) Heat transfer in a high-pressure LEC puller takes place mainly by transmission, although convection in the pressurizing gas, or conduction through the crucible pedestal or seed holder are not negligible. Figure 4 shows a simplified sketch of a puller and the places where the three heat transfer mechanisms are dominant. Major heat transfer models of CZ crystal growth are listed in Table I (Derby and Brown, 1986), according to the main assumptions that were made for each study. Earlier work started from a simple relationship between pull rate and crystal radius to heat transfer models, which were applied to the thermal stress calculation. Up-to-date

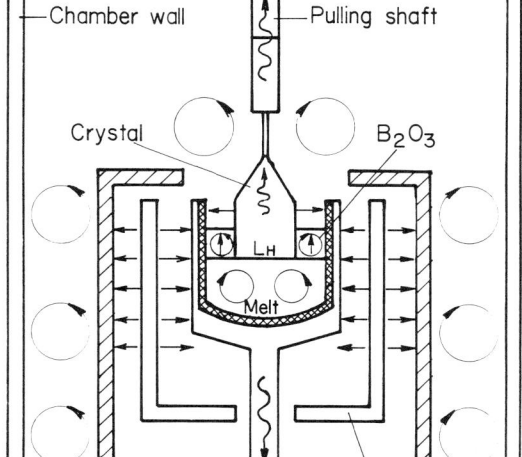

FIG. 4. Simplified sketch of a LEC puller showing the heat-transfer mechanisms.

TABLE I

CLASSIFICATION OF MAJOR THERMAL MODELS FOR CZOCHRALSKI CRYSTAL GROWTH AFTER
J. J. Derby and R. A. Brown, (1986).

	One-dimension fixed interface	Two dimension axisymmetric calculations		
		fixed interface	melt/solid interface	Melt/solid interface meniscus and crystal radius
Radius-pull rate relation	Kim et al. (1983)[a]			
Crystal only	Van der Hart et al. (1983)[b] Rea (1981)[b]	Brice (1968) Jordan et al. (1980a)[e] Duseaux (1983)		
Melt only		Mihelcic et al. (1982)[c,d] Crochet et al. (1983)[a,d]		
Melt and Crystal			Arizumi and Kobayashi (1972)[b]	Derby et al. (1985)[a]
System oriented		Williams and Reusser (1983)		Derby and Brown (1986)[e]

[a] Empirical correlation
[b] includes detailed radiation calculations
[c] includes detailed fluid flow calculations
[d] includes time dependence
[e] includes liquid encapsulated Czochralski growth

works are system-oriented heat transfer models in order to obtain a fully automated process.

1. DISLOCATION GENERATION AND PROPAGATION

Dislocation generation in LEC-grown crystals can appear from different sources, which include excessive thermal stress, native defects, nonstoichiometry, and dislocation propagation from highly dislocated seed.

a. Thermal Stress

During LEC growth, the crystal is subjected to "thermally induced stresses" due to the large thermal gradient near the solid-liquid interface. The excessively cooled crystal periphery produces thermal contraction, and, consequently, the periphery is in tension. putting the hot core in compression. As soon as this thermal stress resolved in the $\{111\}$, $\langle 110 \rangle$ primary slip system exceeds a critical value, the critical resolved shear stress (CRSS), the crystal experiences plastic flow, and the excessive thermal stresses are released by crystallographic glide. Value of the CRSS at the melting point have been extrapolated from results obtained in uniaxial compression experiments at temperatures ranging from 0.5 to 0.8 T_M, the reported value at InP melting point is $6 \cdot 10^5$ Pa, according to Muller *et al.* (1985).

Heat transfer/thermal stress models have been developed by several authors (see Table I) in order to predict the stresses in a growing crystal. For example, Jordan *et al.* (1981) presented dislocation patterns that agree fairly well with the experimental observations obtained from chemical etching or x-ray topographs.

As a consequence of these works, there is now an agreement on the possible routes used to obtain low dislocation density InP substrates:

(1) Decreasing the thermal gradient, and, consequently, the thermal stress inside the crystal to a value lower than the CRSS. In order to achieve stable growth and to get a single crystal, however, a minimum thermal gradient is required.
(2) Increasing the CRSS by an "impurity-hardening" effect, via doping by a selected impurity.

b. Native Defects

Duseaux and Jacob (1982) have reported that the thermal stresses are not always sufficient to explain the dislocation behavior in LEC-grown GaAs. Comparison between their calculations and the experimental results showed that the dislocation nucleation during growth cannot be related directly to the CRSS. They suppose that the CRSS strongly depends on the initial state of the crystal and particularly upon temperature and native defects concentration and mobility.

c. Nonstoichiometry

The influence of stoichiometry upon the dislocation density in LEC-grown GaAs was reported by Brice, (1966). This earlier work has been confirmed in more recent studies on GaAs grown by the horizontal bridgman (HB) method. As reported by Parsey *et al.* (1982), the dislocation density shows a minimum value for an arsenic source temperature of 617 °C, which corresponds to a stoichiometric melt composition.

Lagowski et al. (1984) tried to assess the relative importance of the thermal stress and nonstoichiometry in dislocation formation. They established a relationship between the Fermi energy and the dislocation density and observed that the dislocation density was decreased by a weak donor doping (10^{17} cm^{-3}) and increased by acceptor doping. Nonstoichiometry-related dislocations are created upon condensation of excessive vacancies (V_{GA}) into dislocation loops.

Although the effects of nonstoichiometry and native defects upon dislocation generation have not yet been demonstrated for LEC-grown InP, it seems reasonable to think that they can exist together with the predominant thermal stress mechanism.

d. Dislocation Propagation from Seed

Another important source of dislocation in LEC-grown InP is dislocated seed. Figure 5 is an enlarged picture of the seed-crystal interface and shows that dislocations from the seed are propagating into the growing crystal. There are two possible ways to avoid such an effect: using dislocation-free seeds, and using the so-called "necking" technique. This method will be explained later (see section III 2d).

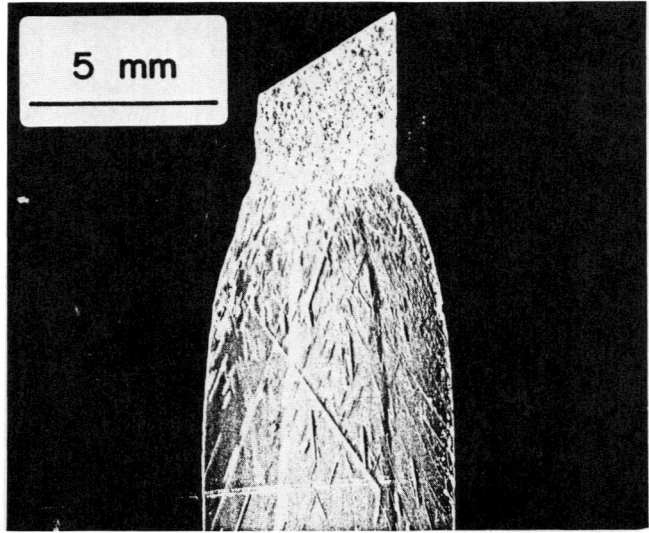

FIG. 5. Enlarged picture of the seed crystal interface. Dislocations from the seed are propagating into the crystal.

e. Dislocation Velocity

Dislocation motion in a crystal requires a displacement energy to overcome the intrinsic lattice resistance. This energy is called the "Peierls barrier." For low-dislocation density, the dislocation moves primarily along the $\langle 110 \rangle$ direction, where the Peierls barrier is minimum.

Compound semiconductors (GaAs, InP, etc.) crystallize in the sphalerite (zinc-blende) structure. They differ from the elemental semiconductors, which crystallize in the diamond structure, by the composition of the $\{111\}$ layers, alternatively composed of atoms of each species. Figure 6 shows the stacking of such a $\{111\}$ plane.

The two different glides that are possible include: (1) the glide along two narrowly-spaced $\{111\}$ planes (i.e., 2-II planes). This set of dislocations is called "glide dislocations." (2) the glide along the widely-spaced $\{111\}$ plane (i.e., I-2 planes). This set of dislocations is called "shuffle dislocations."

Moreover, a distinction must be made: if the extra half plane ends with In atoms, we get α-dislocations in the shuffle set or A-dislocations in the glide set. If the extra half plane ends with P atoms, we get β-dislocations in the shuffle set or B-dislocations in the glide set.

Dislocation velocity is assumed to be proportional to the concentration of double kinks in the material. Hirth *et al.* (1982) discussed the dislocation velocity. They proposed a model where the dislocation velocity is proporational to the concentration of kinks, their velocity, and the driving force τ^m. The velocity can be expressed by:

$$v = A\, \tau^m \exp -\frac{E}{kT},$$

where E is the activation energy, τ is the resolved shear stress, m is constant, and $m \approx 1$ to 2.

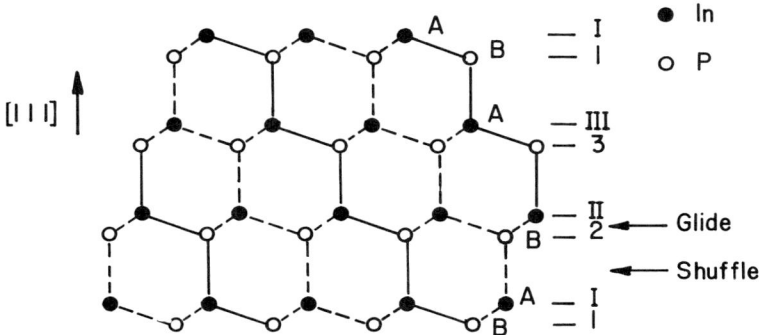

FIG. 6. Stacking of the $\{111\}$ layers, projected on the $\{1\bar{1}0\}$ plane.

2. Dislocation Evaluation

Dislocation evaluation is generally achieved by optical microscopy of chemically etched surfaces, or transmission x-ray topography (TXRT). Other techniques are also available, for example, polarized infrared microscopy (PIM) or electrochemical etching under illumination.

a. Dislocation Evaluation on Chemically Etched Surfaces

Pits and hillocks are commonly observed on etched semiconductors. A pit forms at a point where the dissolution rate is enhanced; therefore, the most important etch pits are those associated to the points of emergence of the dislocations. Numerous reasons have been proposed to explain the enhanced dissolution rate; which include:

(1) The elastic strain field around the dislocation leads to a higher activity, and
(2) The higher rate might be associated with a higher impurity concentration at the dislocation. The most commonly used etches are summarized in Table II. In our opinion, the $A - B$ etch seems to be the most convenient, although it takes the longest time. There are two possibilities to calculate the dislocation density: counting the etch pits for a given surface, and measuring

TABLE II

DISLOCATION ETCHES FOR InP

Composition		Utilization	Etch rate $\mu m\ min^{-1}$
Abrahams and Buiocchi (1965)			
H_2O	2 ml		0.40 for {111} P
$AgNO_3$	8 ml	60°C	0.65 for {100} P
CrO_3	1 g	30 min.	0.60 for {110} P
HF	1 ml		
Huber (1975)			
H_3PO_4	85% 2 vol	24°C	0.30 for {111} P
HBr	47%	2 min.	0.40 for {110} P
Chu et al. (1982)			
HNO_3	1 vol	24°C	
HBr	3 vol	10 to 20 sec.	
Tohno et al. (1981)			
HF	5 vol	25°C	
HBr	1 vol	1 to 2 min.	

the dislocation length for a given volume removed by the etch. A careful comparison between these two methods has been made by Suchet (1987) for $A - B$ etched GaAs substrates, and the discrepancy has been found to be less than 10%.

Prior to etching, the samples are mechano-chemically polished in a 1% solution of bromine in methanol. Usually, up to 100 μm of material are removed in order to avoid effects due to residual work damage.

b. Transmission X-ray Topography (TXRT)

TRXT is usually carried out by the Lang method with MoKα radiations, although some researchers used CuKα or AgKα radiations. X-ray topographs are recorded on ILFORD L4 or K5 nuclear plates. Influences on quality and resolution in TXRT have already been discussed (Lang, 1978). From geometrical consideration of the x-ray source, as well as length spread and resolution of the photographic emulsion, the theoretical resolution ranges between 1.5 and 3 μm. The practical resolution, however, is far more important. If we assume, for example, a dislocation mean length of 50 μm, a crude estimation to distinguish between two similar defaults gives a minimum surface equal to 10×60 μm^2. As a consequence, TXRT is only available when the dislocation density is lower than $\approx 10^5 \cdot$cm^{-2}. Many researchers use the ILFORD L4 nuclear plates, but we think that ILFORD K5 nuclear plates are more convenient to use. Both types have about the same resolution, but processing time for K5 type is three times shorter than for L4 type (Bartels, 1985). Samples for TXRT are mechano-chemically polished (both sides) in bromine-methanol solution, and thinned down to 300 μm. Up to 100 μm per face are removed in order to avoid effects due to residual work damage.

III. Different Alternatives to Reduce the Dislocation Density in InP

It has been demonstrated from the thermoelasticity models that there are two possible routes to decrease the dislocation density in InP by (1) lowering the thermal gradient in order to minimize the thermal stress, and (2) by increasing the CRSS via a lattice hardening mechanism.

In this section, we shall present a survey of the state-of-the-art ways to achieve this.

3. Low Thermal Gradient Bulk-Growth Techniques

The gradient freeze techniques, either horizontal Bridgman (HB), horizontal gradient freeze (HGF), or vertical (VGF) have been widely used in the pioneering ages of III-V compounds bulk growth. Recent improvements, including new multizone furnaces and the use of microprocessors for process control, together with the inherent property of these techniques to achieve very low thermal gradients, have renewed the crystal grower's interest.

a. Horizontal Gradient-Freeze Technique

Horizontal growth systems may roughly be divided into two groups: (1) moving furnace systems (The HB technique belongs to this first group), and (2) gradient-freeze systems, where the temperature gradient moves.

Parsey and Thiel (1985) presented a newly designed furnace, the Mellen "electrodynamic gradient" (EDG) furnace. This apparatus emulates the heat flow characteristic of the HB-Stockarger or GF crystal-growth techniques without requiring motion of the vessel, the growth ampoule, or the furnace. The key features of the system are the structure of the heating element and the computer control to produce the required temperature profile. Figure 7a shows a cross section of an EDG element. An EDG element consists of

—a heating element made of Fe alloy or molybdenum rod,
—a set of zirconia insulating disks which act as mechanical supports, and electrical isolation between each adjacent EDG element,
—aluminium or bronze heat-extraction plates,
—connections to pressurized flowing water heat exchanger system, and
—thermocouple port.

Figure 7b show the thermal profiles that can be generated by the EDG furnace.

The maximum load of a compound semiconductor is 1.5 kg, and crystals of GaAs, InAs, CdTe, and GaSb have been grown in the apparatus either from in situ synthesized melt or from polycrystal materials. Thermal gradients as low as $3.5 \, °C/cm^{-1}$ have been achieved in this apparatus.

The major drawback of these gradient freeze-growth methods is the poor yield of single crystals when growing along $\langle 100 \rangle$ direction. Consequently, crystals are usually growing along $\langle 111 \rangle$ or $\langle 110 \rangle$ direction.

b. Vertical Gradient-Freeze Technique

Monberg et al. (1987) have presented a VGF method for the growth of III-V compounds. Seeded 50 mm in diameter InP crystals have been produced in a pyrolytic boron nitride (PBN) crucible. The thermal environment has been specially designed to achieve a planar liquid-solid interface.

FIG. 7. Mellen electrodynamic gradient furnace, after Parsey et al. © 1985 by Bell Telephone Laboratories, Incorporated, reprinted by permission. (a) sketch of an EDG element, and (b) thermal profiles that can be generated by the EDG furnaces.

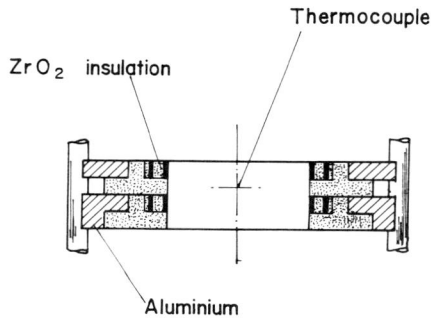

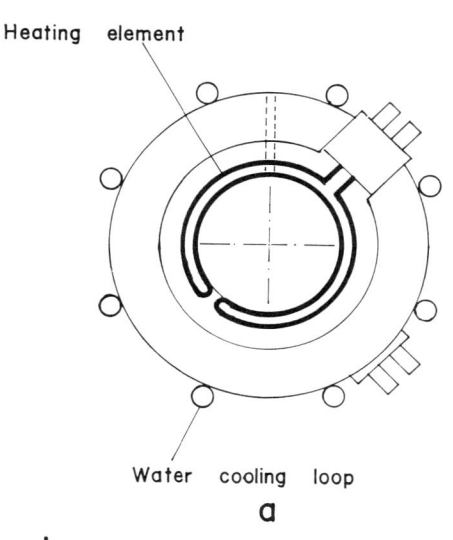

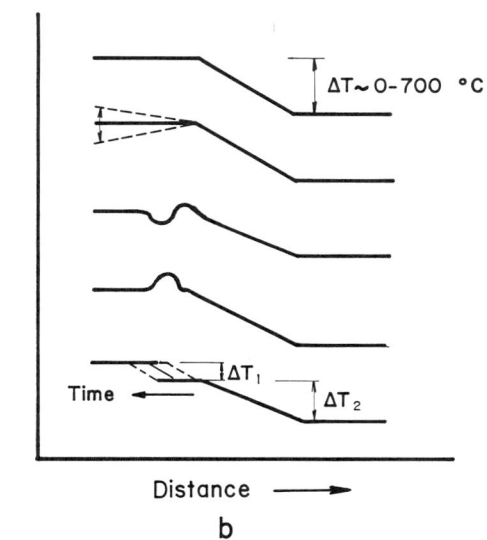

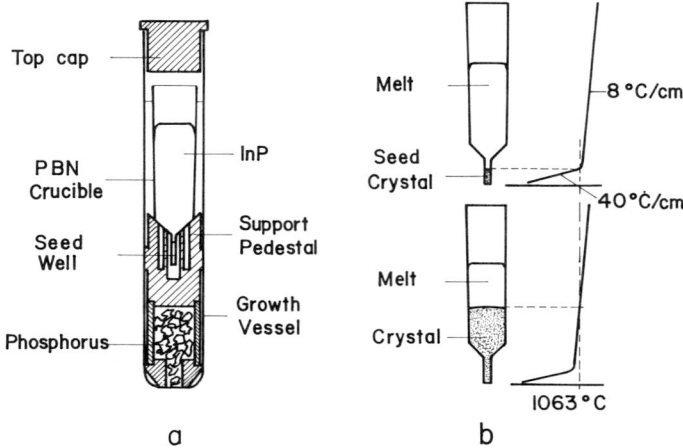

FIG. 8. VGF growth of InP: (a) PRN growth vessel, and (b) thermal profile for seeded VGF growth. (After Monberg, E. M., Gault, N. A., Simchock, F., and Dominguez, F. (1987). *Crystal Growth* **83**, reprinted with permission from Elsevier Pub. B.V.)

Figure 8a shows a scheme of the VGF equipment. Radial and axial thermal gradients have been decreased with respect to a conventional LEC process. Measured temperature gradients are $\approx 40\ °C/cm$ in the seed region, and $\approx 8\ °C/cm$ over the 50 mm in diameter region of the crystal (Fig. 8b). Consequently, the resulting level of stress in the crystal is lower than the critical resolved shear stress and quite dislocation-free material (without slip bands) are obtained, even for a very low level of doping.

Low dislocation density (EPD $\leq 500\ cm^{-2}$), 750 g weight, 50 mm diameter, $\langle 111 \rangle$ seeded InP crystals have been grown by this method, even for S-doping level as low as $2 \times 10^{17}\ cm^{-3}$. Another interesting feature of this method is stoichiometry control, as crystals are pulled under controlled phosphorus vapor pressure.

Single crystal growth, however, is only achieved for the $\langle 111 \rangle$ direction at the moment.

4. HIGH PRESSURE LIQUID ENCAPSULATED CZOCHRALSKI (HPLEC) TECHNIQUES

HPLEC is used largely to grow standard quality (6EPD $\geq 10^4\ cm^{-2}$) InP ingots. Major advantages of this technique are the possibility to pull along the $\langle 001 \rangle$ direction with a good yield of single crystals, and in situ synthesis from the elements In and P when using the injection-tube technique (see III2c).

The most important drawbacks of the HPLEC method are the large thermal gradient and the inhomogenous temperature distribution of the method, which make it quite impossible to grow high-quality, large-diameter ingots. Several different routes have been explored in order to minimize the LEC thermal gradient. They will be reviewed next.

a. Lowering the Thermal Gradient in HPLEC

Key parameters to control the LEC thermal gradient are boric oxide height over the melt, thermal shield, gas nature and pressure, and multiheater furnaces.

(1) *Boric oxide height over the melt.* Shinoyama et al. (1981) have measured the axial temperature distribution in a crucible for various InP and B_2O_3 charges. Measurements were performed with silica encased thermocouples. Using a 100 mm diameter silica crucible covered with a 24 mm B_2O_3 layer, a vertical thermal gradient as low as 38°C/cm was achieved (Fig. 9).

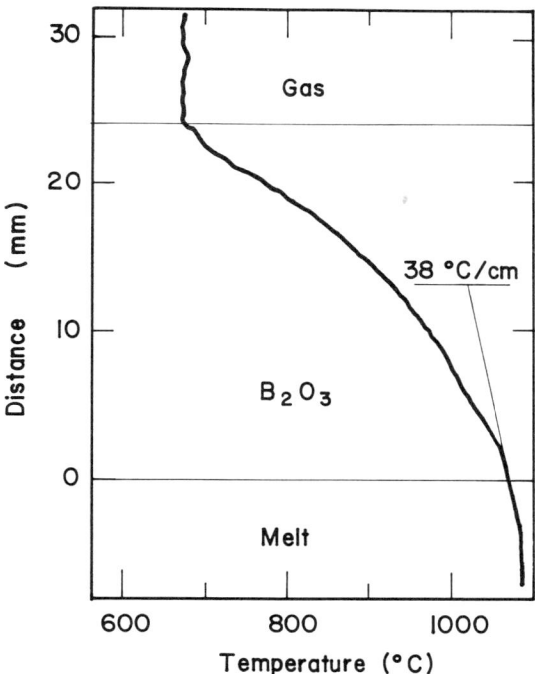

FIG. 9. Temperature distribution along the axis in a crucible of 100 mm diameter. Boric oxide height is 24 mm, after Shinoyama et al. (1981), IOP Publishing Limited.

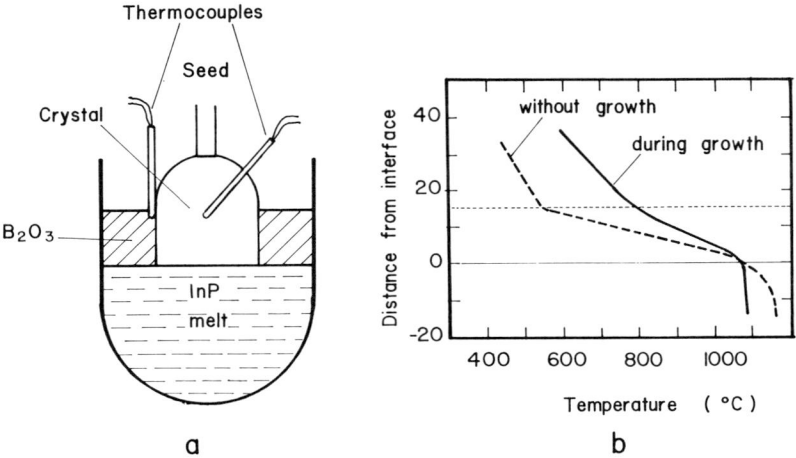

FIG. 10. Temperature measurements during InP LEC growth: (a) schematic arrangement of the thermocouples, and (b) results of temperature measurements versus distance from S-L interface during growth (——) and without growth (-----). (After Muller, G., Volko, J., and Tomzig, E. (1983). *J. Crystal Growth* **64**, reprinted with permission from Elsevier Pub. B.V.)

Muller et al. (1983) have also reported on the vertical thermal gradient dependance on the B_2O_3 height. In their experiments, temperature measurements are performed during the growth process at center and periphery of the growing crystal (Fig. 10a). The temperature profiles obtained during growth differ markedly from that obtained without growth (Fig. 10b).

(2) *Thermal shield.* Katagiri et al. (1985) used a thermal baffle in order to decrease the axial temperature gradient. Growth experiments have been performed in a Metals Research Melbourn high-pressure puller. The thermal baffle was made of graphite. Figure 11 shows axial temperature profile measures with and without thermal baffle.

The temperature gradient near the InP boric oxide interface is decreased from $100°C/cm^{-1}$ for conventional LEC method to $40°C/cm^{-1}$ when using the thermal baffle. As a result, three inch diameter S-doped InP ingots exhibit a very metallic luster, which demonstrates that there is almost no surface decomposition due to phosphorus loss, or to B_2O_3 spalling. These crystals show very large dislocation-free areas: approximately 17 cm^2 in surface, for a S-doping level equal to $5 \cdot 10^{18}$ cm^{-3} to 39 cm^2 for a S-doping level equal to $8 \cdot 10^{18}$ cm^{-3} (surface of a three inch diameter wafer is ≈ 44 cm^2).

(3) *Gas nature and pressure.* Jordan (1980b) has already discussed the influence of gas nature and pressure upon the cooling of GaAs crystals. It was

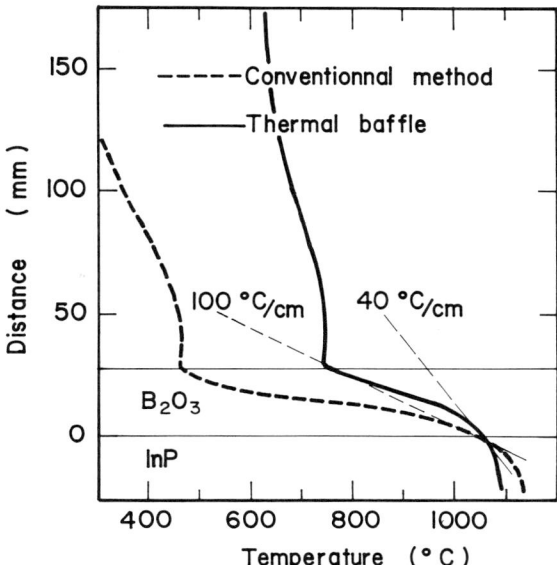

FIG. 11. Axial temperature profile in the crucible with (——) and without (-----) thermal baffle, after Katagiri et al. (1985), IOP Publishing Limited.

found that the convection heat transfer coefficient K depends on the square root of the gas pressure p, expressed in atmospheres, and

$$K = \left[\frac{MC_p}{K_a \mu_a}\right]^{1/4},$$

where M is the gas mass, C_p the heat capacity, K_a the thermal conductivity, and μ_a the viscosity.

Therefore, it is obvious that the lighter gas molecules are the most effective heat conveyor.

(4) *Multiheater furnaces.* A multiheater furnace has been used by Shimizu et al. (1986) to decrease the temperature gradient in LEC by optimizing the power distribution between the main heater and the sub heater. Figure 12 shows the dependence of axial temperature profiles on the rates of power between sub and main heater.

It should be noted that these different techniques to decrease the thermal gradient in HPLEC (increasing the boric oxide height, using thermal shield, or multiheater furnaces) lead to the same results: vertical thermal gradient is decreased from 100–150°C cm^{-1} for conventional LEC method down to 40°C cm^{-1} for improved thermal environment.

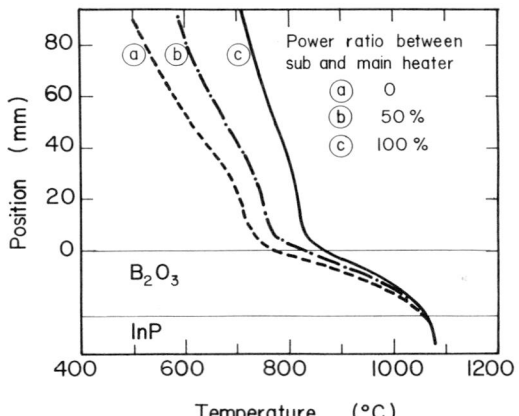

FIG. 12. Dependance of the axial temperature profile on the ratio of power between subheater and main heater. (After Shimizu, A., Nishine, S., Morioka, M., Fujita, K., and Amai, S. (1986). *Proc. 4th Intl. Conf. on Semi-Insulating III-V Materials*, Hakone Japan, reprinted with permission from OHMSA Ltd.)

b. *Minimizing the Thermal Convection Inside the InP Melt*

Striation patterns in a crystal have been related to changes in microscopic growth rates resulting from random convection currents inside the melt, or radial thermal asymmetry. These striations are associated with large variations in dopant concentrations. In order to suppress these growth striations, we have to decrease the thermal oscillation in the melt. This can be achieved in the following ways: (1) magnetic field, (2) crucible and crystal rotation speed, and (3) baffle plate in the melt.

(1) *Magnetic field.* The application of magnetic fields to the melt during growth was first applied for silicon. It has been extended to GaAs crystal growth by Terashima *et al.* (1983). Both vertical and horizontal magnetic fields have been used. Figure 13, after Osaka *et al.* (1984), gives a sketch of the

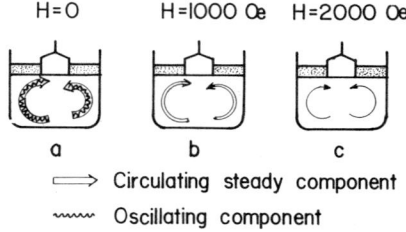

FIG. 13. Thermal convection components inside a melt with and without VMF applied, after Osaka *et al.* (1984).

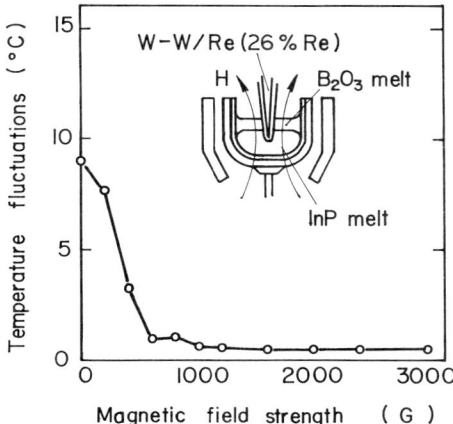

FIG. 14. Influence of a magnetic field on the temperature fluctuations inside an InP melt. Measurements are performed at center of the crucible, 15 mm under the melt surface. (After Miyairi, H., Inada, T., Eguchi, M., and Fukuda, T. (1986). *J. Crystal Growth* **79**, reprinted with permission from Elsevier Pub. B.V.)

thermal convection components in a melt with and without VMF applied. Miyari *et al.* (1986) have reported on InP growth using VMF. Figure 14 shows the temperature fluctuations in a 1 kg load InP melt measured at the center of the crucible, 15 mm under the surface. Temperature fluctuation is reduced to less than 0.3°C as soon as a magnetic field higher than 1000 G is applied. Another interesting feature of magnetic fields is the possibility to produce an homogenous impurity distribution along the growth axis of the crystal, as reported by Ozawa *et al.* (1987). Decreasing the strength of the magnetic field in a programmed manner during the growth (Fig. 15) as a function of the fraction solidified g results in a constant impurity concentration.

(2) *Crucible and crystal rotation rate.* Stirring the melt vigorously is another approach for decreasing the thermal convection inside the melt. Iseler (1984) has performed experiments to determine how the growth striations are affected by the rates of seed and crucible rotation. Figure 16 shows the temperature fluctations obtained at two different crucible rotation speeds. Measurements were performed at the center of the crucible with a silica-encased thermocouple inserted into the melt just below the B_2O_3 layer.

(3) *Baffle plate inside the InP melt.* The use of a baffle plate inside the melt has been proposed by Nishizawa from Gakei Electric Manufacturing Co.,

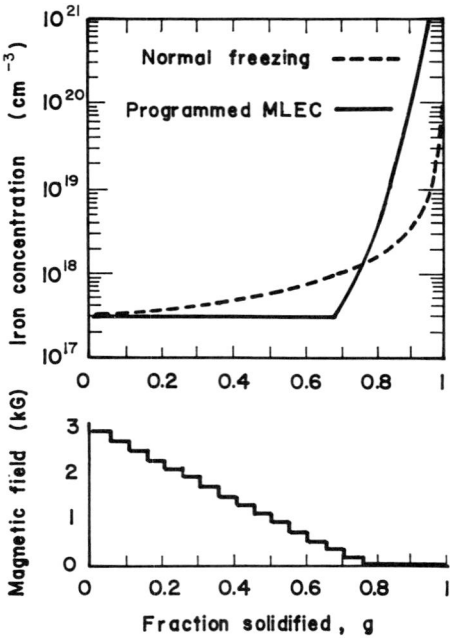

FIG. 15. Programmed MLEC growth of Fe-doped InP. Decreasing the magnetic field in a programmed manner results in a constant doping concentration, after Ozawa et al. (1987).

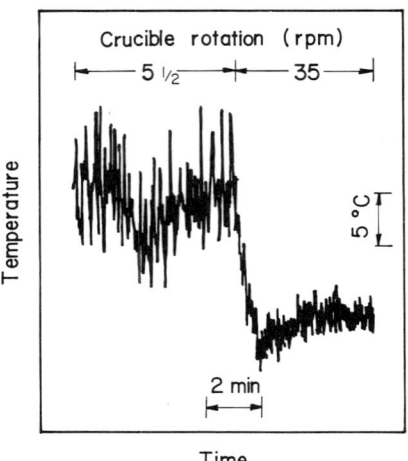

FIG. 16. Crucible rotation rate influence on the temperature fluctuation inside the InP melt, after Iseler (1984). Reprinted with permission from *Journal of Electronic Materials*, vol. 13, no. 6 (1984), a publication of the Metallurgical Society, Warrendale, Pennsylvania.

Ltd., Japan (European patent no 173764) for GaAs growth. The use of a baffle plate is reported to decrease the temperature fluctuations inside the melt. The baffle plate is held in a predetermined position below the surface of the melt during the growth of the single crystal.

Other interesting features of the newly designed puller developed by Gakei include the following: (1) the lift and rotation mechanisms for crystal and crucible are magnetically driven; therefore, most of the seals are suppressed and leaks are minimized; and (2) the resistance heater is made from molybdenum, which avoides carbon contamination from the graphite heater.

c. Influence of Stoichiometry

(1) The phosphorus injection technique. Synthesis of InP is rather difficult to achieve, due to the high phosphorus pressure at the melting point: 27.5 atm. at 1335 K, according to Bachmann *et al.* (1974). Synthesis is usually carried out in a horizontal, double furnace, gradient-freeze system. However, this method gives indium excess at the last-to-freeze part of the ingot, and a premelt is necessary to get stoichiometric material. Farges (1982) has proposed a method for the in situ synthesis and growth of indium phosphide in a LEC puller using the phosphorus vapor injection technique (Fig. 17). Up to 1 kg of poly InP has been grown successfully, and the feasibility of this in situ synthesis and growth of InP in a single run has been demonstrated. Crysta Comm. Inc. (USA) is routinely manufacturing polycrystals using the P-injection technique. According to Antypas (1987a), it is possible to obtain either In-rich or P-rich poly. depending on the phosphorus excess in the phosphorus reservoir. This method has been upscaled and up-to-date equipment allows for the production of up to 10 kg of poly InP (Dowling *et al.* 1986). Fig. 18 is a sketch of the newly designed puller with phosphorus injection facility developed by MCP (GB).

(2) The liquid phosphorus encapsulated Czochralski method. Inada *et al.* (1987 a; b) have presented a new method for the in situ synthesis and growth of InP through the use of liquid P instead of P gas. High purity indium and phosphorus are loaded in a specially designed container (see Fig. 19a). The upper part of the container is cooled so as to condense the phosphorus vapor and to produce liquid white phosphorus which flows down onto the In melt, reacts with it at about 1000–1100°C and produces InP melt. After the synthesis has been completed, the InP melt is covered with a liquid layer, the thickness of which depends upon the initial P excess. Vertical thermal gradient inside this P layer has been measured to be $27°C/cm^{-1}$ (Fig. 19b). Small crystals, 20 mm in diameter and 100 g in weight, have been successfully pulled along the $\langle 111 \rangle$ direction using this new method.

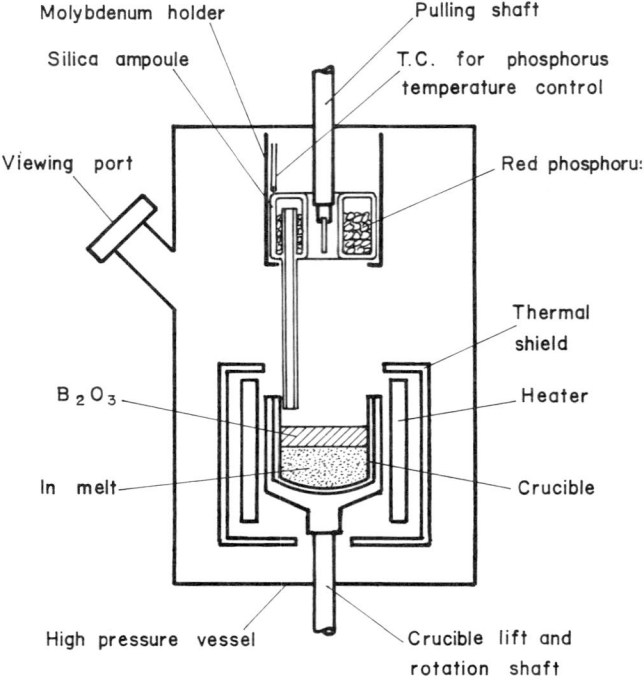

FIG. 17. InP synthesis and growth in a LEC puller using the phosphorus injection technique.

d. *The Necking Technique*

Dash (1958) was the first to use the "necking" technique in order to suppress the dislocation propagation from highly dislocated seed in CZ-grown crystals.

Taking into account the fact that dislocations propagate in the $\{111\}$ $\langle 110 \rangle$ crystallographic glide system, the original feature of the necking method consists in slowly decreasing the diameter of the growing crystal (the "neck") in such a way that dislocations cannot propagate in the material.

Application to the growth of III-V compounds has been proposed by Grabmaier *et al.* (1972) for GaAs, and by Tohno *et al.* (1983) for InP. Fig. 20 shows an example of the necking efficiency for crystals pulled along $\langle 111 \rangle$ or $\langle 001 \rangle$ direction. It should be noticed that keeping the diameter constant during 8 to 12 mm, depending on the pulling direction, is sufficient to suppress the dislocation propagation.

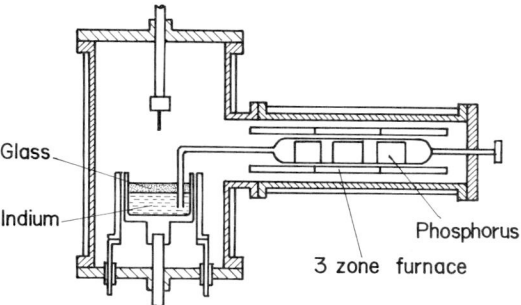

FIG. 18. Scheme of a new puller with the phosphorus injection facility. Pulling capacity is 11 kg for polycrystal, after Dowling et al. (1986).

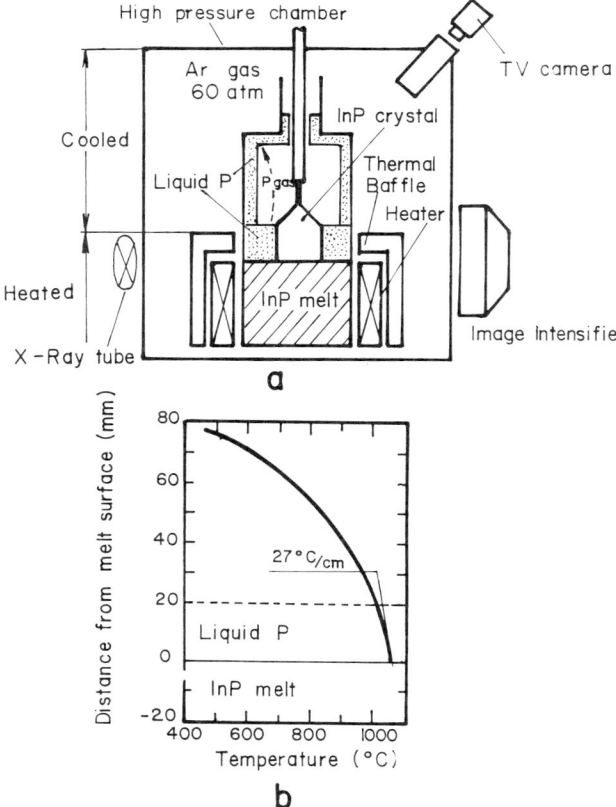

FIG. 19. The liquid phosphorus encapsulated Czochralski method for the direct synthesis and growth of InP: (a) scheme of the apparatus, and (b) vertical temperature profile. After Inada et al. (1987a, b).

FIG. 20. Necking efficiency when pulling along the $\langle 111 \rangle$P and $\langle 001 \rangle$ direction.

5. Doping to Suppress the Dislocation

"Impurity hardening" to increase the CRSS can be achieved by doping from appropriate elements. Choice of the doping elements depends upon the device application, which requires either *n*- or *p*-type conductive, or semi-insulating subtrates. Several impurities have been found to be effective in increasing the CRSS.

a. Doping with Electrically Active Impurities

Seki *et al.* (1976; 1978) have first experimented on the effectiveness of doping to decrease the dislocation density in LEC-grown InP. Quite dislocation-free substrates were obtained when the doping concentration exceeds the mid 10^{18} cm^{-3}. They suggested that the relevant mechanism was the strength of the bonds formed between the impurities atoms and the host crystal atoms.

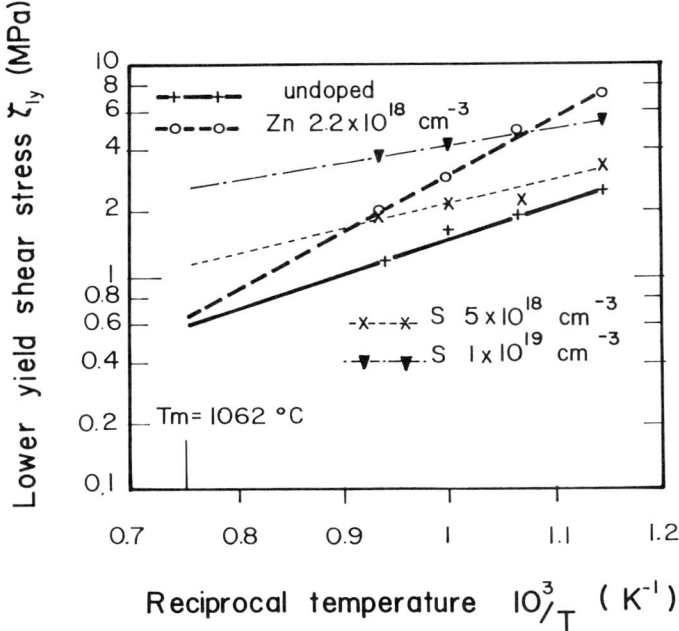

FIG. 21. Temperature dependance of the CRSS for S-doped and Zn-doped InP crystals.

The most effective impurity was found to be Zn, followed by S and Te. Works by Muller et al. (1986) have demonstrated the influence of Zn and S doping on the CRSS of InP. Figure 21 shows the temperature dependance of the CRSS for these two impurities.

The minimum sulphur doping concentration necessary to get rid of dislocations when using the standard growth conditions, i.e., vertical thermal gradient about $150°C.cm^{-1}$, is $5.10^{18}.cm^{-3}$ at seed side as illustrated in Fig. 22. Due to the low sulphur segregation coefficient, 0,50 as reported by Mullin et al. (1972), the sulphur concentration increases from $5 \cdot 10^{18}$ cm^{-3} at seed side to over 10^{19} cm^{-3} at bottom side. Unfortunately, such high doping levels induce detrimental effects in the material such as (1) At $\lambda = 1.3$ μm, the I.R. absorption is proportional to the carrier concentration (Fig. 23). Therefore, in the case of heavily S-doped material, this absorption will reduce the device quantum efficiency, and low S-doped substrates ($n < 3.10^{18}$ cm^{-3}) should be very useful.

(2) Heavily doping concentrations change the lattice parameter, which can induce strains in the epilayers grown on the substrates. Multiple x-ray diffraction (MXD) experiments have been performed using the method

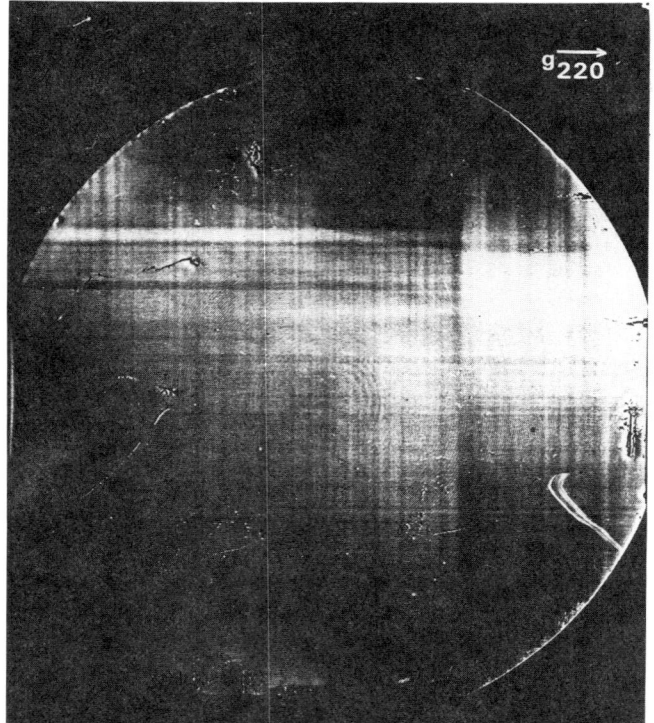

| 1 cm

FIG. 22. TXRT performed at seed side on quite dislocation-free S-doped InP crystal. S-concentration at seed side is $5.5 \; 10^{18}$ cm^{-3}.

described by Bartels (1983) to measure the lattice parameter variation related to S doping (Fig. 24). It should be noted that the lattice parameter variation $\Delta a/a_o$ with respect to undoped material exceeds $5 \cdot 10^{-5}$ when the sulphur concentration is higher than 10^{19} cm^{-3}.

b. *Doping with Isoelectronic Impurities*

Jacob (1982; 1983) first reported that isoelectronic dopant decreases the dislocation density in GaAs and InP. Sb, Ga and As doping experiments were done, and it was concluded that Ga was the most effective to obtain dislocation-free InP single crystals.

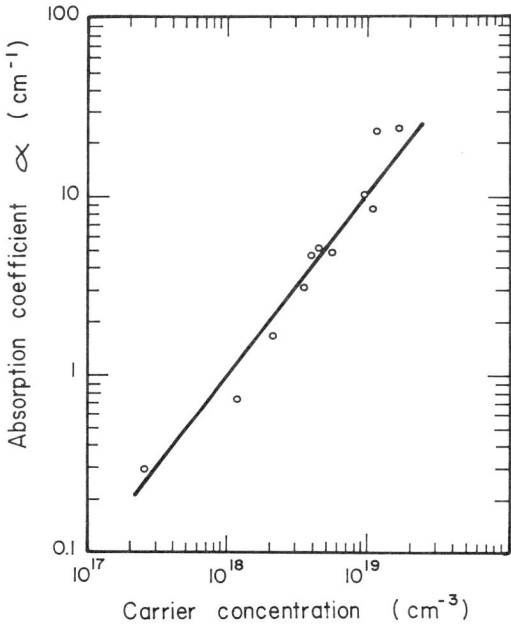

FIG. 23. Absorption coefficient at $\lambda = 1.3$ μm as a function of the free carrier concentration.

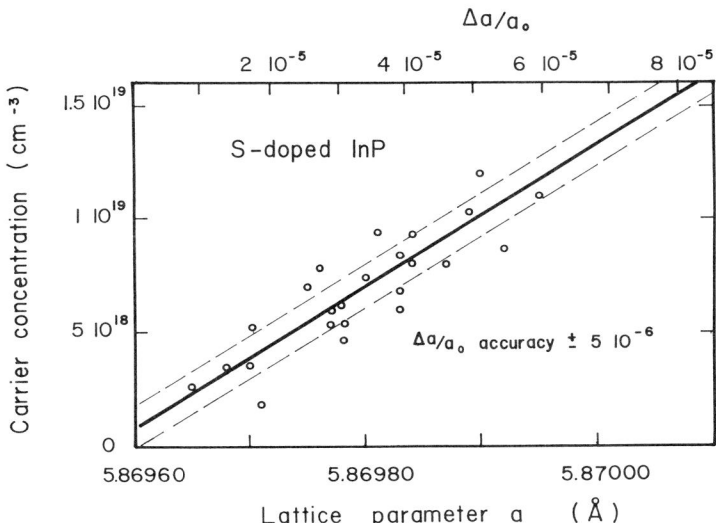

FIG. 24. Lattice parameter variation in S-doped InP, deduced from MXD measurements at 25 °C.

Tohno et al. (1984) have calculated the distribution coefficient of Ga in InP to be 3.35. Consequently, only the ingot's top parts are affected by the Ga doping and doping levels higher than 5.10^{19}cm^{-3} are necessary to reduce significantly the dislocation density from seed to tail end of the ingot. Unfortunately, the single crystal yield decreases drastically when the Ga concentration is higher than 10^{20}cm^{-3}. Moreover, the lattice parameter variation $\Delta a/a_o$ with respect to undoped InP equals 4×10^{-4} when Ga doping is about $5 \times 10^{19} \text{cm}^{-3}$ (Fig. 25). Such high lattice parameter variations may induce a strain in the epilayers grown on Ga-doped substrates, which can be detrimental for the device's reliability.

d. Multidoping Experiments

Tohno et al. (1984) and Katsui et al. (1986) have proposed isoelectronic multidoping experiments from (Ga + Sb) impurities in order to decrease the dislocation from seed to tail end. Using impurity with segregation coefficient higher than one ($k_{Ga} = 3.35$) and impurity with coefficient lower than one ($k_{Sb} = 0.12$) results in a uniform doping concentration all along the pulled

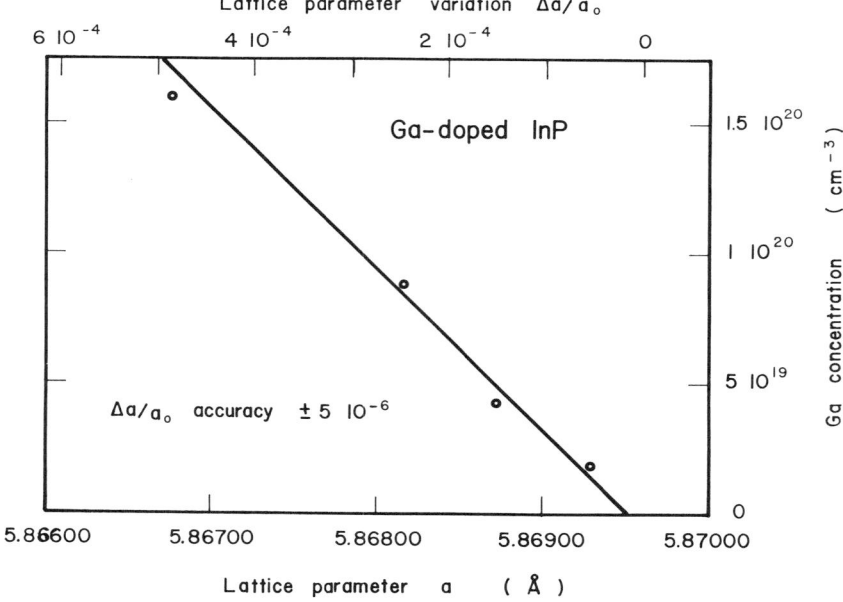

FIG. 25. Lattice parameter variations in Ga-doped InP, deduced from MXD measurements at 25 °C.

ingot. As an example, from a doping concentration of $1.7\ 10^{19}$ cm^{-3} of Ga and $1.5\ 10^{19}$ cm^{-3} of Sb, they reported that a resulting (Ga + Sb) concentration above $8 \cdot 10^{18}$ cm^{-3} is maintained all over the growth of the ingot.

Shimizu *et al.* (1986) experimented with the multidoping from Ga + As elements to decrease the dislocation density in Fe-doped InP. They calculated the Ga segregation coefficient to be 2.8 ± 0.2 and As segregation coefficient to be 0.57 ± 0.03. From a Ga concentration of $2 \cdot 10^{19}$ cm^{-3} and As concentration of $6.4 \cdot 10^{19}$ cm^{-3}, they obtained large dislocation-free areas over a two inch diameter wafer.

6. STATE OF THE ART IN LEC-GROWN DISLOCATION-FREE INP

State of the art LEC-grown dislocation-free InP subtrates are usually obtained from a combination of the before-mentioned techniques, for example, pulling in low thermal-gradient conditions from a slightly doped melt.

a. Combination of Several Techniques to Improve The Crystalline Quality

(1) *Low thermal gradient plus doping.* We have seen that low thermal gradient can be achieved in the following ways: by increasing the boric oxide quantity, by using multiheater furnaces, by using a thermal shield, and so on.

The first solution leads to difficulties in control of the crystal diameter. The second solution can give good results if used together with computer-controlled furnaces. As a matter of fact, Antypas (1987b) reported to have succeeded in growing dislocation-free S-doped InP crystals using multiheater furnace and thermal shield over the melt. S doping in order to get dislocation-free substrates was consequently decreased down to $\approx 3 \cdot 10^{18}$ cm^{-3}.

(2) *Necking plus doping.* Farges *et al.* (1986) have used the necking technique for lightly S-doped crystals. They reported that, when pulling along the $\langle 111 \rangle P$ direction, a large facet often occurs, depending on the growth parameters. The sulphur concentration variation between the faceted core of the ingot and the nonfaceted periphery, has been measured by I.R. absorption at $\alpha = 1.3$ μm. Sulphur concentration in the facet is about 50% larger than at the nonfaceted rim of the ingot, which results in a blockade of the dislocations that propagate from the crystal periphery toward the crystal core (Fig. 26). The same effect has been obtained when pulling in the $\langle 001 \rangle$ direction after growing a thin neck (Fig. 27); although the yield of single crystals is fairly poor in this latter case.

Using the necking plus doping technique, the sulphur concentration to get dislocation-free InP has been lowered to 2 to $3 \cdot 10^{18}$ cm^{-3} at seed side.

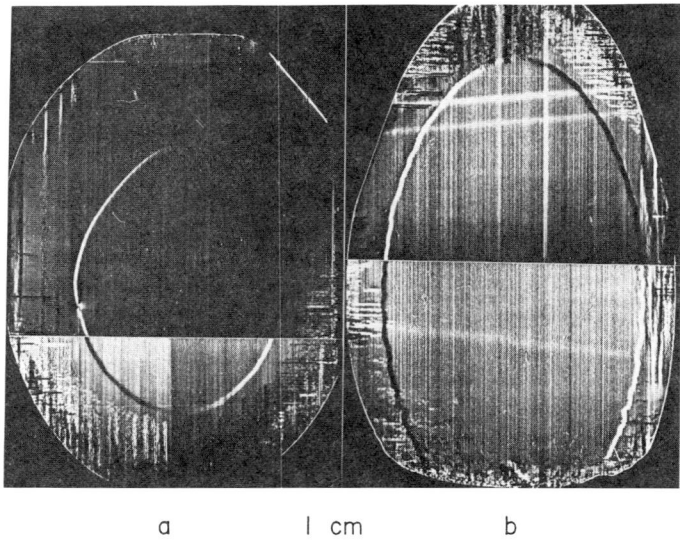

FIG. 26. TXRT performed on {001} slices cut along a ⟨111⟩ pulled ingot after first growing a thin neck. S-concentration is measured at center of the wafer. (a) S-concentration at seed side is $2.7 \cdot 10^{18}$ cm^{-3}. (b) S-concentration at bottom side is $7.8 \cdot 10^{18}$ cm^{-3}.

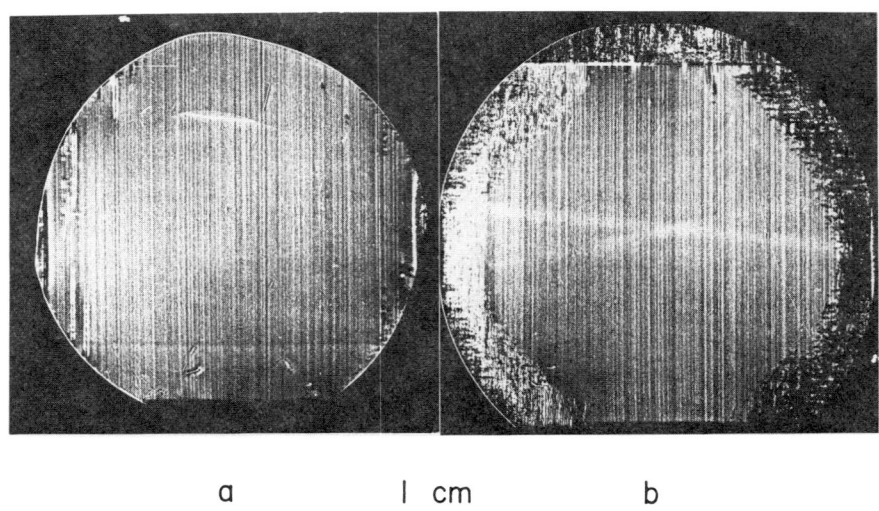

FIG. 27. TXRT on {001} slices cut along a ⟨001⟩ pulled ingot after first growing a thin neck. (a) S-concentration at seed side is $2 \cdot 10^{18}$ cm^{-3}. (b) S-concentration at middle side is $2.6 \cdot 10^{18}$ cm^{-3}.

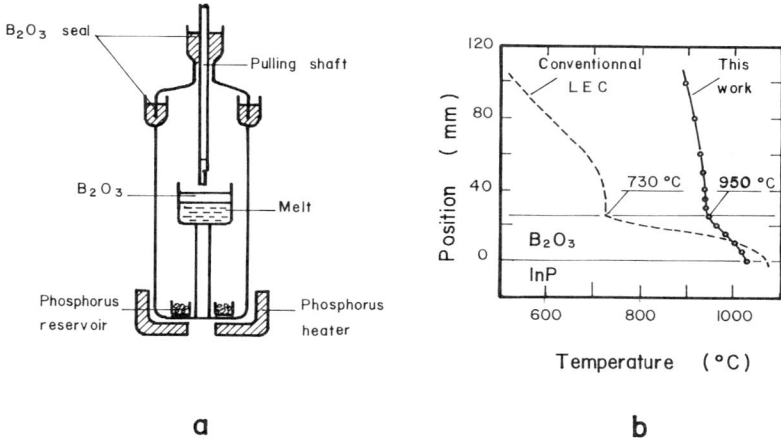

FIG. 28. Modified LEC method for the growth of InP under controlled vapor phosphorus pressure: (a) scheme of the apparatus, and (b) temperature measurements versus distance from InP melt interface. After Tada et al. (1987), IOP Publishing Limited.

(3) *Low thermal gradient plus vapor phosphorus controlled pressure.* Tada et al. (1987) have reported on the growth of low dislocated Fe-doped InP using a multiheater furnace and a modified LEC puller. Figure 28a shows a scheme of the modified LEC method under phosphorus pressure. The major advantage of this method is to provide a very low vertical thermal gradient [Fig. 28b) without dissociation of the InP crystal surface due to the high surface temperature of boric oxide. Low dislocation density ($<10^3$ cm^{-2}) Fe-doped semi-insulating InP crystals have been achieved in this way.

IV. Conclusion

It is now largely accepted that excessive thermal stresses are responsible for the dislocation generation and propagation in growing crystals. An advantage of the gradient-freeze techniques is very low thermal gradients. These techniques have gained a renewed interest due to recent improvements. New HGF furnace design and computer control result in a very low and easily adjustable temperature gradient. Reproducible growth of large-diameter III-V crystals by the VGF technique has been reported. In both cases, the melt stoichiometry can be controlled by heating a volatile component reservoir to maintain the proper vapor pressure over the melt. Advantages of these techniques are that they can achieve a low thermal gradient and, consequently, low dislocated materials are easily available. The major drawback at the moment is the poor yield of single crystals when pulling along the $\langle 001 \rangle$ direction.

In LEC-grown material, two routes have been proposed to decrease the dislocation density (Jordan et al. 1985c), which include decreasing the thermal gradient, and increasing the CRSS by impurity hardening.

However, the single crystal yield is badly affected in the first case. Decreasing the thermal gradient results in poor diameter control and low single-crystal yield. Increasing the CRSS by impurity hardening requires sulphur doping in the $5 \cdot 10^{18}$ to 10^{19} cm^{-3}, which is the necessary level to get dislocation-free material when pulling in standard LEC conditions.

Unfortunately, a high doping level induces several disadvantages, such as a large lattice constant, which gives strain in the crystal and makes slicing difficult, and a large optical absorption, which makes these substrates unsuitable for some device applications.

Therefore, there is a need for low-dislocation, low-doped InP substrates. Recent works have shown that a combination of the two proposed routes (low thermal gradient, impurity hardening) is effective in significantly reducing the dislocation density. General trends to pull state of the art LEC-grown dislocation-free material are the use of multiheater furnaces. sometimes in combination with a thermal shield over the melt, together with a slight doping for an impurity hardening effect.

The use of a magnetic field is only significant to reduce irregular striation effects, but it should be noted that this can be achieved by carefully adjusting the seed and crucible rotation rate.

Other works have demonstrated the possibility of pulling InP under liquid phosphorus encapsulation (Inada et al, 1987) or under controlled vapor phosphorus pressure (Tada et al., 1987). In both cases, vertical temperature gradients have been significantly reduced with respect to conventional LEC method. Besides the promising results, which have already been obtained via these techniques, we think that they present potentialities to further decrease the dislocation density.

REFERENCES

Abrahams, M. S., and Buiocchi, C. J. (1965). Etching of Dislocation on the Low-index Face of GaAs. *J. Appl. Phys.* **36**, 2855–2863.

Antypas, G. (1987a). Private communication.

Antypas, G. (1987b). Private communication.

Arizumi, T., and Kobayashi, N. (1972). Theoretical studies of the temperature distribution in a crystal being grown by the Czochralski method. *J. Crystal Growth* **13/14**, 615–618.

Bachman, K. J., and Buehler, E. (1974). Phase equilibria and Vapor Pressures of pure Phosphorus and of the Indium/Phosphorus System and their implications regarding crystal growth of InP. *J. Electrochem. Soc.* **121** (6), 835–846.

Bartels, W. J. (1983). Characterization of thin layers on perfect crystals with a multipurpose high resolution X-ray diffractometer. *J. Vacuum Sci. Technol.* **B1**, 338–345.
Bartels W. J. (1985). Private communication.
Brice, J. C., and King, G. D. (1966). Effect of the arsenic pressure on dislocation densities in melt-grown gallium arsenide. *Nature* **209**, 1346.
Brice, J. C. (1968). Analysis of the temperature distribution in pulled crystals. *J. Crystal Growth* **2**, 395–401.
Chu, S. N. G., Jodlauk, C. M., and Ballman, A. A. (1982). New dislocation etchant for InP. *J. Electrochem. Soc.* **129** (2), 352–354.
Cohen, L. G., and Lin, C. (1977). Pulse delay measurements in the zero material dispersion wavelength region for optical fibers. *Appl. Optics* **16**, 3136.
Crochet, M. J., Wouters, P. J., Geyling, F. T., and Jordan A. S. (1983). Finite-element simulation of Czochralski bulk flow. *J. Crystal Growth* **65**, 153–165.
Dash. W. C. (1958). *In* "Growth and perfection of crystals." (R. M. Doremus, B. W. Roberts, and D. Turnbull, ed.) Wiley, New York.
Derby, J. J., Brown, R. A., Geyling, F. T., Jordan, A. S., and Nikolakopoulos, G. A. (1985). Finite element analysis of a thermal capillarity model for liquid encapsulated Czochralski growth. *J. Electrochem. Soc.* **132**, 470–482.
Derby, J. J., and Brown, R. A. (1986). Thermal capillarity analysis Czochralski and liquid encapsulated Czochralski crystal growth. *J. Crystal Growth* **74**, 605–624.
Dowling, D. J., Wardill, J. E., Brunton, R. A. Crouch, D. A. E., and Thompson, J. P. (1986). The large-scale synthesis and growth of polycrystalline InP. *In:* "Proc. of the 3rd NATO Workshop on Materials Aspects of InP," Harwichport, USA, Sept. 86.
Duseaux, M. (1983). Temperature profile and thermal stress calculations in GaAs crystals growing from the melt. *J. Crystal Growth* **61**, 576–590.
Duseaux, M., and Jacob, G. (1982). Formation of dislocations during liquid encapsulated Czochralski growth of GaAs single crystals. *Appl. Phys. Lett.* **40** (9), 790–793.
Farges, J. P. (1982). A method for the "in-situ" synthesis and growth of indium phosphide in a Czochralski puller. *J. Crystal Growth* **59**, 665–668.
Farges, J. P., Schiller, C., Bartels, W. J. (1986). Growth of large diameter dislocation-free indium phosphide ingots. *J. Crystal Growth* **83**, 159–166.
Grabmaier, B. C., and Grabmaier, J. G. (1972). Dislocation-free GaAs by the liquid encapsulation technique. *J. Crystal Growth* **13/14**, 635–639.
Hirth, J. P., and Lothe, J. (1982). *In:* "Theory of Dislocations," second edition, Wiley, New York.
Huber, A., and Linh, N. T. (1975). Revelation metallographique des défauts cristallins dans InP. *J. Crystal Growth* **29**, 80–84.
Inada, T., Fujii, T. Eguchi, M., and Fukuda, T. (1987a). New direct synthesis technique for indium phosphide using liquid phosphorus. *J. Crystal Growth* **82**, 561–565.
Inada, T., Fujii, T. Eguchi, M., and Fukuda, T. (1987b). Technique for the direct synthesis and growth of indium phosphide by the liquid phosphorus encapsulated Czochralski method. *Appl. Phys. Lett.* **80** (2), 86–88.
Iseler, G. W. (1984). Advances in LEC Growth of InP Crystals. *J. of Electronics Mat.* **13** (6) 989–1011.
Jacob, G. (1982). How to decrease defect densities in LEC S.I. GaAs and InP crystals. *In:* "Proc. of the second Conf. on Semi-insulating III–V Materials," Evian, France. (Makram-Ebeid, S., Tuck, B., eds.) Shira Publishing Ltd.
Jacob, G., Duseaux, M., Farges, J. P., Van Den Boom, M. M. B., and Roksnoer, P. J. (1983). Dislocation-free GaAs and InP crystals by isoelectronic doping. *J. Crystal Growth* **61**, 417–424.

Jordan, A. S., Caruso, R., and Von Neida, A. R. (1980a). A thermoelastic analysis of dislocation generation in pulled GaAs. *Bell System Tech. J.* **59**, 593–637.

Jordan, A. S. (1980b). An evaluation of the thermal and elastic constants affecting GaAs crystals growth. *J. Crystal Growth* **49**, 631–642.

Jordan, A. S., Caruso, R., Von Neida, A. R., and Nielsen, J. W. (1981). A comparative study of thermal stress induced dislocation generation in pulled GaAs, InP and Si crystals. *J. Appl. Phys.* **52**, 3331–3336.

Jordan, A. S. (1985a). Estimated thermal diffusivity, Prandtl number and Grasshof number of molten GaAs, InP and GaSb. *J. Crystal Growth* **71**, 551–558.

Jordan, A. S. (1985b). Some thermal and mechanical properties of InP essential to crystal growth modelling. *J. Crystal Growth* **71**, 559–565.

Jordan, A. S., Brown, G. T., Cockayne, B., Brasen, D., and Bonner, W. A. (1985c). An analysis of dislocation reduction by impurity hardening in the liquid encapsulated growth of $\langle 111 \rangle$ InP. *J. Appl. Phys.* **58** $\langle 11 \rangle$, 4383–4389.

Katagiri, K., Yamakazi, S., Takagi, A., Oda, O., Arraki, H., and Tsuboya, I. (1985). LEC growth of large diameter InP crystals doped with Sn and with S. *In*: "Proceedings of the 12th Int. Symposium on GaAs and Related Compounds," Karuizawa, Japan, 67–72.

Katsui, A., and Tohno, S. (1986). Growth and characterization of isoelectronically double-doped semi-insulating InP (Fe) single crystals. *J. Crystal Growth* **79**, 287–290.

Kim, K. M., Kraw, A., Smetana, P., and Schwuttke, G. H. (1983). Computer simulation and controlled growth of large diameter Czochralski silicon crystals. *J. Electrochem. Soc.* **130**, 1156–1160.

Lagowski, J., Gatos, A. C., Auyama, T., and Lin, D. G. (1984). On the dislocation density in melt-grown GaAs. *In*: "Proc. of the third Conf. on Semi-insulating III–V Materials," Kah-Nee-Ta, USA (D. C. Look and J. S. Blakemore; eds.), Shiva Publishing Ltd., 60–67.

Lang, A. R. (1978). *In*: "Diffraction and Imaging Techniques in Material Science, Vol. II." North Holland Publishing Co., 647–654.

Martin, G. M., Deconinck, P., Duseaux, M., Maluenda, J., Nagel G., Nöhnert, K. Crochet, M. J., Dupret, F., and Nicodème, P. (1987). Improvements in GaAs materials for IC's application. *In*: "Proc. of the Fourth Annual ESPRIT Conference," 28–30 Sept. 1987, Brussels, Belgium, North Holland Publishing Co.

Mihelcic, M., Schrock-Pauli, C., Wingerath, K., Wenzl, H., Uelhoff, W., and Van den Hart, A. (1982). Numerical simulation of free and forced convection in the classical Czochralski method and in CACRT. *J. Crystal Growth* **57**, 300–317.

Miura, S., Hamaguchi, H., and Mikawa, T. (1987). High-speed GaInAs monolithic PIN/FET receiver. *In*: "Proc. of the 13th European Conference on Optical Communication," Helsinki, Finland, Vol. 1, 66–69.

Miya, T., Terunuma, Y., Hosaka, T., and Miyashita, T. (1979). Ultimate low-loss single mode fibre at 1.55 μm. *Elect. Lett.* **15**, 106–108.

Miyairi, H., Inada, T., Eguchi, M., and Fukuda, T. (1986). Growth and properties of InP single crystals grown by the magnetic field applied LEC method. *J. Crystal Growth* **79**, 291–295.

Monberg, E. M., Gault, N. A., Simchock, F., and Dominguez, F. (1987). Vertical gradient freeze growth of large diameter low defect density indium phosphide. *J. Crystal Growth* **83**, 174–183.

Muller, G., Völkl, J., and Tomzig, E. (1983). Thermal analysis of LEC InP growth. *J. Crystal Growth* **64**, 40–47.

Muller, G., Rupp, R., Völkl, J., and Blum, W. (1985). Deformation behaviour and dislocation formation in undoped and doped (Zn, S) InP Crystals. *J. Crystal Growth* **71**, 771–778.

Mullin, J. P., Royle, A., Straughan, B. W., Tufton, P. J., and Williams, E. W. (1972). Crystal growth and properties of group IV doped indium phosphide. *J. Crystal Growth* **13/14**, 640–646,

Nishizawa, M. Gakei Manufacturing Co., Ltd. 18-9 Senjya-Sakurazi 2-chome, Adachi-ku, Tokyo, Japan.

Osaka, J., and Hoshikawa, K. (1984). Homogeneous GaAs grown by vertical magnetic field applied LEC method. *In*: "Proc. of the third Conf. on Semi-insulating III–V Materials," Kah-Nee-Ta, USA, 126–13. (Look, D. C., Blakemore, J. S., eds.) Shiva Publishing Ltd.

Ozawa, S., Kimura, T., Kobayashi, J., and Fukuba, T. (1987). Programmed magnetic field applied liquid encapsulated Czochralski crystal growth. *Appl. Phys. Lett.* **50** (6), 329–331.

Payne, D. N., and Gambling, W. A. (1975). Zero material dispersion in optical fibres. *Elect. Lett.* **11**, 176–178.

Parsey, J. M., Jr., Nanishi, Y., Lagowski, J., and Gatos, H. C. (1982). Bridgman-type apparatus for the study of growth property relationships: arsenic vapor pressure–GaAs property relationship. *J. Electrochem. Soc.* **129**, 388–393.

Parsey, J. M., Jr., and Thiel, F. A. (1985). A new apparatus for the controlled growth of single crystals by horizontal Bridgman techniques. *J. Crystal Growth* **73**, 211–230.

Rea, S. N. (1981). Czochralski silicon pull rate limits. *J. Crystal Growth* **54**, 267–274.

Seki, Y., Matsui, J., and Watanabe, H. (1976). Impurity effect on the growth of dislocation-free InP single crystals. *J. Appl. Phys.* **47** (7), 3374–3376.

Seki, Y., Watanabe, H., and Matsui, J. (1978). Impurity effect on grown-in dislocation density of InP and GaAs crystals. *J. Appl. Phys.* **49** (2), 822–828.

Shimuzu, A., Nishine, S., Morioka, M., Fujita, K., and Amai, S. (1986). Low dislocation crystal growth of semi-insulating InP through multi-heater LEC technique and co-doping of Ga and As. *In*: "Proc. of the Fourth Int. Conf. on Semi-Insulating III–V Materials," Hakone, Japan (Hukimoto, H., Miyazawa, S., eds.) 41–46.

Shinoyama, S., Tohno, S., and Uemura, C. (1981). Growth of dislocation-free single crystals by the LEC technique. *In*: "Proc. of the Ninth Int. Conf. on GaAs and Related Compounds," OISO, Japan.

Suchet, P. (1987). Nouvelle thèse. Institut National Polytechnique de Lorraine, France.

Suchet P., M. (1987). Effects of dislocations on threshold voltage of GaAs field-effect transistors. *J. Appl. Phys.* **62** (3), 1097–1101.

Tada, K., Tatsumi, M., Nakagawa, M., Kawase, T., and Akai, S. (1987). Growth of low-dislocation density InP by the modified Czochralski method in the atmosphere of phosphorus vapor pressure. *In*: "Proc. of the 14th Int. Symp. on GaAs and Related Compounds," Heraklion, Greece.

Terashima, K., Katsumata, K. Orito, F., Kikuta, T., and Fukuda, T. (1983). Electrical resistivity of undoped GaAs single crystals grown by magnetic field applied LEC technique. *Jap. J. Appl. Phys.* **22**, L-325-327.

Tohno, S., Kubota, E., Shinoyama, S., Katsui, A., and Uemura, C. (1987). Isoelectronic double doping effects on dislocation density of InP single crystals. *Jap. J. Appl. Phys.* **23**, (2), 272–274.

Ueda, O., Umebu, I., Yamakoshi, S., and Kotani, T. (1982). Nature of dark defects revealed in InGaAsP/InP double heterostructure light emitting diode aged at room temperature. *J. Appl. Phys.* **53** (4), 2991–2997.

Van den Hart, A., and Velhoff, W. (1981). Macroscopic Czochralski growth. *J. Crystal Growth* **51**, 251–266.

Williams, G., and Reusser, W. E. (1983). Heat transfer in Silicon Czochralski crystal growth. *J. Crystal Growth* **64**, 448–460.

CHAPTER 2

High Purity InP Grown by Hydride Vapor Phase Epitaxy*

M. J. McCollum[†] and G. E. Stillman

CENTER FOR COMPOUND SEMICONDUCTOR MICROELECTRONICS
MATERIALS RESEARCH LABORATORY AND COORDINATED SCIENCE LABORATORY
UNIVERSITY OF ILLINOIS AT URBANA–CHAMPAIGN
URBANA, ILLINOIS

	LIST OF ACRONYMS	37
I.	INTRODUCTION.	38
II.	GROWTH	40
	1. *Growth Reactions*	40
	2. *Growth System*	41
	3. *Substrate Preparation*	44
	4. *Growth Procedure*	47
	5. *Methods for Enhancing Purity*	48
III.	CHARACTERIZATION OF HIGH PURITY INDIUM PHOSPHIDE . . .	53
	6. *Hall Effect*	54
	7. *Photothermal Ionization Spectroscopy*	56
	8. *Photoluminescence*	59
	9. *Magneto-photoluminescence*	59
	10. *Constant Capitance-Deep Level Transient Spectroscopy* . .	59
IV.	CONCLUSIONS AND SUMMARY.	66
	ACKNOWLEDGMENTS	66
	REFERENCES.	67

List of Acronyms

APMOCVD	atmospheric-pressure metalorganic chemical vapor deposition
CC-DLTS	constant capacitance-deep level transient spectroscopy
C-VPE	chloride vapor-phase epitaxy
GSMBE/CBE	gas source molecular beam epitaxy/chemical beam epitaxy
H-VPE	hydride vapor-phase epitaxy
LPE	liquid-phase epitaxy
LPMOCVD	low-pressure metalorganic chemical vapor deposition

* Work supported in part by the National Science Foundation under contracts ECS 82-09090. DMR 86-12860, and by the Joint Services Electronics Program under contract N00014-84-C-0149. The Center for Compound Semiconductor Microelectronics Engineering Research Center is funded by the National Science Foundation under contract CDR 85-22666.

[†] Present address: National Institute of Standards and Technology, Boulder, Colorado.

MBE	molecular beam epitaxy
MFC	mass flow controller
MOCVD	metalorganic chemical vapor deposition
MPL	magneto-photoluminescence
PL	photoluminescence
ppm	parts per million
PTIS	photothermal ionization spectroscopy
VPE	vapor-phase epitaxy

I. Introduction

High purity crystal growth and the study of high purity electronic materials is extremely important to many aspects of electronic devices. Certainly, those devices that require high purity layers benefit directly, but all devices benefit from improved material quality and increased process understanding and control. The study of high purity epitaxial growth leads to greater process control through increased understanding of system thermodynamics and reaction chemistry. Critical growth procedures relating to such areas as substrate preparation, reactor design, maintenance and operation can be evaluated so that unnecessary steps can be avoided and important procedures optimized. Material characteristics such as background impurity concentrations and identification, morphology, and luminescence properties help researchers to identify critical growth procedures and parameters as well as contamination sources. This benefits the state of the art of crystal growth by giving system designers and source material producers information on what areas are best for targeting improvement efforts. Thus, all epitaxial growth and resulting electronic devices see improvement and benefit from the study of high purity epitaxial crystal growth.

Avalanche photodiodes and p-i-n detectors are devices that benefit directly from material purity (Stillman et al., 1982a). The low capacitance required for p-i-n detectors implies large depletion widths, which translates into a requirement for high purity material. Avalanche photodiodes also require low dark current, which translates into high-quality low-defect density material. There has been an expanding interest in these detectors and related lasers realized in the InGaAsP/InP material system. This lattice-matched material system has a bandgap that ranges from about 0.9 to 1.65 μm, which makes it suitable for use in the 1.3 and 1.55 μm low-loss region of today's silica fibers.

High purity epitaxial indium phosphide has been grown by liquid-phase epitaxy (LPE), chloride vapor-phase epitaxy (C-VPE), hydride vapor-phase epitaxy (H-VPE), atmospheric-pressure metalorganic chemical vapor deposition (APMOCVD), low-pressure metalorganic chemical vapor deposition

TABLE I

SOME RECORD 77 K MOBILITY EPITAXIAL LAYERS GROWN BY VARIOUS TECHNIQUES[1]

	1977	1980	1981	1982	1983	1984	1985	1986	1987	1988	1990
APMOCVD		Fukui and Horikoshi 36, 4.4, 31.			Bass, et al. 55, 4.7, 18.	Bass and Young 74, 5.2, 8		Chen, et al. 131, 5.1, —			
LPMOCVD					Razeghi and Duchemin 60, 5.4, 15.		DiForte-Poisson, et al. 147, 5.0, 3.4		Thrush, et al. 264, —, —	Razeghi, et al. 150, 6.0, 0.3	Bose, et al. 305, 4.9, 0.5
LPE		Eastman 94, —, —									
GSMBE/CBE								Kawaguchi, et al. 105, —, 9.1			
C-VPE	Fairman, et al. 140, 6.06, 1.3		Fairhurst, et al. 121, 5.16, 2.2		Taylor and Anderson 130, —, 0.6						
H-VPE			Zinkiewicz, et al. 56, —, 2	Roth, et al. 71, 3.27, 5.2					Iwata and Inoshita 92, —, 1.3	McCollum, et al. 125, 4.9, 3.3	

[1]The three numbers under each reference are: 77 K mobility (10^3 cm^2/V-s), 300 K mobility (10^3 cm^2/V-s), $N_D - N_A$ at 77 K (10^{14} cm^{-3}), respectively. For example, Fairman, et al. refers to Fairman, et al. (1977). A sample with 77 K mobility of 140,000 cm^2/V-s at carrier concentration 1.3×10^{14} cm^{-3}. The 300 K mobility was 6,060 cm^2/V-s.

(LPMOCVD), and gas source molecular beam epitaxy/chemical beam epitaxy (GSMBE/CBE). Molecular beam epitaxy (MBE) has not proved to be successful in the growth of high purity indium phosphide due to the difficulty of handling elemental phosphorus.

Table I summarizes some of the highest 77 K electron mobilities obtained in samples grown by each of the techniques mentioned. It is interesting that the C-VPE sample with 140,000 cm^2/V-s 77 K mobility reported by Fairman et al. (1977), and the LPE sample with 94,000 cm^2/V-s 77 K mobility reported by Eastman (1980) have not since been exceeded in mobility by these growth methods. In each of the other methods, except for GSMBE/CBE, which is still quite young, there has been an increase in highest reported mobilities with time. There have been a number of recent reports by different researchers of very high purity InP grown by both APMOCVD and LPMOCVD. Thrush et al. (1987) have reported a sample with mobility at 77 K of 264,000 cm^2/V-s. Another sample grown by Thrush was reported by Bose et al. (1990) to have a 77 K mobility of 305,000 cm^2/V-s and peak mobility of 420,000 cm^2/V-s at 54 K. Donor and acceptor concentrations were 8.8×10^{13} cm^{-3} and 3.7×10^{13} cm^{-3} respectively. This is the highest purity InP, reported to date, grown by any method. This is also higher than the highest 77 K mobility for GaAs of 210,000 cm^2/V-s reported by Shastry et al. (1988). Skromme et al. (1984) have observed that C is not incorporated as a residual acceptor in H-VPE or LPE grown InP. This very high purity MOCVD InP may indicate that C is not incorporated into InP grown by that method.

As indicated by Table I, H-VPE had lagged significantly behind other growth methods in the demonstration of its capability to produce and control high purity indium phosphide, but indium phosphide with purity comparable to that grown by most of the other methods has now been demonstrated and controlled (McCollum et al., 1988). The growth of InP of higher purity than previously reported is described in this chapter. The remainder of this chapter will contain discussions of this high purity indium phosphide grown by hydride vapor-phase epitaxy. Characterization data for this material will be presented.

For a general discussion of H-VPE growth techniques, the reader is referred to Olsen and Zamerowski (1979), Olsen (1983), and Beuchet (1985).

II. Growth

1. Growth Reactions

The kinetics and thermodynamics of hydride vapor-phase epitaxial InP have been discussed in detail by Jones (1982). A brief and simplified summary of the transport processes involved in vapor-phase epitaxy and the differences and similarities between hydride and chloride VPE will be presented next. The actual processes are probably much more complex than the simple reactions described here.

In both hydride and chloride VPE, pure metals are transported by chlorine from HCl. For indium phosphide the reaction is

$$HCl + In \rightarrow \tfrac{1}{2}H_2 + InCl. \tag{1}$$

In C-VPE the HCl is formed by

$$PCl_3 + \tfrac{3}{2}H_2 \rightarrow 3HCl + \tfrac{1}{4}P_4. \tag{2}$$

As can be seen by Eqs. (1) and (2), the ratio of P to In (V/III ratio) in C-VPE is only alterable by the efficency of Eq. (1). If the conversion of HCl to InCl is nearly 100%, this ratio is 1/3. H-VPE, in contrast, uses separate sources for P(PH_3) and In(HCl + Eq. (1)), and the (V/III) ratio can be varied over a wide range since the partial pressures of P and In can be controlled independently. The samples of this study were grown with a V/III ratio of 3.1/1.

In C-VPE, the P source is PCl_3 by way of Eq. (2). In H-VPE the P source is cracked PH_3. It is evident that although C-VPE and H-VPE use different source materials, the chemistry is very similar. This will be discussed more in Section 5.

2. Growth System

The reactor used for the growth of high purity InP should be as simple as possible and free of sources of unwanted impurities. However, the reactor used in this study was originally designed for the growth of doped binary, ternary, and quaternary multilayered structures and has been described by Zinkiewicz *et al.* (1981). The original reactor tube is shown in Fig. 1. This tube was modified as described below for the growth of high purity InP.

The diethylzinc + H_2 input line was capped off since it was not needed. The Ga source was removed to keep Ga contamination from being an issue in the InP growth. HCl readily transports metals to the growing layer and to nearly everywhere else in the reactor. If Ga is left in the reactor tube, it may contaminate the In melt and also the grown InP. The AsH_3 source was removed from the system. The O_2 + Ar flow was included with the HCl + H_2 over the In source boats to investigate the effect of 0_2 injection on Si incorporation in high purity InP as will be discussed next.

This vertical reactor tube is unique among hydride reactors. Virtually all other hydride reactors are horizontal and use a large and open In melt (Olsen, 1983). In this system, the In source material is held in three vertically

InGaAsP VPE Growth System

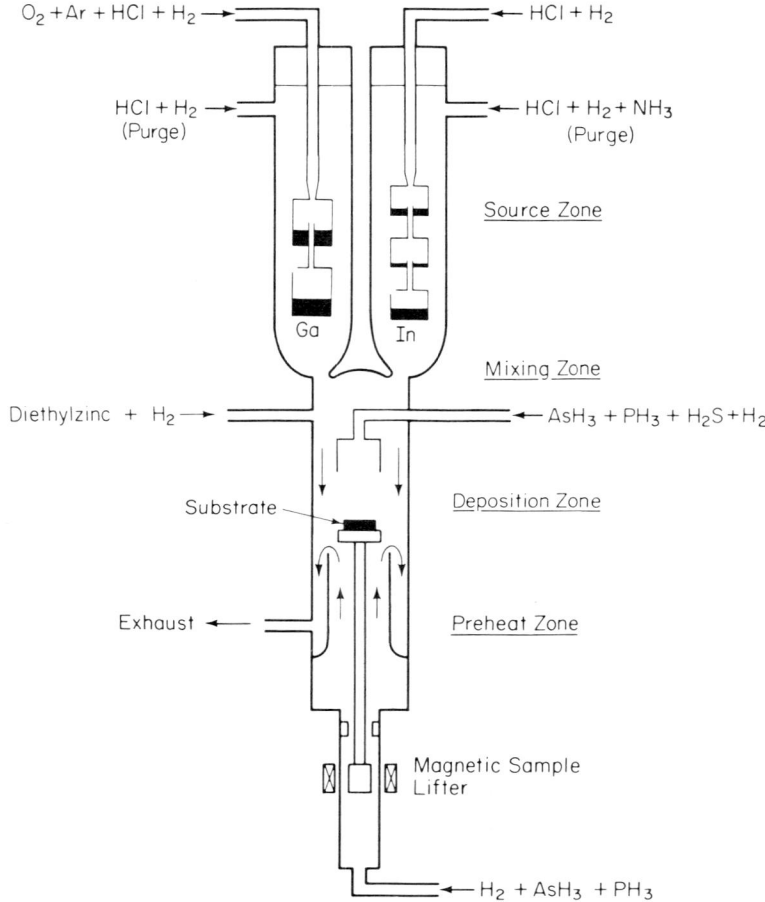

FIG. 1. University of Illinois InGaAsP quaternary H-VPE growth system before changes were made for growth of high purity InP.

cascaded quartz boats. These boats confine the HCl and other In source gas flows to provide intimate contact with the In melt. In horizontal reactors, an open In melt is usually used, which may make it difficult to obtain complete reaction of the HCl with the In (Eq. 1). Figure 2 shows the modified growth tube.

Some reactor tubes must be completely cooled to load or remove a sample. This may represent a very long process due to the thermal mass often

InP VPE Growth Systems

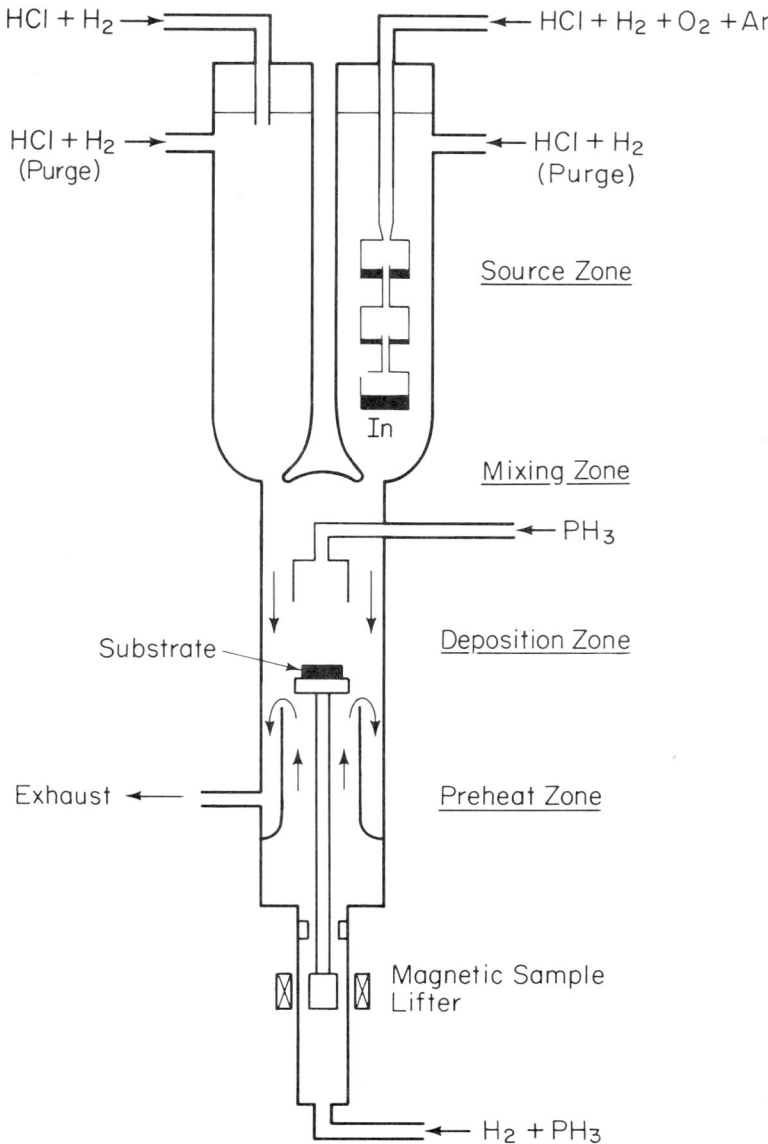

FIG. 2. High purity H-VPE growth system after modifications.

associated with a hot walled reactor. To avoid this delay in time, some reactors have included a load lock. The reactor of Fig. 2 can be opened near the bottom end while it is at growth temperatures. Three large thumb screws, holding an O-ring seal, are loosened, and the entire magnetic sample lifter assembly can be removed and lowered about two inches. Substrates are easily positioned on the lowered pedestal. A nitrogen flow is initiated around the outside of the reactor tube so that nitrogen surrounds the open loading area to prevent room air from entering the reactor. A peristaltic pump is used in the exhaust system to maintain the flow through the reactor. The reactor must be cooled only for replenishing the indium source and for maintainance or repair. The exhaust system is shown schematically in Fig. 3. The exhaust path is through the dump tube, then through bubbler 1, liquid nitrogen cold traps no. 1 and no. 2, bubbler 2, then is exhausted. The back flow is a hydrogen flow that keeps the cool end of then reactor tube side arm from becoming plugged. Although this works very effectively, the tall alternate bubbler will become activated in the event of an exhaust blockage. The peristaltic pump is used only during the substrate loading procedure.

All of the gas source flows are controlled with mass flow controllers and positive shutoff is provided with air actuated valves. The gas source flow schematic is shown in Fig. 4.

The PH_3 is a 10% mixture in H_2 purchased from Phoenix Research and contained in a stainless steel cylinder. HCl from Airco (five-nines plus) was supplied in a carbon steel cylinder. These gases were used directly as supplied without additional purification. The indium used in this study was Johnson Matthey Puratronic A1A, with a silicon concentration of less than 0.04 ppm according to mass spectroscopic measurements.

The preheat, deposition, mixing, and source zone temperatures were 680° C, 680° C, 760° C, and 790° C, respectively. The 790° C source zone temperature is significantly lower than that used by other researchers. Usui and Watanabe (1983) have reported a gettering of impurities at source zone temperatures as high as 920° C. We have not found this higher temperature helpful. This may indicate that the sources used by Usui and Watanabe were of lesser purity so that a gettering effect was realized. In contrast, we have found that residual impurity concentration increases with increasing source zone temperature.

3. SUBSTRATE PREPARATION

Substrate surface orientation and preparation are very important to epitaxial layer quality in H-VPE. The growth rate is dependent on crystal orientation. The preferred substrate orientation is (100) misoriented from 2° to 6° → ⟨110⟩. This will produce a much smoother grown surface than an

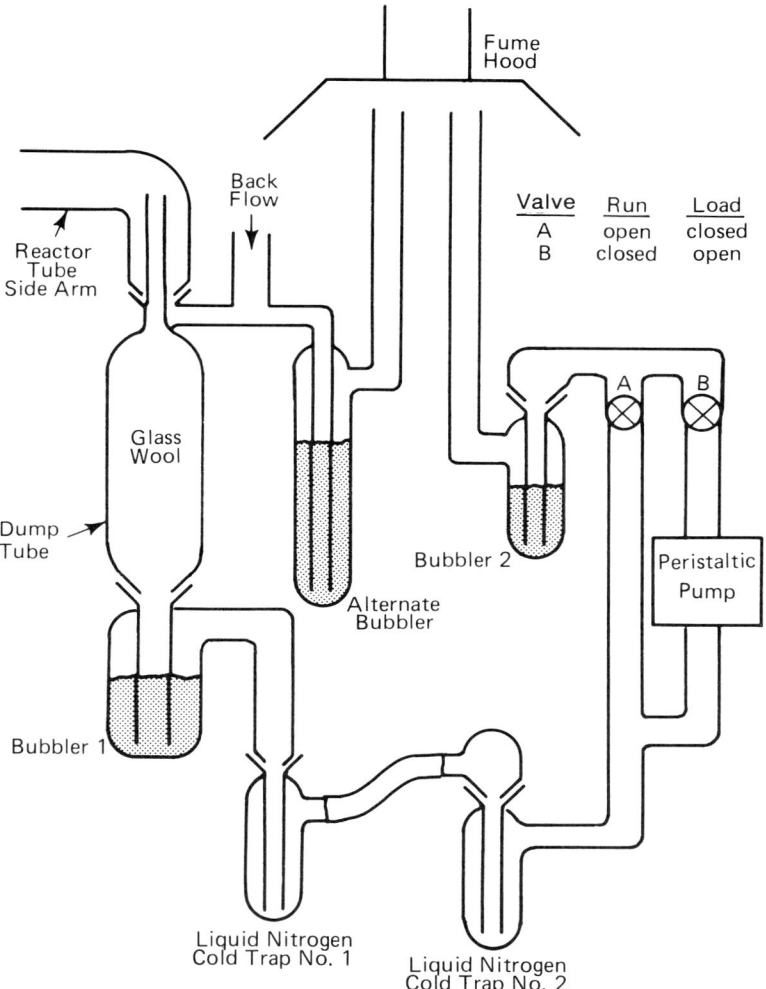

Fig. 3. Exhaust system.

accurately oriented (100) substrate as discussed by DiLorenzo (1972). An accurately oriented (100) surface will produce a large density of hillocks (Roth *et al.*, 1982).

It is very important in all methods of crystal growth to have a clean substrate surface, and cleaning procedures sometimes are very complex. Cleaning procedures often include organic solvents, an $H_2SO_4 : H_2O : H_2O_2$ etch, bromine-methanol, and KOH or some combination of these. In H-VPE,

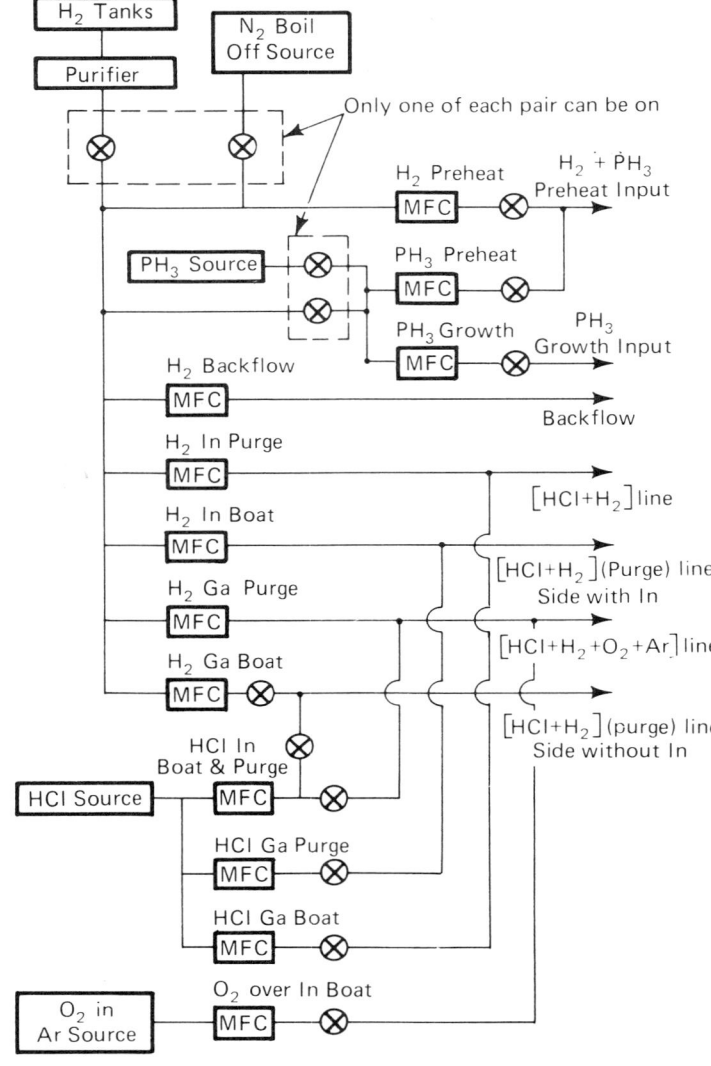

Fig. 4. Gas source flow system.

this lengthy procedure can be simplified for prepolished substrates and eliminated if substrates are polished just prior to growth.

The hydride technique is unique among all epitaxial growth methods in that HCl is used for column III metal transport and is controlled separately from the column V source. The HCl diluted with H_2 can be used to etch the substrate in situ prior to growth. This produces a clean surface within the growth chamber environment which is excellent for crystal growth.

If unpolished substrates are used, they can be loaded into the growth reactor directly after a chemical/mechanical polish with a bromine-methanol soaked rotating polishing pad. All cleaning can be eliminated between polishing and sample loading. Both surfaces of the substrate should be polished to remove crystal damage induced by wafer sawing. If saw damage is not removed from the backside of the substrate, it can propagate through to the growth surface.

If prepolished substrates are used, a proper surface is realized by degreasing with organic solvents, followed by a two minute etch in a two percent bromine-methanol solution. This procedure takes about five minutes as compared to the one hour or more required by unnecessary, extensive cleaning procedures.

4. Growth Procedure

After a substrate is loaded, it resides in a hydrogen atmosphere where a PH_3 overpressure is introduced before the substrate is heated to growth temperatures. In the hydride system shown in Fig. 2, this preheat position is on the growth pedestal in the bottom of the reactor tube. This pedestal can be rotated, raised and lowered by means of a magnetically coupled lifter. Hydrogen flows up through the bottom of the reactor and through the preheat zone. After these flows have had sufficient time to equilibrate, the substrate is lifted to the preheat zone. In this position the substrate is at growth temperature in H_2 with a PH_3 partial pressure of 76 Torr. Two minutes later the pedestal is raised to the deposition zone, and the substrate is etched for two minutes. The substrate is etched with an HCl partial pressure of 2.4 Torr in the H_2 flow. The substrate is then returned to the preheat position, where it is protected from thermal damage at growth temperature by a 76 Torr PH_3 partial pressure. The substrate is kept in this position while the growth gas flows are set and allowed to equilibrate. The substrate is then returned to the deposition zone for 25 minutes of growth. Table II summarizes the growth flows.

To reduce PH_3 use, the PH_3 flow up through the preheat zone is replaced with H_2 from two minutes after the beginning of growth until three minutes before the end of growth. With PH_3 flow reestablished in the preheat zone,

TABLE II

SUMMARY OF GROWTH FLOWS AND PARTIAL PRESSURES

(Torr)	Without O_2 Flow (sccm)	Partial Pressure	With O_2 Flow (sccm)	Partial Pressure (Torr)
HCl In boat	2.4	5.4	2.4	5.2
PH_3 growth (10% in H_2)	7.5	16.7	7.5	16.4
Ar In boat (sets O_2)	0	0	8	17.5
O_2 In boat (998 ppm in Ar)	0	0	.008	0.02
H_2 Ga purge	100	223.3	100	218.1
H_2 In boat	63	140.6	63	137.4
H_2 from PH_3	67.5	150.7	67.5	147.3
H_2 In purge	100	223.3	100	218.1
H_2 total	330.5	737.9	330.5	720.9
TOTAL	340.4	760.0	348.4	760.0

the pedestal is lowered to that zone. The PH_3, HCl, and O_2/Ar (if used, O_2 injection is explained in Section 5) flows that were used during growth are turned off at this time. The sample is then lowered to the bottom of the reactor and allowed to cool. The PH_3 flow in the preheat zone is then replaced with H_2, and the PH_3 and O_2/Ar tank valves are closed. The reactor is allowed to idle for an hour while it is purged with H_2 to remove the PH_3. The sample is then removed.

After the sample is removed, the temperatures in the reactor are raised to 800° C in all four zones, and HCl is used to etch the reactor tube. For etching tube deposits, the HCl is introduced through the purge inputs and not over the In source. This procedure takes 30 minutes. The system is then returned to an idle mode of 600° C with H_2 flowing through all ports until the next growth.

5. Methods for Enhancing Purity

Cairns and Fairman (1968) were the first to report the effect of $AsCl_3$ concentration on electrical properties and growth rates in the C-VPE growth of GaAs. DiLorenzo and Moore (1971) confirmed this effect and also modeled the reactions. They concluded that the increased $AsCl_3$ caused the HCl partial pressure to be increased. The increased HCl suppressed the incorporation of column IV impurities but did not suppress column VI impurities that were introduced as hydrides, because the column IV elements (silicon and germanium) have stable chlorides at epitaxial growth temperatures and can chemically interact with the HCl, whereas the column VI elements (selenium and sulfur) do not. For high purity InP growth by H-VPE, the column IV impurity that needs to be minimized is silicon, since it is the dominant donor in H-VPE grown material, probably due to reactions between HCl and the quartz reactor tube.

The reactions between hydrogen, quartz, and HCl, which would lead to the formation of volatile chlorosilanes $SiClH_3$, $SiCl_2H_2$, $SiCl_3H$, $SiCl_4$, and SiO, have been summarised by DiLorenzo and Moore (1971). These reactions for growth of GaAs by C-VPE, are listed next.

$$H_2(g) + SiO_2(s) \rightarrow SiO(g) + H_2O(g) \tag{3}$$

$$3H_2(g) + HCl(g) + SiO_2(s) \rightarrow SiClH_3(g) + 2H_2O(g) \tag{4}$$

$$2H_2(g) + 2HCl(g) + SiO_2(s) \rightarrow SiCl_2H_2(g) + 2H_2O(g) \tag{5}$$

$$H_2(g) + 3HCl(g) + SiO_2(s) \rightarrow SiCl_3H(g) + 2H_2O(g) \tag{6}$$

$$4HCl(g) + SiO_2(s) \rightarrow SiCl_4(g) + 2H_2O(g) \tag{7}$$

$$4Ga(l) + SiO_2(s) \rightarrow Si(s) + 2Ga_2O(g). \tag{8}$$

Equations (4)–(7) would also apply to the growth of InP by H-VPE. The general form of Eqs. (4)–(7) is

$$(4-n)H_2(g) + nHCl(g) + SiO_2(s) \rightarrow SiCl_nH_{(4-n)}(g) + 2H_2O(g), \tag{9}$$

where n is an integer 1, 2, 3, or 4. Equation (9) applies to transport of SiO_2 by HCl. Equation (3) describes the transport of SiO_2 by H_2. These reactions represent paths for SiO_2 to be transported from the reactor to the growing sample. Reaction (8) represents a different path. Here the Si is incorporated into the column III melt. For InP growth this reaction would be

$$4In(l) + SiO_2(s) \rightarrow Si(s) + 2In_2O(g). \tag{10}$$

This path is set apart as different from those represented by Eq. (3) and Eq. (9) due to the reactant paths in a hydride reactor tube. There are two main paths for reactant gases to reach the growth surface in a hydride reactor, one over the metal source and the other directly over the substrate. When HCl flows over the metal source, the metal in the form of a chloride will be transported to the sample. The other path, around the metal source or over the substrate, is usually the path for the column V hydride during growth and HCl during pregrowth etch.

Several different gases have been used to try to suppress these volatile chlorosilanes by introducing the gases into one of these two substrate-bound growth flow routes. Stringfellow and Ham (1977) introduced NH_3 into the

AsH$_3$ flow over the substrate in the H-VPE growth of GaAs and found they were able to reduce N_D-N_A from about 5×10^{15} cm^{-3} to below 10^{14} cm^{-3}.

Sun et al. (1983) introduced O$_2$, NH$_3$, or additional HCl into various H-VPE InP growth runs. It is not clear whether these were injected over the In or over the substrate, but since HCl is additional only if injected over the substrate, the O$_2$ and NH$_3$ were probably injected over the substrate as well. Since injection of additional HCl had little effect in this experiment, it was concluded that column VI impurities were dominant. Background N_D-N_A concentrations were as high as 10^{16} cm^{-3}. Injection of O$_2$ and NH$_3$ also had little effect on background carrier concentrations. Liquid nitrogen temperature electron mobilities were as high as 6,270 cm^2/V-s for the purest O$_2$ doped sample and 14,000 cm^2/V-s for NH$_3$ doped sample compared to 13,700 cm^2/V-s for the best undoped sample.

Jürgensen et al. (1984) discussed the effect of additional HCl injection on growth rate and wall deposition in H-VPE grown InP.

Usui and Watanabe (1983) injected O$_2$ over the substrate and were able to reduce background carrier concentration and increase mobility. They were able to reduce N_D-N_A from about 10^{16} cm^{-3} with an assumed residual O$_2$ concentration of 0.05 ppm to less than 10^{15} cm^{-3} with O$_2$ concentration of about 1.2 ppm. Their best sample had a mobility at 77 K of 40,000 cm^2/V-s for a carrier concentration of 5×10^{14} cm^{-3} and a mobility at 300 K of 3,480 cm^2/V-s for a carrier concentration of 8×10^{14} cm^{-3}. Figure 5 shows

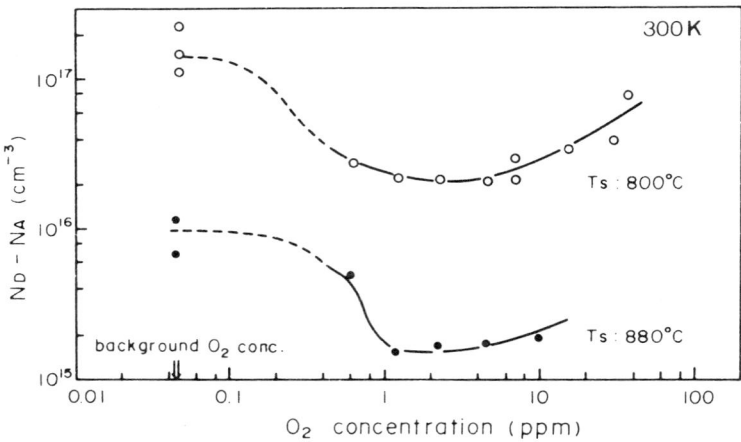

FIG. 5. Carrier concentration dependence on input oxygen concentration into the growth region, when the In source temperature is 800 °C and 850 °C. (Reprinted with permission from *Journal of Electronic Materials* **12**, 891–902, a publication of The Metallurgical Society, Warrendale, Pennsylvania 15386 USA.)

2. HIGH PURITY InP BY HYDRIDE VAPOR PHASE EPITAXY

their carrier concentration dependence on input O_2 concentration at two different In source temperatures.

Although none of the samples in the previous discussion are of very high purity, the work just mentioned by Usui and Watanabe is somewhat promising as a novel way to reduce the background concentration in H-VPE grown InP.

Silicon incorporation was significantly suppressed in H-VPE GaAs by injecting O_2 over the Ga melt (Lee et al., 1989a). Photothermal ionization spectroscopy was used to confirm the reduced Si incorporation. A sample grown without intentional O_2 injection had a 77 K mobility of 91,000 cm^2/V-s for a carrier concentration of 2.9×10^{14} cm^{-3}. The 77 K mobility of a sample grown with an O_2 concentration of 300 ppm was increased to 112,000 cm^2/V-s for a decreased carrier concentration of 2.3×10^{14} cm^{-3}. Another sample grown with an O_2 concentration of 630 ppm had a 77 K mobility of 92,000 cm^2/V-s for a carrier concentration of 1.9×10^{14} cm^{-3}. The PTIS data shown in Fig. 6, indicates a decreasing Si peak as O_2 concentration is increased. The samples were grown with the reactor described in Section 2 and shown in Fig. 1. According to Roth (1989), the O_2 had no effect if it was injected over the substrate rather than over the Ga. This would indicate that Eq. (8) produces less Si when O_2 is present. This is similar to the mole fraction effect of HCl in reducing Si incorporation via Eq. (9) but through a different reaction. The path of the injected O_2 is important in the suppression of Si incorporation in H-VPE GaAs.

The injection of O_2 has recently been used to suppress Si incorporation in the H-VPE growth of InP. Iwata and Inoshita (1987) injected the O_2 with the PH_3 through a quartz tube, which bypassed the indium metal source. This was apparently the same reactor used by Usui and Watanabe (1983) of NEC Corporation. They were able to confirm by PTIS that the Si donor concentration was suppressed by the injected O_2. A sample grown without O_2 added had a 77 K mobility of 78,000 cm^2/V-s with a carrier concentration of 4.7×10^{14} cm^{-3}. A sample grown with an injected O_2 concentration of 0.65 ppm had a 77 K mobility of 92,000 cm^2/V-s with a carrier concentration of 1.3×10^{14} cm^{-3}. PTIS indicated that in the sample grown without O_2, Si was the dominant donor (4.2×10^{14} cm^{-3}) with S also present (1.7×10^{14} cm^{-3}). In the sample grown with O_2 injected over the substrate, the [Si] was reduced to the level of the [S] (1.5×10^{14} cm^{-3}). The photothermal ionization spectra are shown in Fig. 7. It is interesting that this effect was seen for H-VPE of InP with $[O_2] = 0.65$ ppm when the O_2 was injected over the substrate, bypassing the In melt, while Roth (1989) observed this effect only when the O_2 was injected over the Ga melt with $[O_2] = 300 - 630$ ppm for H-VPE growth of GaAs.

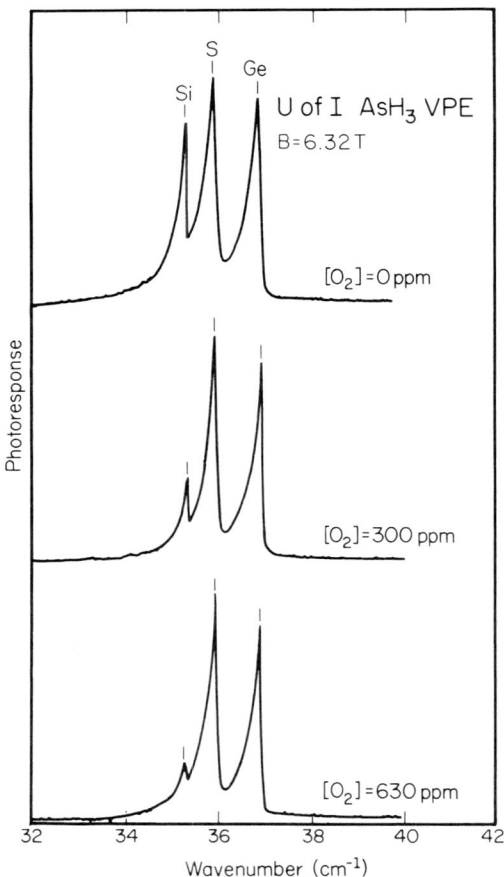

FIG. 6. Photothermal ionization spectra of three H-VPE GaAs samples grown with different O_2 concentrations indicating that Si incorporation is reduced by the O_2. [From Lee et al. (1989a).]

In still another study (McCollum et al. 1988), O_2 was injected over the In melt during the H-VPE growth of high purity InP. Suppression of Si incorporation was not observed in this study. In this work, 11 consecutively grown samples yielded the highest purity H-VPE InP reported thus far. Electron mobilities at 77 K of 91,000–125,000 cm^2/V-s and carrier concentrations of 3–6×10^{14} cm^{-3} were routinely grown. These results will be reviewed in the following section.

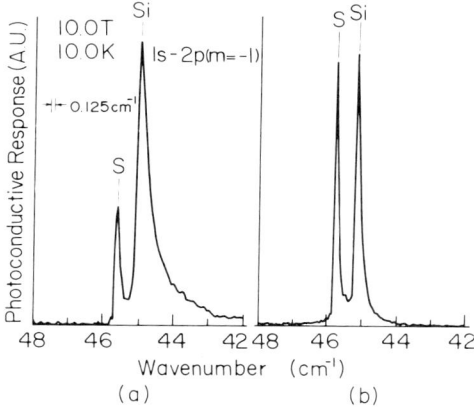

FIG. 7. $1s - 2p_-$ transition spectra for a sample grown (a) without O_2 injection and (b) with injected O_2 concentration of 0.65 ppm. [From Iwata and Inoshita (1987).]

III. Characterization of High Purity Indium Phosphide

Determining the purity of high purity semiconductor layers is neither a simple nor a trivial matter. Hall-effect measurements at room temperature and liquid nitrogen temperature have become the accepted method to determine carrier concentration and mobility of high purity samples. Hall measurements at these two temperatures are straightforward and can be obtained quickly and easily. Even though the peak mobility occurs around 60 K, the 77 K mobility is reported as a figure of merit for high purity epitaxial layers. Caution, however, must be taken with the analysis of high purity samples that are apparently much higher than previously reported.

Examples of anomalously high mobility in GaAs have been described by Wolfe and Stillman (1971). Inhomogeneity in epitaxial layers was modeled in this work, and inhomogeneous GaAs samples were intentionally produced in which the apparent 300 K mobility was increased from 7,400 to 24,000 cm²/V-s. Apparent 77 K mobility was increased from 150,000 to 740,000 cm²/V-s. These high mobilities would, of course, be recognized as exceeding theoretical limits in GaAs. The concern here, however, is that one may measure an InP sample to have a realizable, yet anomalously high, mobility. Several steps can be taken to increase confidence in Hall measurements. Examples of normal and anomalous characteristics for mobility versus temperature, resistivity versus inverse temperature, and magnetic field dependence of Hall factor are given by Wolfe and Stillman (1975), and Rode et al. (1983). These can be used to help consider the 'normality' of a sample.

Gross inhomogenity can be detected by simply cleaving the measured sample into four pieces and then measuring each of the four pieces if the original sample is large enough. Each of these four measurements should be in agreement and should be the same as the original piece, accounting for size and contact errors in the smaller Hall samples. X-ray analysis on various portions of the layer can also be helpful in detecting inhomogeneities.

Kim (1989) has described a method for measurement of the Hall mobility of thin, fully depleted layers by illuminating the sample under test at low temperatures. Although this has been shown to be a useful technique for measuring fully depleted high purity layers, care must be taken. Doping spikes and charges at the substrate/epitaxial layer interface can give erroneous mobility measurements. Verification in this case is best accomplished by growing a thick sample under the same growth conditions so that dark Hall measurement can be made.

The best way to verify the purity of epitaxial layers is to measure them by more than one technique and then to consider the ensemble of data. The remainder of this section will review the material characterization data for the high purity epitaxial indium phosphide mentioned earlier. These include fixed temperature Hall-effect measurements, variable temperature Hall-effect measurements, photothermal ionization spectroscopy (PTIS), photoluminescence (PL), magneto-photoluminescence (MPL), and constant capacitance-deep level transient spectroscopy (CC-DLTS). Data for high purity hydride indium phosphide grown with oxygen injected over the In source and without oxygen will be given. These samples are the highest purity H-VPE InP known to date. For additional discussion on high purity material and characterization techniques, the reader is referred to Stillman *et al.* (1982b).

6. HALL EFFECT

Hall-effect measurements were made by the Van der Pauw technique. Geometry is important to Hall measurements. If the sample to be tested is larger than a half inch on a side and square, contact can be made to the four corners. Epitaxial layers used for Hall measurements must be grown on insulating substrates. The traditional-cut cloverleaf Hall sample works very well but is difficult and time consuming to make. If contacts to a square sample are very small compared with the sample and very near the corners of the sample, one will obtain accurate measurements. The contacts are made by alloying a small tin or indium/tin spot on the sample corner in a hydrogen ambient for two minutes at 300° C. In-coated Pt wires are then set into these ohmic spots and connected to a mounting that goes into the Hall system. For carrier-concentration determinations the thickness of the layer must be known and is usually measured with an optical microscope after cleaving and

TABLE III

Mobilities and Carrier Concentrations

Sample	μ_{300} (10^3 cm^2/V-s)	μ_{77} (10^3 cm^2/V-s)	n_{77} (10^{14} cm^{-3}) Depletion Corrected	O_2
IPMM049	4.7	94	5.8	No
IPMM050	4.9	97	—	No
IPMM052	5.0	92	6.0	No
IPMM053	4.9	92	6.8	No
IPMM054	4.9	111	4.6	No
IPMM055	4.8	91	3.9	No
IPMM056	4.9	119	3.3	Yes
IPMM057	4.9	125	3.3	Yes
IPMM058	4.9	109	3.4	Yes
IPMM060	4.8	108	3.9	Yes
IPMM061	4.8	106	4.0	No

staining. The information obtained in these measurements is resistivity, Hall coefficient, N_D-N_A, and electron mobility.

Hall measurements at 300 K and 77 K are easy to obtain and provide quick feedback to the crystal grower. Table III gives a listing of mobilities at 300 K and 77 K and depletion corrected carrier concentration at 77 K for 11 samples grown by the author. Four of these were grown with oxygen ($[O_2] = 20$ ppm), and 7 were grown without oxygen. An increase in mobility achieved with O_2 injection is apparent from these data. It is probably more prudent to suggest that there was a steady increase in the 77 K mobilities in growths 49 through 57. Analysis by other techniques will help to indicate that the O_2 had little or no effect on material purity.

Repeated Hall measurements performed on high purity InP over several months have demonstrated a peculiar behavior in some samples grown by H-VPE, LPE, and MOCVD. An increase in 77 K mobility and decrease in carrier concentration has been observed by Lee et al. (1989b), Stillman et al. (1982b), Roth et al. (1982), and Eastman (1980). Roth reported an increase in 77 K electron mobility from 71,260 cm^2/V-s to 95,920 cm^2/V-s over a 3 month period in one sample stored at room temperature. This 'aging effect' has also been observed in the samples of this study. The Hall mobilities reported here were all taken shortly after growth and immediately after contact alloy, before any increase was realized. The samples return to their original mobility and carrier concentration when they are reheated in the alloy system.

It is significant to note that the overall average of the 77 K mobilities of the 11 samples is 104,000 cm^2/V-s and that the highest mobility of any of the samples is 125,000 cm²/V-s. Each of the layers had a uniform thickness of about 14 microns. The average growth rate was 0.57 microns per minute or 34 microns per hour. All of the samples exhibited good surface morphology.

A sample with O_2 and one without O_2 were measured by variable temperature Hall to consider the difference due to O_2 injection over the In source in hydride VPE. The variations of carrier concentration with temperature for a sample grown without O_2, IPMM054, and a sample grown with O_2, IPMM056, are shown in Fig. 8a and Fig. 8b respectively. The open dots are actual data, and the line through them is a theoretical fit based on donor concentration, N_D, acceptor concentration, N_A, and the impurity ionization energy. By making this fit on the sample data, the donor concentration, N_D, acceptor concentration, N_A, and donor activation energy, E_D can be calculated. These values for donor concentration and acceptor concentration can be used with the ionization energy and scattering mechanisms to calculate the theoretical variation of the mobility over temperature. This is shown in the plots of log mobility versus log temperature of Fig. 9a for sample IPMM054 grown without O_2 and Fig. 9b for sample IPMM056 grown with O_2. The open dots represent actual data. The mobility curves for each scattering process used in this calculation are also shown and labelled. Piezoelectric scattering was also considered in the calculation but lies above the region shown. The slight differences in the mobilities between those of Fig. 9 and Table III are probably due to the 'aging' mentioned previously. These calculations fit well to the data for sample IPMM056, and generally fit to the data for sample IPMM054 and indicate that the samples are of high purity. The drop in mobility at very low temperatures in each of these samples and the lack of fit below 10 K in sample IPMM054 are not understood at this time.

7. Photothermal Ionization Spectroscopy

Photothermal ionization spectroscopy is done at near liquid He temperature leaving very few electrons in the conduction band. Photothermal ionization takes place when an electron in a donor ground state absorbs a far infrared photon and is raised to one of the donor excited states. This bound electron may then be excited to the conduction band by absorption of one or more phonons, contributing to electrical conductivity. The slight difference between donor ground states can be resolved under a high magnetic field, which splits the threefold degenerate $2p$ excited state into the $2p_-$, $2p_o$, and $2p_+$ states. The reader is referred to Stillman *et al.* (1977) for additional discussion on PTIS.

A plot of photoresponse versus wave number for the $1s - 2p_-$ and $1s - 2p_+$ transitions for sample IPMM054 (without O_2) and sample IPMM056 (with

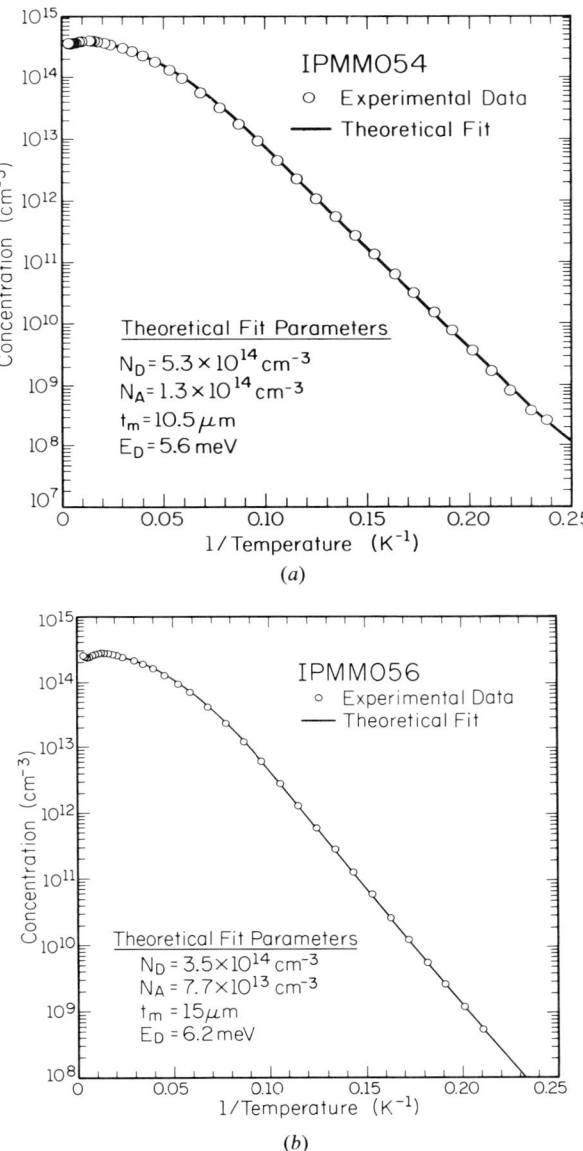

FIG. 8. Variation of carrier concentration with inverse temperature for samples (a) IPMM054 grown without O_2 and (b) IPMM056 grown with O_2 injection concentration of 20 ppm. The circles represent experimental data. Solid lines represent a theoretical fit to the data to determine donor concentration, acceptor concentration, and impurity activation energy. "t_m" is layer thickness. [(a) is from Kim (1988). (b) is from McCollum et al. (1988).]

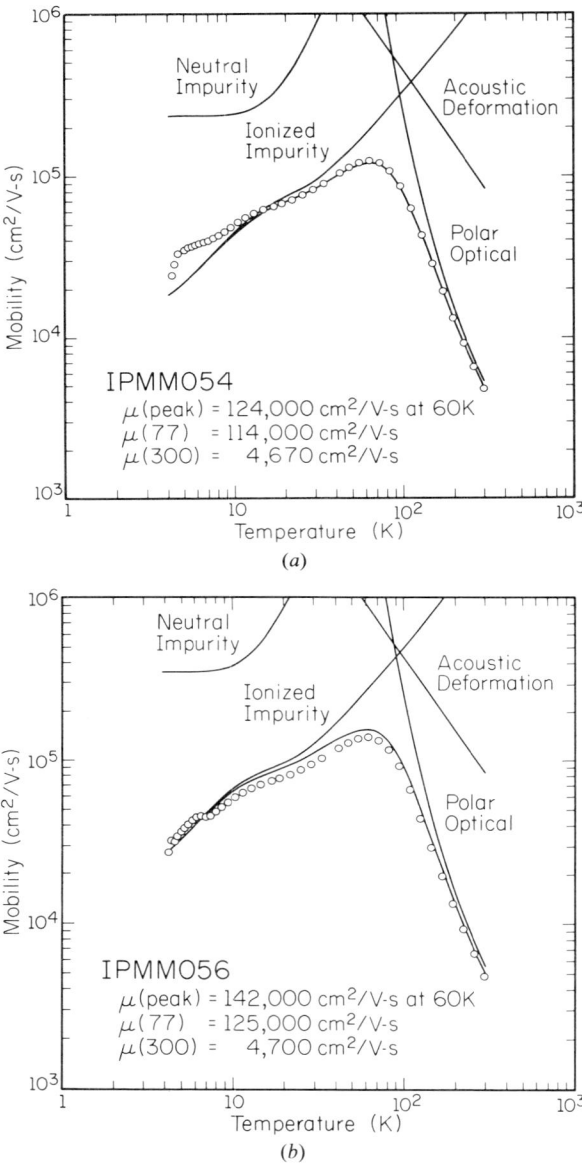

FIG. 9. Variation of mobility with temperature for samples (a) IPMM054 grown without O_2 and (b) IPMM056 grown with O_2 injection concentration of 20 ppm. The circles represent experimental data, and the solid lines are from theoretical calculations. The mobility as limited by each of the individual scattering mechanisms is also shown. [(a) is from Kim (1988). (b) is from McCollum et al. (1988).]

O_2) are shown in Fig. 10. The magnetic field strength was 6.32 T. The peaks are identified for Si, S, and Ge, as expected for hydride InP. Although the FWHM of the Si line and the height of the Ge line may be slightly reduced in the spectrum with O_2, these data are not significantly different. The reduction of Si to the level of S as was reported by Iwata and Inoshita (1987) for InP, shown in Fig. 7, and Lee et al. (1989a) for GaAs, shown in Fig. 6, is not seen here. Iwata and Inoshita reported this reduction for InP grown with O_2 injected downstream from the In source in a horizontal reactor. Lee et al. (1989a) reported this for GaAs grown with O_2 injected over the Ga boats in the reactor tube used in this study. The Si level in the O_2 doped samples of this InP study do not appear to be significantly reduced from the Si level found in the undoped samples.

8. Photoluminescence

Sample IPMM054 (without O_2) and IPMM057 (with O_2) were measured by low temperature photoluminescence at 1.7 K. The samples were optically excited with 5145 Å radiation from an argon ion laser. The spectrum of exciton recombination for the samples is shown in Fig. 11. As can be seen, the data from the two samples are virtually indistinguishable. The very sharp and distinct peaks of the $(D°, X)_{n=1}$ lines indicate samples of high purity. The FWHM of the major peaks were 0.08 meV for both samples.

9. Magneto-photoluminescence

Magneto-photoluminescence measurements were made on a sample grown without O_2, IPMM054, and a sample grown with O_2, IPMM056 at a magnetic field of 9.0 T. The spectra are shown in Fig. 12. The sharp peaks again indicate the high purity of the samples. The 'two-electron' satellite spectra, shown in Fig. 13, were then recorded for each of the two samples with a dye laser tuned near the bandgap of IPMM054 and to the strongest principal exciton recombination line for IPMM056. (This resonant excitation is the reason for the lower noise level in the data for IPMM056). The donor species are identified according to Skolnick et al. (1984a) and Skolnick et al. (1984b). The Si concentration is only slightly lower in the sample grown with O_2 injected over the In melt. In both Figs. 13a and 13b, the relative peak heights for Si, S, and Ge appear to be about the same as in the PTIS data of Figs. 10a and 10b. The peaks labelled a, b, and c in Fig. 12b are the same as the peaks labelled β, α, and γ in Fig. 12a.

10. Constant Capacitance-Deep Level Transient Spectroscopy

The deep levels in H-VPE grown InP have been reported by both Sun et al. (1983), and Plano (1988). Sun measured diffused $p^+ - n$ junctions and

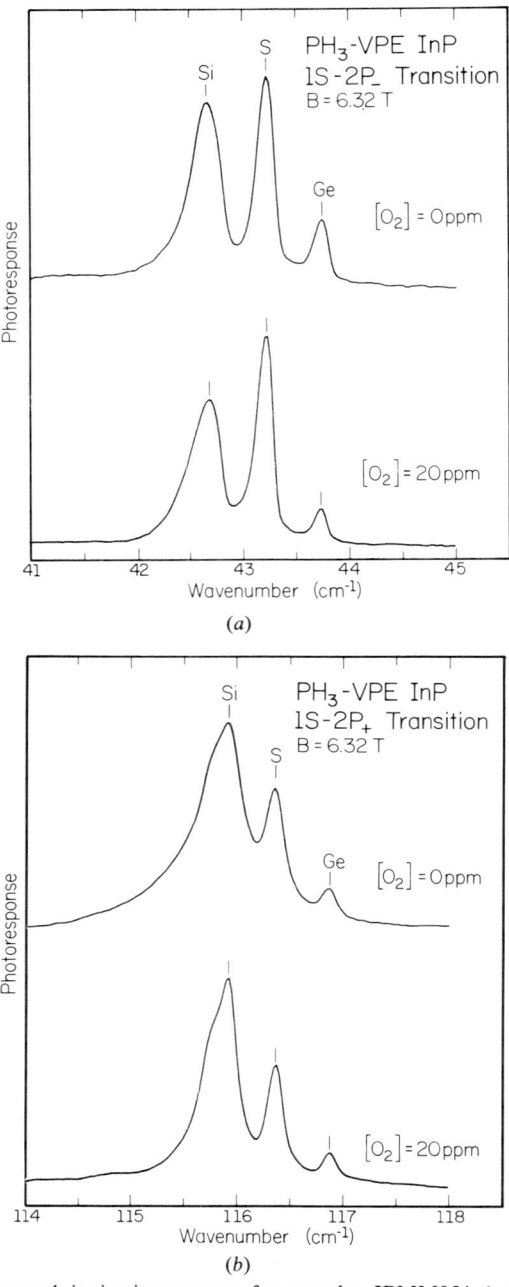

FIG. 10. Photothermal ionization spectra for samples IPMM054 (top) and IPMM056 (bottom) of the (a) $1s - 2p_-$ and (b) $1s - 2p_+$ transitions. Si, S, and Ge are identified in this sample. [(a) and (b) are from Lee (1988).]

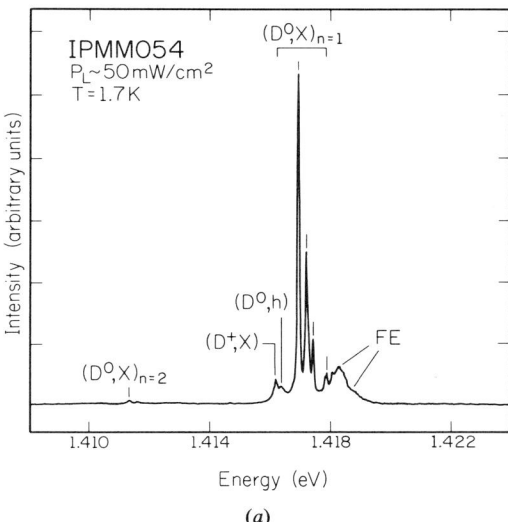

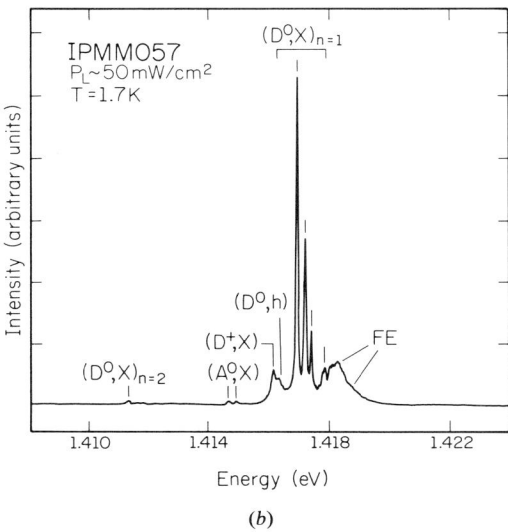

FIG. 11. Photoluminescence spectra of exciton recombination for samples (a) IPMM054 grown without O_2 and (b) IPMM057 grown with O_2 injection concentration of 20 ppm. The strongest peak has a full width half maximum of 0.08 meV in both samples. [(a) is from Bose (1988). (b) is from McCollum et al. (1988).]

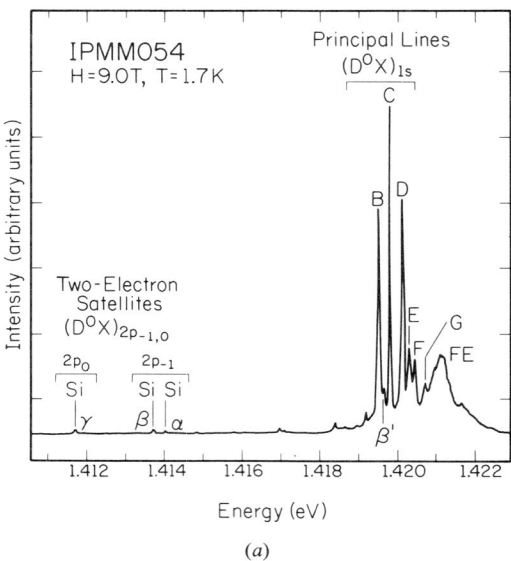

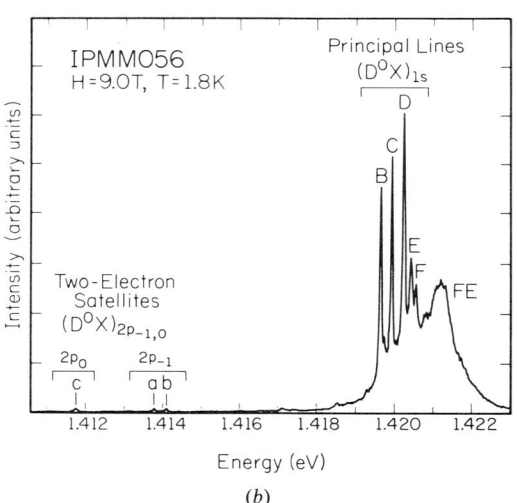

FIG. 12. Magneto-photoluminescence spectra taken under 9.0 T. (a) IPMM054 (without O_2). (b) IPMM056 (with O_2 injection). [(a) and (b) are from Bose (1988).]

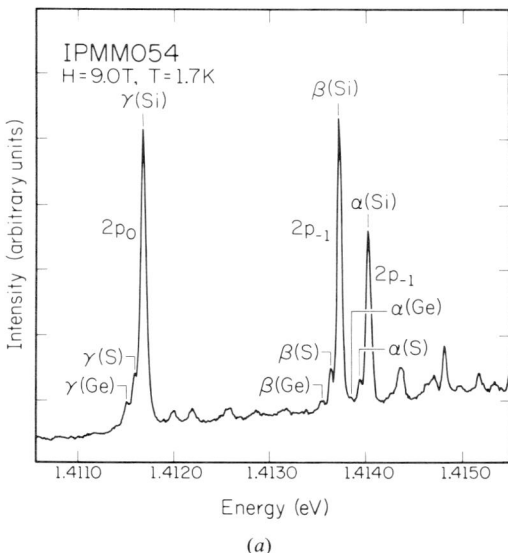

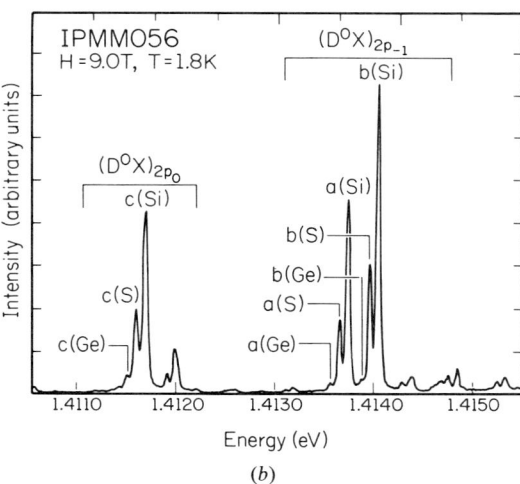

FIG. 13. Magneto-photoluminescence spectra of "two-electron" satellites for samples (a) IPMM054 taken with the dye laser tuned near the bandgap, and (b) IPMM056 taken with the dye laser tuned to the principal line D in Fig. 12(b). The b(Si) line in (b) has a full width at half maximum of 0.046 meV. Peaks are sharp and well resolved. Si, S, and Ge donors are identified in this sample. [(a) is from Bose (1988). (b) is from McCollum et al. (1988).]

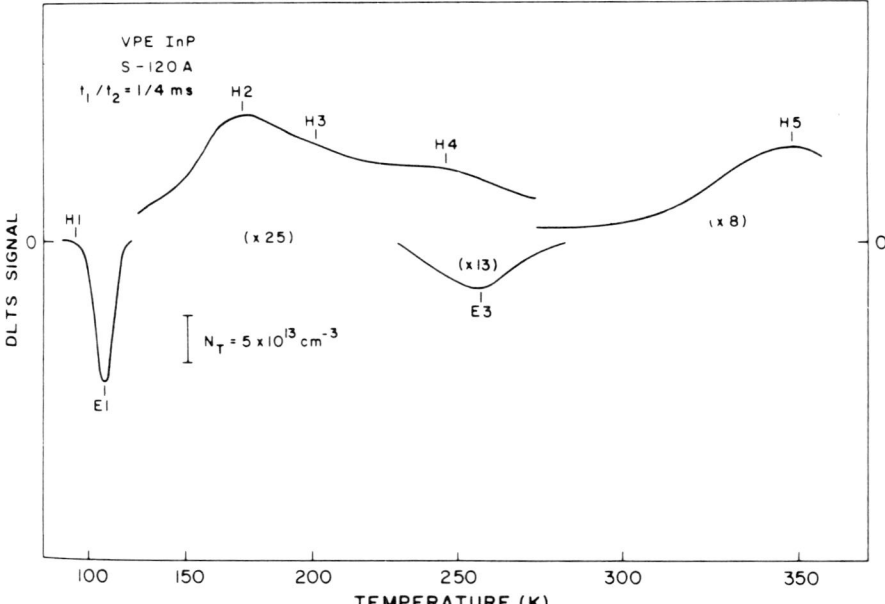

FIG. 14. DLTS spectrum for H-VPE material. (Reprinted with permission from Elsevier Science Publishers, Sun, S. W., Constant, A. P., Adams, C. D., and Wessels, B. W. (1983). Defect Centres in High Purity Hydride VPE Indium Phosphide. *J. Crystal Growth* **64**, 149–157.)

observed two electron traps and five hole traps. The DLTS spectrum is shown in Fig. 14. Table IV lists the energies and capture cross sections of these seven traps.

In the work by Plano (1988), *n*-type InP was measured using majority carrier diodes. Samples were grown on Si doped substrates as described in

TABLE IV

ELECTRON AND HOLE TRAPS IN UNDOPED AND DOPED VPE InP. (Reprinted with permission from Elsevier Science Publishers, Sun, S. W., Constant, A. P., Adams, C. D., and Wessels, B. W. (1983). Defect Centres in High Purity Hydride VPE Indium Phosphide. *J. Crystal Growth* **64**, 149–157.)

Trap	E (eV)	$\sigma_\infty \exp(\alpha/k)$ (cm^2)	T (τ = 2.2 ms)
E1	$E_c - 0.14$	2.4×10^{-16}	108
E3	$E_c - 0.51$	1.8×10^{-13}	255
H1	—	—	90
H2	$E_v + 0.44$	7.6×10^{-16}	171
H3	$E_v + 0.43$	1.0×10^{-13}	198
H4	$E_v + 0.29$	1.7×10^{-18}	238
H5	$E_v + 0.53$	2.3×10^{-17}	349

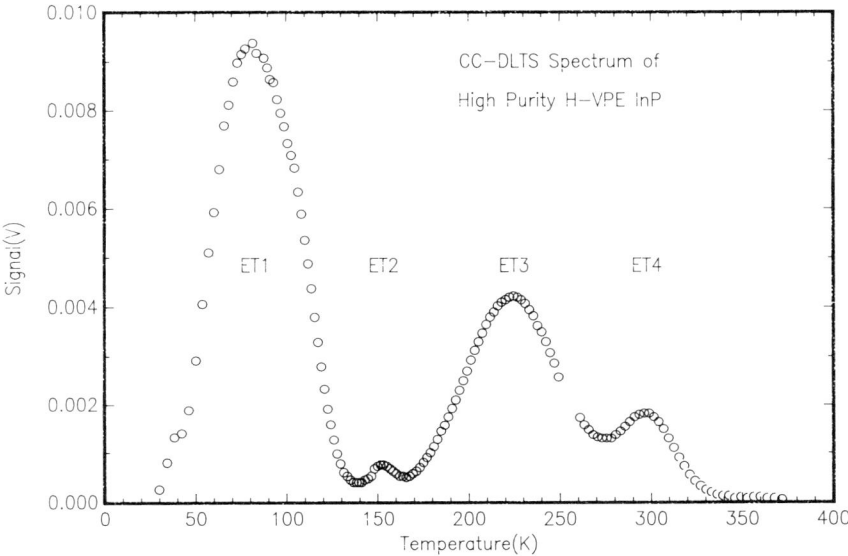

FIG. 15. CC-DLTS spectrum of a high purity H-VPE InP sample (IPMM068). The sample was pulsed from zero bias to 3 V reverse bias and a rate window of 14.25/s was used. [From Plano (1988).]

Section 4. Samples were grown simultaneously on semi-insulating, Fe-doped, substrates so that Hall measurements could be made. Constant capacitance-deep level transient spectroscopy (CC-DLTS) was used to measure the electron traps. Four electron traps were observed and are depicted in the CC-DLTS spectrum shown in Fig. 15. The energy values, capture cross sections, and trap concentrations are listed in Table V. The only trap that is common to both of these studies is labelled ET1 (0.10 eV) by Plano and E1 (0.14 eV) by Sun *et al.* The energies of these traps are not identical but their capture

TABLE V

ELECTRON TRAPS IN HIGH PURITY H-VPE InP.
[From Plano et al. (1988).]

Trap	Energy (ev)	Capture Cross Section (cm^2)	Trap Concentration (cm^{-3})
ET1	$E_c - 0.10$	2.4×10^{-16}	1×10^{14}
ET2	$E_c - 0.35$	3.4×10^{-13}	1×10^{13}
ET3	$E_c - 0.35$	2.6×10^{-17}	1×10^{14}
ET4	$E_c - 0.69$	1.7×10^{-13}	5×10^{14}

cross sections are equal, indicating that they are probably the same trap. The Fe-doped substrates grown simultaneously with the sample of Fig. 15 (IPMM068) had a 77 K mobility of 79,000 cm^2/V-s for a depletion corrected carrier concentration of 2.3×10^{14} cm^{-3}. (This sample was not part of the study discussed in Sections 6–9.)

IV. Conclusions and Summary

High purity InP has been routinely grown as indicated by 11 consecutive samples with an average 77 K mobility of 104,000 cm^2/V-s. The mobility at 77 K of 125,000 cm^2/V-s reported here is significantly higher than the highest 77 K mobility previously reported for InP grown by H-VPE. The high purity of these samples is also indicated by PTIS, PL, MPL, and variable temperature-Hall measurements. CC-DLTS data identified four electron traps in high purity H-VPE InP. Silicon is the dominant donor in both the O_2-doped samples and the undoped samples. In the experiments of McCollum et al. (1988), the injection of oxygen over the indium source ($[O_2] = 20$ ppm) had little, if any, effect on impurity incorporation. When O_2 is injected over the substrate, bypassing the In source ($[O_2] = 0.65$ ppm), Iwata and Inoshita (1987) have been able to reduce Si incorporation in H-VPE InP. Roth (1989) and Lee et al. (1989a) have seen the opposite effect in O_2 injection in H-VPE GaAs. They found that O_2 injected over the substrate had no effect, and the O_2 injected over the Ga source in GaAs growth ($[O_2] = 300$ and 630 ppm) was effective in suppressing Si incorporation in H-VPE GaAs. The $[O_2]$ and actual growth conditions differ in each experiment. The mechanisms here are not well understood at this point and are apparently different in the H-VPE growth of GaAs and InP.

The hydride technique is capable of growing high purity InP at high growth rates with good morphology with a minimum of substrate wet chemical cleaning due to the etch available within the H-VPE system. It is now a promising technique for devices requiring reproducible high purity InP layers.

Acknowledgments

We would like to thank K. Kuehl, A. Wilson, H. Russell, and C. Henderson for technical assistance and B. L. Payne, R. F. MacFarlane, M. K. Suites, and R. T. Gladin for help in manuscript preparation. Thanks are owed to B. Lee (PTIS), M. H. Kim (variable-temperature Hall), S. S. Bose (PL and MPL), and M. A. Plano (CC-DLTS) for making data available for this manuscript. M. A. Haase, A. D. Reed, and T. J. Roth are acknowledged for helpful discussions on crystal growth.

References

Bass, S. J., and Young, M. L. (1984). High Quality Epitaxial Indium Phosphide and Indium Alloys Grown Using Trimethylindium and Phosphine in an Atmospheric-pressure reactor. *J. Crystal Growth* **68**, 311–318.

Bass, S. J., Pickering, C., and Young, M. L. (1983). Metal Organic Vapour Phase Epitaxy of Indium Phosphide. *J. Crystal Growth* **64**, 68–75.

Beuchet, G. (1985). Halide and Chloride Transport Vapour-phase Deposition on InGaAsP and GaAs. *In "Semiconductors and Semimetals*, Vol. 22", Part A (W. T. Tsang, ed.), 261–298. Academic Press, Orlando.

Bose, S. S., (1988). Unpublished Data.

Bose, S. S., Szafranek, I., Kim, M. H., and Stillman, G. E. (1990). Residual Acceptor Impurities in Undoped High Purity InP Grown by Metalorganic Chemical Vapor Deposition. *Appl. Phys. Lett.* **56**, 752–754.

Cairns, B., and Fairman, R. (1968). Effects of the $AsCl_3$ Concentration of Electrical Properties and Growth Rates of Vapour Grown Epitaxial GaAs. *J. Electrochem. Soc.* **115**, 327C–328C.

Chen, C. H., Kitamura, M., Cohen, R. M., and Stringfellow, G. B., (1986). Growth of Ultrapure InP by Atmospheric Pressure Organometallic Vapour Phase Epitaxy. *Appl. Phys. Lett.* **49**, 963–965.

Di Forte-Poisson, M. A., Brylinski, C., and Duchemin, J. P. (1985). Growth of Ultrapure and Si-doped InP by Low Pressure Metalorganic Chemical Vapour Deposition. *Appl. Phys. Lett.* **46**, 476–478.

DiLorenzo, J. V. (1972). Vapour Growth of Epitaxial GaAs: A summary of Parameters Which Influence the Purity and Morphology of Epitaxial Layers. *J. Crystal Growth* **17**, 189–206.

DiLorenzo, J. V., and Moore, G. E. (1971). Effects of the $AsCl_3$ Mole Fraction on the Incorporation of Germanium, Silicon, Selenium, and Sulfur into Vapour Grown Epitaxial Layers of GaAs. *J. Electrochem. Soc.* **118**, 1823–1830.

Eastman, L. F. (1980). High Purity InP Grown by Liquid Phase Epitaxy. *In "Proc. 1980 NATO Sponsored InP Workshop"*, June 1980, unpublished.

Fairhurst, K., Lee, D., Robertson, D. S., Parfitt, H. T., and Wilgoss, W. H. E. (1981). A Study of Vapour phase Expitaxy of Indium phosphide. *J. Mater. Sci.* **16**, 1013–1022.

Fairman, R. D., Omori, M., and Fank, F. B. (1977). Recent Progress in the Control of High-purity VPE InP by the $PCl_3/In/H_2$ Technique. *Conf. Ser.-Inst. Phys.* **33B**, 45–54.

Fukui, T., and Horikoshi, Y. (1980). Properties of InP Grown by Organometallic VPE Method. *Jpn. J. Appl. Phys.* **19**, L395–397.

Iwata, N., and Inoshita, T. (1987). Spectroscopic Evidence that Oxygen Suppresses Si Incorporation into Vapour Phase Epitaxial InP. *Appl. Phys. Lett.* **50**, 1361–1363.

Jones, K. A. (1982). A Thermodynamic Analysis for the Growth of InP Using the Hydride Technique. *J. Crystal Growth* **60**, 313–320.

Jürgensen, H., Grundmann, D., Heyen, M., and Balk, P. (1984). Effect of Additional HCl Injection on Growth and Doping in the $In\text{-}PH_3\text{-}HCL\text{-}H_2$ System. *J. Crystal Growth* **70**, 123–126.

Kawaguchi, Y., Asahi, H., and Nagai, H. (1986). Gas Source MBE Growth of High-quality InP. *Conf. Ser-Inst. Phys.* **79**, 79–84.

Kim, M. H. (1988). Unpublished data.

Kim, M. H. (1989). Photo–Hall Studies of Compound Semiconductors. Ph.D. thesis, University of Illinois at Urbana–Champaign, 74–110.

Lee, B. (1988). Unpublished data.

Lee, B., Arai, K., Skromme, B. J., Bose, S. S., Roth, T. J., Aguilar, J. A., Lepkowski, T. R., Tien, N. C., and Stillman, G. E. (1989a). Spectroscopic Studies of the Influence of Oxygen Partial

Pressure on the Incorporation of Residual Silicon Impurities in Vapor-Phase Epitaxial Gallium Arsenide. *J. Appl. Phys.* **66**, 3772–3786.

Lee, B., Kim. M. H., McCollum, M. J., and Stillman, G. E. (1989b). Incorporation of Residual Donor Impurities in High Purity Indium Phosphide. *In* "SPIE Vol. 1144 Indium Phosphide and Related Materials for Advanced Electronic and Optical Devices," 39–47.

McCollum, M. J., Kim, M. H., Bose, S. S., Lee, B., and Stillman, G. E. (1988). High Purity Epitaxial Indium Phosphide Grown by the Hydride Technique. *Appl. Phys. Lett.* **53**, 1868–1870.

Olsen, G. H. (1983). Vapor Phase Epitaxy of III-V Compound Optoelectronic Devices. *In* "*Proceedings of the Symposium on III–V Opto-Electronics Epitaxy and Device Related Processes*" (V. G. Keramidas, and S. Mahajan, eds.), pp. 231–251. Electrochemical Society, Pennington, NJ.

Olsen, G. H., and Zamerowski, T. J. (1979). Crystal Growth and Properties of Binary, Ternary and Quaternary (In,Ga)(As,P) Alloys Grown by the Hydride Vapor Phase Epitaxy Technique. *In* "*Prog. Crystal Growth Charact.*," Vol. 2, 309–375.

Plano, M. A. (1988). Unpublished data.

Razeghi, M., Navrel, P., Defour, M., and Omnes, F. (1988). Very high purity InP Epilayer Grown by Metalorganic Chemical Vapor Deposition. *Appl. Phys. Lett.* **53**, 117–119.

Razeghi, M., and Duchemin, J. P. (1983). Growth and characterization of InP Using Metalorganic Chemical Vapor Deposition at Reduced Pressure. *J. Crystal Growth* **64**, 76–82.

Rode, D. L., Wolfe, C. M., and Stillman, G. E. (1983). Magnetic-field Dependence of the Hall Factor for Isotropic Media. *J. Appl. Phys.* **54**, 10-13.

Roth, T. J. (1989). High Purity Epitaxial Growth and Characterization of III–V Compound Semiconductors. Ph.D. thesis, University of Illinois at Urbana–Champaign, 71–74.

Roth, T. J., Skromme, B. J., Low, T. S., Stillman, G. E., and Zinkiewicz, L. M. (1982). Growth and Characterization of High Purity H_2-In-HC1-PH_3 Vapor Phase Epitaxy (VPE) InP. *In* "*SPIE Vol. 323 Semiconductor growth Technology*", 36–44.

Shastry, S. K., Zemon, S., Kenneson, D. G., and Lambert, G. (1988). Control of Residual Impurities in Very High Purity GaAs Grown by Organometallic Vapor Phase Epitaxy. *Appl. Phys. Lett.* **52**, 150–152.

Skolnick, M. S., Dean, P. J., Taylor, L. L., Anderson, D. A., Najda, S. P., Armistead, C. J., and Stradling, R. A. (1984a). Identification of Germanium and Tin donors in InP. *Appl. Phys. Lett.* **44**, 881–883.

Skolnick, M. S., Dean, P. J., Grover, S. H., and Kuphal, E. (1984b). Donor Identification in Liquid Phase Epitaxial Indium Phosphide. *Appl. Phys. Lett.* **45**, 962–964.

Skromme, B. J., Stillman, G. E., Oberstar, J. D., and Chan, S. S. (1984). Photoluminescence Identification of the C and Be Acceptor Levels in InP. *J. Elect. Materials* **13**, 463–491.

Stillman, G. E., Cook, L. W., Bulman, G. E., Tabatabaie, N., Chin. R., and Dapkus, P. D. (1982a). Long-wavelength (1.3- to 1.6-μm) Detectors for Fiber-optical Communications. *IEEE Trans. on Elec. Dev.* **ED-29**, 1355–1371.

Stillman, G. E., Cook, L. W., Roth, T. J., Low, T. S., and Skromme, B. J. (1982b). High-purity Material. *In* "*GaInAsP Alloy Semiconductors*" (T. P. Pearsall, ed.), 121–166. John Wiley and Sons, New York.

Stillman, G. E., Wolfe, C. M., and Dimmock, J. O. (1977). Far-infrared Photoconductivity in High Purity GaAs. *In* "*Semiconductors and Semimetals*", Vol. 12 (R. K. Willardson, and A. C. Beer, eds.), 169–290. Academic Press, New York.

Stringfellow, G. B., and Ham, G. (1977). Hydride VPE Growth of GaAs for FETs. *J. Electrochem. Soc.* **124**, 1806–1811.

Sun, S. W., Constant, A. P., Adams, C. D., and Wessels, B. W. (1983). Defect Centres in High Purity Hydride VPE Indium Phosphide. *J. Crystal Growth* **64**, 149–157.

Taylor, L. L., and Anderson, L. L. (1983). The Growth of Ultra-pure InP by Vapor Phase Epitaxy. *J. Crystal Growth* **64,** 55–59.
Thrush, E. J., Cureton, C. G., Trigg, J. M., Stagg, J. P., and Butler, B. R. (1987). Reactor Design and Operating Procedures for InP based MOCVD. *Chemtronics* **2,** June, 62–68.
Usui, A., and Watanabe, H. (1983). High-purity InP Grown by Hydride VPE Technique with Impurity Gettering by Indium Source and Oxygen. *J. Elect. Materials* **12,** 891–902.
Wolfe, C. M., and Stillman, G. E. (1975). Apparent Mobility Enhancement in Inhomogeneous Crystals. *In "Semiconductors and Semimetals"*, Vol. 10 (R. K. Willardson, and A. C. Beer, eds.) 175–220. Academic Press, New York.
Wolfe, C. M., and Stillman, G. E. (1971). Anomalously High "Mobility" in Semiconductors. *Appl. Phys. Lett.* **18,** 205–208.
Zinkiewicz, L. M., Roth, T. J., Skromme, B. J., and Stillman, G. E. (1981). The Vapour Phase Growth of InP and $In_xGa_{1-x}As$ by the hydride ($In-Ga-AsH_3-PH_3-HCl-H_2$) technique. *Conf. Ser.-Inst. Phys.* **56,** 19–28.

CHAPTER 3

Direct Synthesis and Growth of Indium Phosphide by the Liquid Phosphorus Encapsulated Czochralski Method

Tomoki Inada

METAL RESEARCH LABORATORY
HITACHI CABLE, LTD.
IBARAKI, JAPAN

and

Tsuguo Fukuda

INSTITUTE FOR MATERIALS RESEARCH
TOHOKU UNIVERSITY
MIYAGI, JAPAN

I.	INTRODUCTION.	71
II.	DIRECT SYNTHESIS SYSTEM.	73
	1. *Principle of Synthesis*.	73
	2. *Apparatus for Synthesis*.	75
	3. *Procedure for Direct Synthesis*	76
	4. *Observation of Phosphorus Liquid*.	78
	5. *Evaluation of Synthesized Polycrystals*.	81
III.	DIRECT SYNTHESIS AND CZOCHRALSKI GROWTH.	82
	6. *Apparatus and Growth Procedure*	82
	7. *Single Crystal Growth by LP-CZ Method*.	83
	8. *Evaluation of Grown Single Crystals*	84
IV.	COMPOSITION OF LIQUIDS AND CRYSTALS.	86
V.	CONCLUSION.	90
	ACKNOWLEDGMENTS.	91
	REFERENCES.	91

I. Introduction

There is a strong need for cheap, high-quality indium phosphide substrates for both optical and high-speed devices. It seems unlikely, however, that commercially available substrates will be able to meet the demand, because the cost performance of crystal production is very low. It can be said that the high cost of indium phosphide substrates has hindered their broad use. The reason for the high cost is that crystals are usually grown by a "two-step" process: the synthesis of polycrystal by the horizontal Bridgman (HB) method and single crystal growth by the liquid encapsulated Czochralski (LEC) method. In any two-step operation, the cost performance is basically lower than in a one-step operation. In practice, the source polycrystal for LEC growth, commercially produced by the HB method, is very expensive at present. Moreover, contamination by impurities is inevitable. So the need for low-cost, high quality polycrystalline indium phosphide to be used in LEC growth and/or for a one-step growth technique is strongly being felt.

There have been several papers on compound methods of synthesizing high purity indium phosphide. They are the HB method (e.g., Bonner et al., 1983 and Adamski, 1983), the phosphorus injection method in which phosphorus gas is injected into an indium melt (Farges, 1982 and Hyder and Holloman, 1983) and synthesis in an autoclave (Savage et al., 1984). One thing these methods have in common is that phosphorus reacts with indium as a gas. In the HB method, the pressure of the phosphorus gas must be maintained at 27.5 atm in order to compensate for the vapor pressure of the phosphorus from the indium phosphide melt at the melting point (1062°C). In the phosphorus-injection method, phosphorus gas is injected into the melt while the temperature or the pressure in the phosphorus reservoir is controlled. In the autoclave-synthesis method, an inert gas at an extremely high pressure, as high as around 2000 atm, is used in order to prevent the leakage of phosphorus gas from the container, since the pressure of phosphorus reaches 200–300 atm at 1000–1100°C. Thus, great care must be taken with the conditions, which necessitates the fine control of pressure of the phosphorus, especially so as to avoid the danger of explosion.

With regard to a one-step process, the direct synthesis and LEC-growth method is commonly used for gallium arsenide crystals. Unlike gallium arsenide, however, a similar method has not been developed for indium phosphide. It is quite difficult to synthesize indium phosphide simply by heating a mixture of indium and phosphorus. In the case of gallium arsenide, gallium and arsenic are covered with boric oxide (B_2O_3) during synthesis and the vapor pressure of arsenic at the temperature of synthesis (about 800°C) is a relatively low 30–40 atm. In contrast, in the case of indium phosphide, phosphorus cannot be covered with boric oxide because phosphorus sublimes at a lower temperature (about 416°C for red phosphorus) than the softening temperature of boric oxide (about 500°C). Even if it were possible to cover the phosphorus with an encapsulant and thus prevent sublimation of the phosphorus gas, the very high vapor pressure of phosphorus, which is as high as 200–300 atm during synthesis at about 1000–1100°C, would preclude such a synthesis because it would be very difficult to obtain such a high temperature in an inert gas at high pressure. Another technique is the phosphorus-injection technique. It is basically an LEC method for pulling a single crystal from an "in situ" synthesized melt produced by injecting phosphorus gas into an indium melt. Farges (1982) and Hyder and Holloway (1983) reported the details of this method and the characteristics of the grown crystals. It is presently being used commercially. All the previously mentioned methods commonly use phosphorus gas in the synthesis so as to compensate for the vapor pressure of the phosphorus from the indium phosphide melt.

We have developed a new synthesis system for producing polycrystalline indium phosphide and have also developed a direct synthesis and Czochralski growth system (one-step process) for single crystals, through the ingenious use of liquid phosphorus instead of phosphorus gas. We have named the new system the liquid phosphorus encapsulated Czochralski (LP-CZ) method (Inada et al., 1987). This chapter describes the new systems and the results regarding the crystal growth and the characteristics of the grown crystals. The direct synthesis system is described in Section II. The direct synthesis and Czochralski growth system is described in Section III, with the main focus on single crystal growth and the characteristics of the grown crystals. We have observed an interesting phenomenon during crystal growth in the phosphorus liquid, and this is taken up in Section IV, which describes the composition of the liquids and crystals and the mechanism of growth.

II. Direct Synthesis System

1. PRINCIPLE OF SYNTHESIS

Phosphorus has many allotropes, such as red, white, and black phosphorus, etc. Red phosphorus is the most widely used allotrope for the synthesis of indium phosphide owing to safety considerations. We also used red phosphorus for the new system. Fig. 1 shows the phase transition of phosphorus. Red phosphorus sublimes and becomes a gas when heated over 416°C (sublimation temperature). There are two structures for the gas: a P_4 structure at a

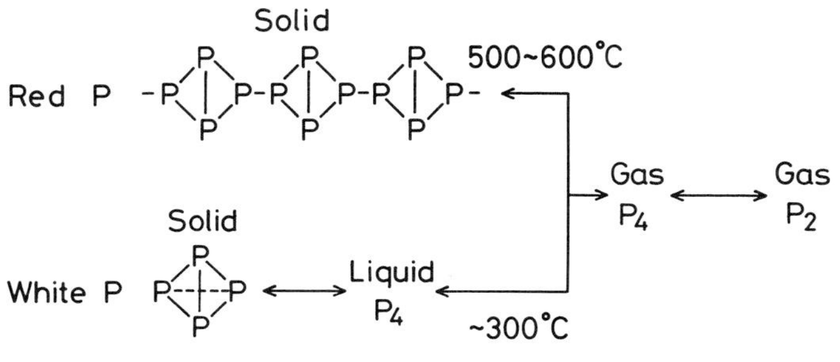

FIG. 1. Phase transition of phosphorus.

lower temperature and a P_2 structure at a higher temperature. The temperature for the structural change varies depending on the ambient pressure. When the gas is cooled quickly at a higher temperature of about 500–600°C, it becomes solid red phosphorus again. The temperature yielding this change depends on the ambient pressure and the cooling conditions. However, when cooled and condensed at a lower temperature, i.e. lower than about 300°C, the gas becomes a liquid, which can further be supercooled to become solid white phosphorus (white phosphorus has a melting point of 44°C). Fig. 2 shows the vapor pressure of liquid phosphorus after Wazer (1958). Because of the rapid formation of red phosphorus, the liquid does not exist in the region indicated by the broken portion of the line in the figure. Wazer, in *Phosphorus and Its Compounds*, mentioned that it is believed that the same liquid is obtained regardless of whether white, red, or black phosphorus is melted or whether the vapor is condensed. Thus the nature of the phase transition of phosphorus is very interesting and depends on the temperature and the pressure. The main feature of our new synthesis system is the use of liquid phosphorus generated from solid red phosphorus.

A schematic outline of the synthesis system is shown in Fig. 3. The starting materials are solid red phosphorus and solid indium. When heated, first the

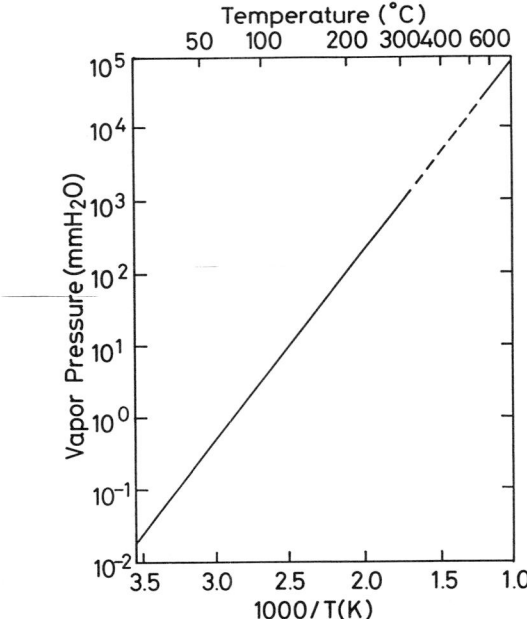

FIG. 2. Vapor pressure of phosphorus as a function of temperature.

3. DIRECT SYNTHESIS OF INDIUM PHOSPHIDE

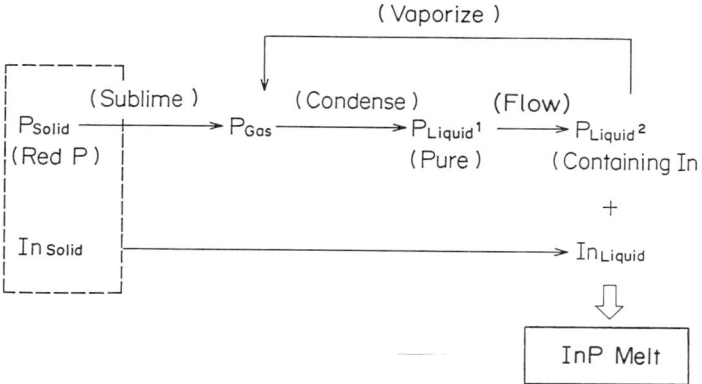

FIG. 3. Principle of the synthesis of indium phosphide using liquid phosphorus and the cyclic phase transition of phosphorus.

indium melts and then the phosphorus sublimes, generating phosphorus gas. Then the gas is cooled to a liquid (P_{liquid}^{1}), which flows down onto the indium melt (P_{liquid}^{2}), and is vaporized again. The difference between (P_{liquid}^{1}) and (P_{liquid}^{2}) is that (P_{liquid}^{2}) seems to contain a certain amount of indium and (P_{liquid}^{1}) doesn't. During this continuous cyclic phase transition of phosphorus from liquid to gas and from gas to liquid, the liquid phosphorus reacts with the indium melt to produce indium phosphide. If there is more phosphorus than that required for stoichiometry, the system produces an indium phosphide melt with a near stoichiometric composition.

2. APPARATUS FOR SYNTHESIS

A cross section of the chamber developed for the new system is shown schematically in Fig. 4. A commercially available MSR-6RA type high-pressure puller (Cambridge Instruments Ltd.) was modified and supplemented with an x-ray transmission image observation system. The chamber can withstand pressure of up to 80 atm. The same modification of the puller has been made by Ozawa and Fukuda (1986) for the "in situ" observation of the solid-liquid interface during the LEC growth of gallium arsenide. The chamber has two windows made of aluminum, through which X-rays can enter and exit. During synthesis, the outside of the container can be observed through a quartz rod with a TV camera, and the inside can be observed with the X-ray transmission image. A quartz container 40 mm in diam. is placed on a susceptor made of graphite in the hot zone of the furnace. The furnace can accomodate a container up to 100 mm in diameter. The container has a chimney 100 mm long and 10 mm in diameter, with an opening 10 mm in

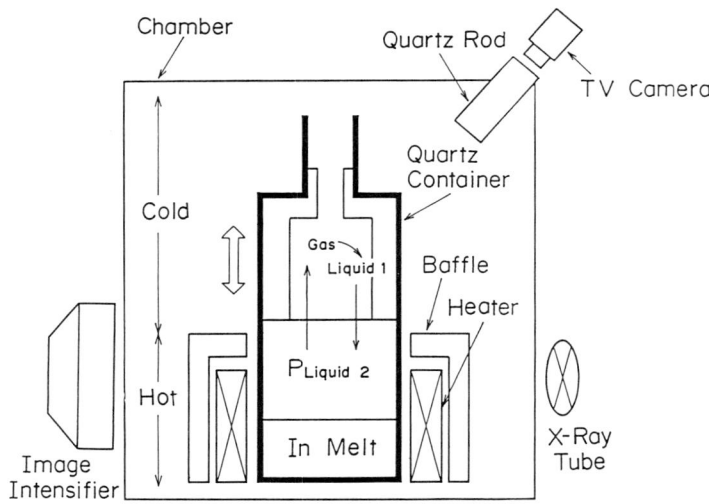

FIG. 4. Schematic cross section of the furnace for the synthesis system.

diameter at the top and is not sealed at all. The container can be moved up and down by means of a lift located in the furnace. The lower part of the container is heated, while the upper part is cooled by employing a thermal baffle made of boron nitride (BN) and alumina (Al_2O_3) as a heat shield and as a reflector for thermal radiation. In addition, an inert gas under high pressure removes heat from the upper part of the container. The temperature of the container is measured with a W-W.Re thermo-couple placed just beneath the crucible.

3. Procedure for Direct Synthesis

High-purity indium and red phosphorus (both 6-9s grade) were placed in the container. An amount of red phosphorus 100 wt% in excess of that needed for stoichiometry was used. The total charge weight of the raw materials was 150–200 g. No inert encapsulant, such as B_2O_3, was used in the system. An Ar ambient was maintained at a pressure of 60 atm throughout the synthesis process.

Initially, the container was placed in the cool zone, which is the upper part of the chamber. Then the graphite heater raised the temperature of the hot zone to 1000–1100°C at a rate of 8°C/min, and the container was then lowered into the middle of this zone at a rate of 150 mm/h. As the temperature in the container rose, first the indium melted at around 250°C and then the phosphorus sublimed at around 500°C. The generation of phosphorus gas was observed visually as a kind of shimmering. The gas cooled and condensed on the inside wall near both the shoulder and neck portions of the container. The liquid phosphorus flowed down onto the indium melt and reacted with it. The reaction was almost complete within a couple of seconds and, as a result, indium phosphide was produced.

The synthesis process showed up clearly in the X-ray transmission images, and they proved to be a very good way to observe this process. Photographs of the images are shown in Fig. 5(a)–(c). In Fig. 5(a) the solid indium can be seen above the red phosphorus in the quartz container in the initial stage. As the container was heated, the indium melted and flowed down to the bottom, and then liquid phosphorus was generated as is shown in Fig. 5(b). At around 1000–1100°C, the volume of the dark region at the bottom suddenly increased in a few seconds as in Fig. 5(c). Considering that the density of molten indium phosphide (5.05 g/cm^3) is lower than that of molten indium (7.31 g/cm^3), this increase in volume means that indium phosphide was synsthesized at that temperature. It was not possible to measure the volume

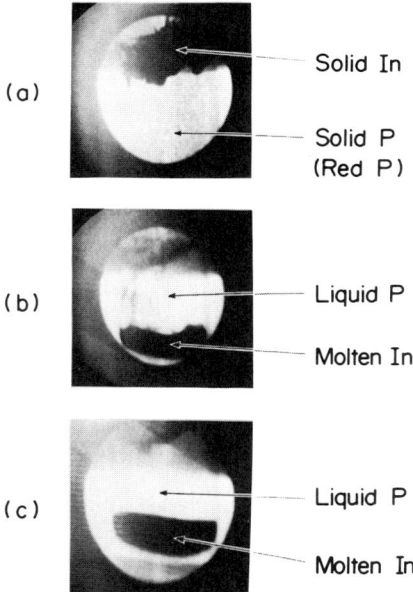

FIG. 5. X-ray transmission images taken during direct synthesis. (a) Solid indium at the top and solid red phosphorus below in the initial stage at room temperature. (b) At about 200–250°C, the indium melts and flows down to the bottom, and at about 400–500°C liquid phosphorus is generated. (c) At 1000–1100°C indium phosphide is synthesized from the liquid phosphorus and the molten indium.

of the synthesized indium phosphide from the x-ray image because the contrast is not sharp enough for this purpose.

The melt was left as it was for an hour in order to allow the synthesis to reach completion and then cooled at a rate of 10°C/min. Polycrystalline indium phosphide was taken out of the container at room temperature. Figure 6 shows a 120 g polycrystalline ingot. In some cases, a very small amount of indium inclusion was observed in the last-to-freeze portion of the crystal. No phosphorus inclusion was observed. The measurement of the lattice parameter of the crystal was performed with an x-ray diffractometer, and it showed that a stoichiometric ingot had been obtained.

The cost performance of the new system seems to be better than that of the widely used HB method, since the running time is shorter. The synthesis itself takes as little as a few seconds, and the total running time is about five hours including the hour when it is left as is, while it usually takes one to two days or more for HB growth.

4. Observation of Phosphorus Liquid

The phosphorus liquid can be observed with the TV camera. The picture in Fig. 7(a) shows the container before heating. Chunks of indium can be seen in

FIG. 6. Photograph of grown polycrystalline indium phosphide.

3. DIRECT SYNTHESIS OF INDIUM PHOSPHIDE

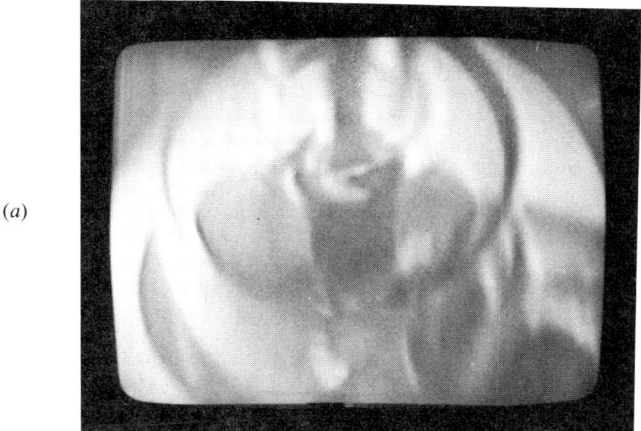

(a)

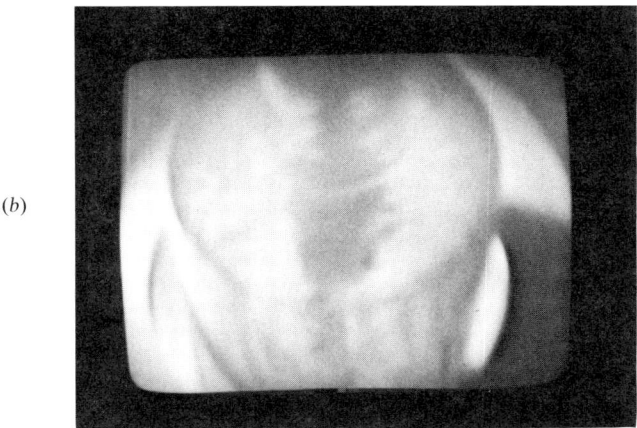

(b)

FIG. 7. Visual images from a TV camera. (a) in the initial stage, (b) flow of liquid phosphorus when heated.

it. When it was heated, phosphorus gas was generated. The gas then liquified and flowed down the inside wall of the container as is shown in Fig. 7(b). The liquid continued to flow during the whole synthesis process.

An interesting discovery was that there were two distinct layers after the synthesis had been completed; that is to say, a layer of liquid phosphorus remained on the molten indium phosphide layer. The amount of phosphorus in this layer seemed to correspond to the amount in excess of the stoichiometric amount with which the container had been charged. In order to find out the thickness of the layer the electrical current passing between an indium phosphide seed crystal and the container was measured by the method schematically shown in Fig. 8. The method is exactly the same as that reported for gallium arsenide growth by Terashima *et al.* (1982). The results are given in Fig. 9, which shows the change in the measured current as a function of the distance from the bottom of the crucible. As the seed crystal was lowered into the melt, the current showed a sharp increase at two positions. One was at the surface of the liquid phosphorus and the other was at the surface of the molten indium phosphide. The positions were observed by means of x-ray transmission images at the same time as the measurement. A value of 20 mm was obtained for the thickness in this experiment. This is smaller, however, than the value calculated for the excess phosphorus in the charge. Since the phosphorus layer became thinner during the cooling process after synthesis, the present cooling system in the furnace does not seem to adequately prevent phosphorus gas from escaping from the container. And in fact, solid red phosphorus was actually observed on the chamber wall and at the neck portion of the container when the crystal was taken out of the chamber after the synthesis procedure. It is supposed that some of the excess phosphorus escaped from the container as a gas during synthesis and/or during cooling, and the gas cooled rapidly to become solid red phosphorus. The problems with the cooling system can be eliminated through appropriate modifications. A water-cooling system around the container and the use of N_2 gas would improve the situation.

Details of the mechanism for the formation of the two distinct layers are given in Section IV.

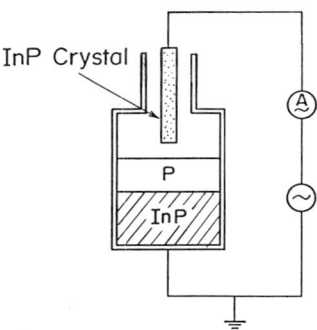

FIG. 8. Method of measuring the thickness of the liquid phosphorus layer by measuring the electrical current.

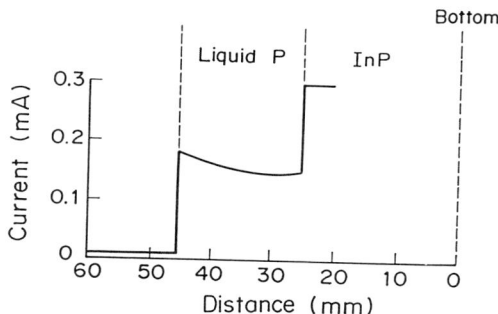

FIG. 9. Changes in current when a seed crystal was lowered into the melt.

5. EVALUATION OF SYNTHESIZED POLYCRYSTALS

The impurity concentrations of crystals were analyzed both by atomic absorption analysis (AA) and by inductively coupled radio frequency plasma atomic emission spectroscopy (ICP). These concentrations are shown in Table I. The concentration of each element was less than the detectable limit except for Si (0.07 ppm) as measured by the AA method.

The electrical properties were measured by Hall method at room temperature. The carrier concentration of the crystal was as low as $5 - 7 \times 10^{15}$ cm^{-3}. The mobility was 3200–3500 cm^2/V-s. The majority carrier in indium phosphide crystal is widely believed to correspond to sulfur originally contained in the indium metal and also to silicon absorbed from the quartz container. But it has not been possible to identify the dominant impurities through this analysis as the concentration of each impurity was less than the detectable limit.

The results of the evaluation show that a high-purity crystal has been obtained, despite the use of a quartz container. For higher purity, the use of pyrolytic BN (pBN) components (crucible and container), instead of the quartz components we used in the system, is more preferable. However,

TABLE I

CONCENTRATION OF IMPURITIES ANALYZED BY ICP AND AA METHOD. THE NUMBERS WITH THE "LESS THAN" SYMBOL ARE THE DETECTABLE LIMIT.

		Si	Pb	Zn	Cu	Mg	Mn	Se	Te
Conc. (ppm)	ICP	<0.1	<0.5			<0.01	<0.05	<0.2	<0.6
	AA	0.07	<0.05	<0.05	<0.05				

experiments in which a pBN crucible (but not container) was employed in the system did not yield a higher purity, and further study is needed. The use of indium metal with a higher purity (commercially available 7-9s grade) has also failed to yield better results.

III. Direct Synthesis and Czochralski Growth

As is mentioned previously, we have observed a separation into two layers: liquid phosphorus and the indium phosphide melt. This phenomenon suggested the possibility that single crystals could be grown from an in situ synthesized melt by the Czochralski technique (one-step process), using the liquid phosphorus layer as an encapsulant to prevent phosphorus evaporation from the melt. If the phosphorus liquid on the molten indium phosphide is stable under an Ar ambient at a pressure of 60 atm, this could compensate for the vapor pressure of phosphorus from indium phosphide melt. Our new LP-CZ (liquid phosphorus encapsulated Czochralski) method is similar to the LEC method, but the greatest difference is that our encapsulant is one of the constituent elements of indium phosphide, while an inert oxide encapsulant is used in the LEC method.

6. APARATUS AND GROWTH PROCEDURE

Figure 10 is a schematic diagram of the growth system. This apparatus is the same as that used for the synthesis. A $\langle 111 \rangle$ single crystal was attached to a molybdenum holder as a seed.

A quartz container 40 mm in diameter was charged with high-purity indium and red phosphorus (both commercially available 6-9s grade). The charge weight of indium was 100 g and that of phosphorus was 54 g. The 54 g of phosphorus includes an amount 100 wt% in excess of that necessary for a stoichiometric composition. Synthesis and growth were carried out under an Ar gas ambient at a pressure of 60 atm.

When the container was heated up to 200-300°C, source solid red phosphorus sublimed to become a gas. The gas then cooled near the neck portion of the container and condensed on the inside wall. This liquid phosphorus flowed down onto the indium melt and reacted with it at about 1000-1100°C. As a result of the reaction, an indium phosphide melt was produced as is described in Section II. The melt was kept as it was for 50 minutes to allow the synthesis to reach completion. During this process the crucible was gradually raised a little way into the upper part to create conditions suitable for seeding. After the seeding process, a $\langle 111 \rangle$ single crystal was grown from the melt, with the liquid phosphorus layer acting as an encapsulant. The whole synthesis and growth process was observed not

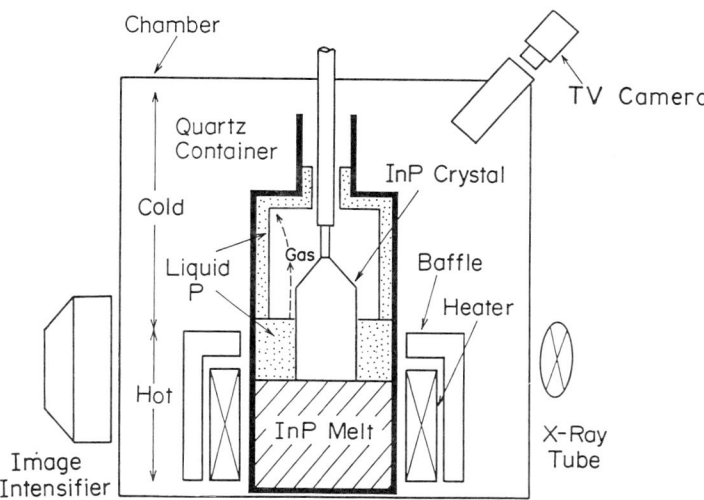

FIG. 10. Schematic diagram of the furnace for the liquid encapsulated Czochralski method.

only with the TV camera but also with the X-ray transmission image observation system.

7. SINGLE CRYSTAL GROWTH BY LP-CZ METHOD

The temperature profile in the container is a very important factor in growth. It was measured in advance with a W-W.Re thermo-couple. The result is shown in Fig. 11. The temperature was measured from the melt surface toward the upper portion of the container. The profile showed a gradual decrease. The vertical temperature gradient near the surface of the melt was as low as 27°C/cm. It was not only lower than the value for conventional conditions, 100–150°C/cm, but also lower than the value of 55°C/cm reported by Shinoyama et al. (1980) as a condition for the growth of dislocation-free crystals. The temperature near the chimney of the container was less than 300°C, which is low enough to produce liquid phosphorus.

The cyclic phase transition of phosphorus, from gas to liquid and from liquid to gas, and the liquid phosphorus layer on the indium phosphide melt were also observed during the growth process just as during the synthesis process. The major role of the liquid phosphorus layer is exactly the same as B_2O_3 usually used in the LEC method: it is an encapsulant that prevents phosphorus evaporation from the melt.

It wasn't easy to observe the growth visually with the TV camera because of the poor transparence of the quartz container after it became coated with

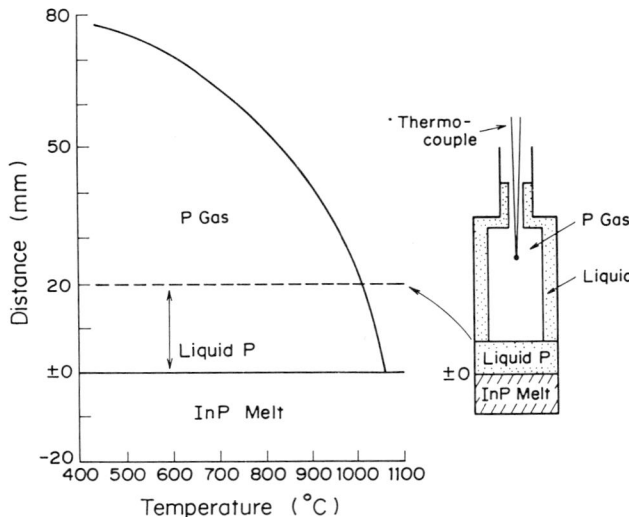

FIG. 11. Temperature profile along the pulling direction. The temperature gradient near the surface of the melt was measured to be 27°C/cm.

liquid phosphorus. On the other hand, the X-ray image was very helpful. It was very informative regarding the generation of polycrystal and changes in the meniscus. Figure 12(a) shows an x-ray transmission image taken during growth at a cooling rate of 1.2°C/h. While a $\langle 111 \rangle$ single crystal was growing, the image confirmed that there was three-fold symmetry, which corresponded to three (110) planes. When it became polycrystal, however, the image became asymmetric.

$\langle 111 \rangle$ single crystals were successfully grown at a cooling rate of 1.2°C/h all during growth. Fig. 12(b) is a photograph of one such crystal. The diameter was about 20 mm. This is a reasonable size for the present system, which has a 40 mm diameter crucible. The surface of the crystal was shiny and did not show any signs of either the dissolution or deposition of phosphorus. The seed crystal wasn't dissociated at all.

8. Evaluation of Grown Single Crystals

Grown crystals were evaluated through Hall measurements at room temperature, through lattice parameter measurements with x-ray diffraction, and through measurement of the etch pit density. Etch pits were revealed with Huber etchant.

The lattice parameter was measured to be 5.8683 Å, which is the same value as that of a commercially available LEC-grown crystal measured at the

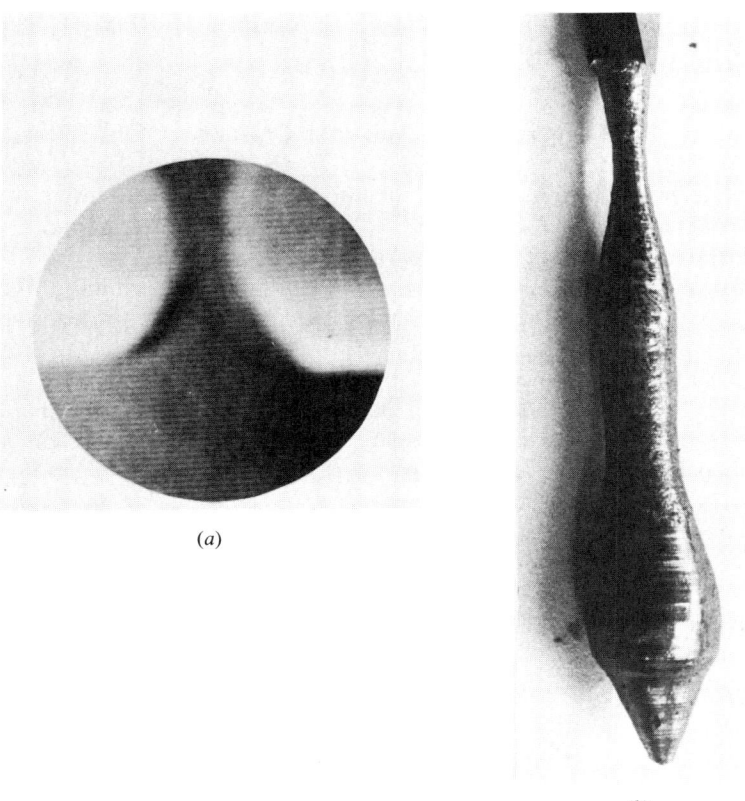

FIG. 12. (a) X-ray transmission image taken during growth of a $\langle 111 \rangle$ single crystal, (b) Photograph of a grown single crystal with a diameter of around 20 mm from 40 mm diameter crucible.

same time. It is likely that the composition of the crystal was near stoichiometric or almost the same as LEC grown crystal. We observed neither indium inclusions nor phosphorus inclusions in any of the crystals.

The carrier concentration was measured to be $5 \times 10^{15} \text{cm}^{-3}$ and the mobility was 3500cm^2/V-s. The purity was as high as that of the synthesized polycrystal described in Section II. The etch pit density was on the order of $10^3 - 10^5 \text{ cm}^{-2}$, which is not very low in spite of the low temperature gradient near the growth interface. This might be because of the high postgrowth cooling rate, 10°C/min, employed in these experiments. Such a high cooling rate is needed for the present system in order to reduce the amount of phosphorus gas escaping from the container. The cooling conditions must be improved for reduction of the dislocation density, and an auxiliary heater would be helpful for this purpose.

IV. Composition of Liquids and Crystals

The mechanism for the occurrence of the interesting phenomenon we have discovered, namely the separation of the liquid phosphorus and the molten indium phosphide, is described in this part. It would seem that knowing the composition of both layers would be very helpful in determining the mechanism, but sampling the liquids is not an easy task. At present the composition can only be speculated on from the experimental results. Two significant results are described next.

One concerns the x-ray absorption. In the x-ray transmission images, there was a clear region above the interface, meaning a very low absorption, and a dark region below, meaning a very large x-ray absorption. The interface between the two layers was comparatively sharp. The atomic weight of the major component of the clear region is thought to be as low as that of phosphorus. The component is probably liquid phosphorus with a slight amount of indium. The atomic weight of the dark region is as large as that of indium phosphide and/or indium.

Another important result is that a quite interesting phenomenon was observed during growth when the cooling rate was 3°C/h. While a single crystal was growing gradually and stably, dendric crystals suddenly grew into the liquid phosphorus layer from both the surface of the crystal and the surface of the melt. As the cooling procedure continued, they grew larger and reached their maximum size, which was equal to the diameter of the container. An x-ray image of the growth of the dendrites is shown in the photograph in Fig. 13(a), and a corresponding schematic drawing is given in

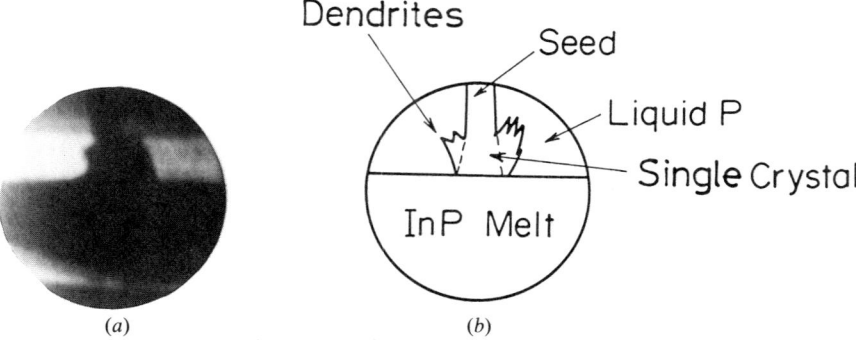

FIG. 13. (a) X-ray transmission image of dendrites; (b) schematic drawing of (a)

Fig. 13(b). Dendrites were observed on the surface of a growing crystal, though this isn't so clear in the photograph. A photograph of actual dendrites is shown in Fig. 14. In the case of Fig. 14, there was a single crystal inside the dendrites. The dendrites were brittle. X-ray diffractometry showed them to be a near stoichiometric compound of both indium and phosphorus. This result suggests that a certain amount of indium was dissolved in the liquid phosphorus layer. The liquid is thought to be supersaturated with indium and supercooled, which would explain why dendric crystals suddenly grow in the liquid. The absolute value of the concentration of the indium dissolved in the liquid phosphorus doesn't seem to be very high because of the relatively low x-ray absorption of the liquid. On the other hand, the dark region most likely consists of near stoichiometric indium phosphide, since the growth of near stoichiometric indium phosphide crystal occurred at nearly the same temperature as the melting point of indium phosphide. There is some possibility, however, that the melt was indium rich judging from the fact that the solidified residual polycrystals in the container included a small amount of indium metal. This possibility has not been confirmed because the evaporation of phosphorus from the melt during the post-growth cooling does occur, and the amount of the loss has not been accurately determined.

Prompted by these speculations, the concentration profile of the indium in the two liquid layers has been proposed, as is shown in Fig. 15. The gradual increase in the concentration of indium in the liquid phosphorus is due to the

FIG. 14. Photograph of grown dendrites. A single crystal was in the dendrites.

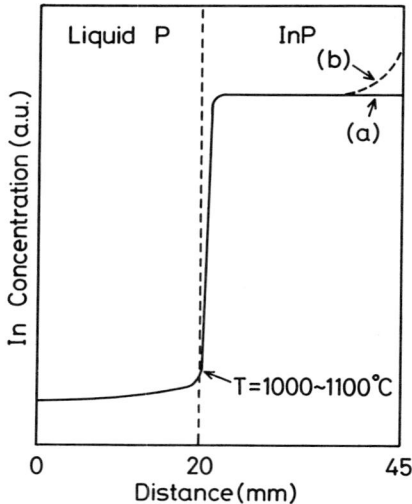

FIG. 15. Schematic curve for the speculated indium concentration in the liquids. (a) and (b) are for the cases without and with gravity segregation, respectively.

increasing solubility of indium in phosphorus as the temperature gradually increases (see Fig. 11). Though the concentration in the liquid seems to be low, more indium might be dissolved in the liquid than that yielding equilibrium solubility. This seems to be the reason why neither the seed crystal nor the single crystal dissolved in the liquid phosphorus layer. The drastic increase in the concentration of indium could be explained by the reactivity between indium and phosphorus, which depends on the liquid interaction parameter of the indium-phosphorus system. According to Ilegems's report (1975), the parameter changes from positive to negative at around 1000–1100°C. This means that the indium and phosphorus tend to react strongly with each other at temperatures above this point. In other words, it means that there is almost no possibility of separation in the liquid phase above this point, because a synthesis reaction occurs. So the separation probably occurs below this point. Based on the temperature profile in Fig. 11, the temperature of the liquid phosphorus was, in fact, below this point. The mean temperature of the liquid is kept lower in our system by the continuous evaporation of phosphorus, i.e., the phosphorus gas extracts the heat of vaporization from the liquid phosphorus. In the lower layer indium-phosphorus structures would be more stable. The solid line ((a) in the figure) shows the case where a near stoichiometric melt exists below. If an indium-rich melt exists in the lower layer, there seems to be a possibility of gravity segregation. The broken line ((b) in the figure) shows the case of gravity segregation.

This also seems to hold true for a gallium-arsenic system, based on our experimental results with the direct synthesis process of the LEC method under 70 atm. The previously mentioned report by Ilegems also states that the liquid interaction parameter in this case becomes negative near 700°C. Actually, the synthesis reaction reproducibly occurs near this temperature, which supports the idea that gallium and arsenic react with each other in the same manner as indium and phosphorus, and this reaction depends on the liquid interaction parameter. This analogy sugests that the synthesis of gallium arsenide under high pressure in the LEC system occurs between the gallium and arsenic in the liquid phase.

The relationship between the liquid interaction parameter and the synthesis reaction might differ depending on the ambient pressure. Thus, though the previous speculations are reasonable, the temperatures themselves have no absolute meaning. The systems aren't under atmospheric pressure: the indium-phosphorus system is under 60 atm, and the gallium-arsenic is under 70 atm.

X-ray diffraction analysis has revealed that the composition of both the dendrites and the single crystal is near stoichiometric. This means that near stoichiometric crystal can be grown from two different compositions of the liquid at different temperatures, that is, from the heavily phosphorus-rich composition in the phosphorus liquid at a temperature lower than the melting point of indium phosphide and from the stoichiometric or possibly slightly indium-rich composition in the indium phosphide melt near the melting point. The widely reported parabolic liquidus line is schematically drawn in Fig. 16. Generally the phosphorus side (broken curve) has not been

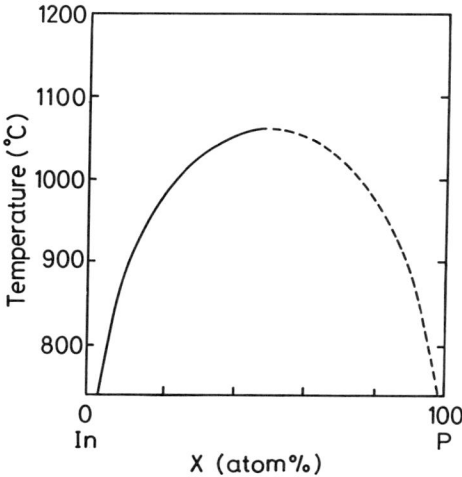

FIG. 16. Parabolic liquidus line for the indium phosphide system.

determined experimentally. The diagram is for equilibrium conditions at atmospheric pressure. It doesn't seem to fit the phenomena described previously, probably because our experiments were not performed at the equilibrium vapor pressure of phosphorus. If we were to draw a diagram for nonequilibrium conditions at a high pressure, the liquidus line would go up to the higher temperature on the phosphorus side, and the curve would become asymmetric. In order to determine a practical liquidus line for our system, further investigations of the composition of the two liquids and the crystals is required.

V. Conclusion

High purity indium phosphide single crystals have been successfully grown by our newly developed LP – CZ technique for the direct synthesis and growth of indium phosphide. This is the first time that liquid phosphorus has been used for both the synthesis and growth instead of phosphorus gas. In the system, liquid phosphorus, which is generated from solid red phosphorus, is used as one of the elements to produce indium phosphide, and it is also used as an "encapsulant" during growth as well as synthesis. Both the synthesized polycrystals and the grown single crystal showed a very high purity with a low carrier concentration and a low impurity concentration.

A very interesting phenomenon, namely the formation of two distinct layers, has been observed. The composition of the upper layer is thought to be liquid phosphorus with a slight amount of indium, where solubility dominates, and that of the lower layer is thought to be near stoichiometric indium phosphide, where reactivity dominates. It hasn't been possible to draw a phase diagram for the present nonequilibrium system. Further investigation of the details of the phase diagram applicable to our experimental conditions is needed for a clearer understanding of the observed phenomena.

This new technique should also be applicable to other materials with the most likely candidates being materials containing phosphorus such as gallium phosphide or indium gallium phosphide mixed alloy, and possibly II–VI compounds containing sulfur or selenium. The technique we have developed is basically a melt-growth method but solution-growth should also be possible. Fig. 17 is a schematic drawing of a solution-growth system. Here the solvent is liquid phosphorus, but it could also be liquid sulfur or liquid selenium. The solute would be gallium, indium, arsenic, zinc, etc. In the conventional solution growth of compound semiconductor materials, the solvent is a II or III metal and the solute is a V or VI element with a high vapor pressure. But it is just the opposite in the proposed system. This system

3. DIRECT SYNTHESIS OF INDIUM PHOSPHIDE

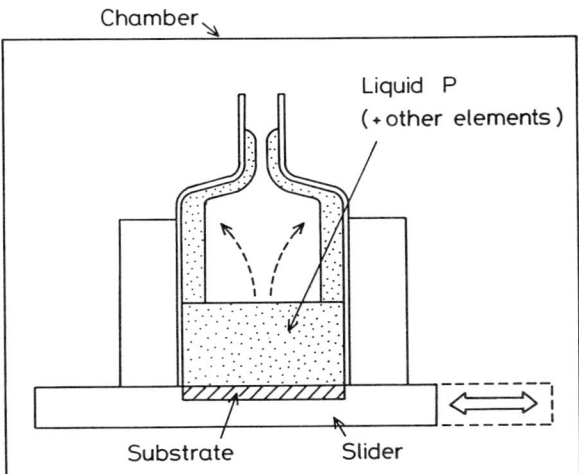

FIG. 17. Schematic drawing of a proposed system for solution growth using liquid phosphorus containing other elements such as indium or gallium.

would be suitable for mixed alloy materials such as indium gallium phosphide arsenide. The most important point would be how to control the solubility of each of the elements.

With regard to industrial applications, our new system would be very useful for the mass production of cheap indium phosphide crystals. The important point is how to improve the cooling system for condensing the phosphorus gas, but actually this wouldn't be very difficult if a water cooling system and an auxiliary heater were used.

Acknowledgments

The authors are indebted to Drs. T. Iizuka, M. Hirano, and I. Hayashi for fruitful and valuable discussions. They would like to express their thanks to Prof. T. Nishinaga and S. Kuma for the valuable discussion on the mechanism. They are also grateful to Messrs. T, Fujii and S. Ozawa for their advice and help in the experiments.

The present research was performed in Optoelectronics Joint Research Laboratory and was supported by the Agency of Industrial Science and Technology, Ministry of International Trade and Industry.

References

Adamski, J. A. (1983). Synthesis of Indium Phosphide. *J. Crystal Growth* **64**, 1–9.
Bonner, K. A., and Temkin, H. (1983). Preparation and Characterization of High Purity Bulk InP. *J. Crystal Growth* **64**, 10–14.

Farges, J. P. (1982). A Method for the "in-situ" Synthesis and Growth of Indium Phosphide in a Czochcralski Puller. *J. Crystal Growth* **59**, 665–668.

Hyder, S. B., and Holloway, C. J., Jr. (1983). In-situ synthesis and Growth of Indium Phosphide. *J. Electron. Mater.* **12**, 575–585.

Ilegems, M. (1975). Phase Studies in III–V, II–VI, and IV–VI Compound Semiconductor Alloy Systems. *Ann. Rev. Mater. Sci.* **5**, 345–371.

Inada, T., Fujii, T., Eguchi, M., and Fukuda, T. (1987). Technique for the Direct Synthesis and Growth of Indium Phosphide by the Liquid Phosphorus Encapsulated Czochralski Method. *Appl. Phys. Lett.* **50**, 86–88.

Inada, T., Fujii, T., Eguchi, M., and Fukuda, T. (1987). New Direct Synthesis Technique for Indium Phosphide Using Liquid Phosphorus. *J. Crystal Growth* **82**, 561–565.

Ozawa, S., and Fukuda, T. (1986). In-situ Observation of LEC GaAs Solid-liquid Interface with Newly Developed x-ray Image Processing System. *J. Crystal Growth* **76**, 323–327.

Savage, R. O., Anthony, J. E., AuCoin, T. R., Ross, R. L., Harsh, W., and Cantwell, H. E. (1984). High Pressure Direct Synthesis of Bulk Indium Phosphide. In "*Semi-Insulating III–V Materials*" (D. C. Look, and J. S. Blakemore, eds.). 171–174. Shiva Publishing Limited.

Shinoyama, S., Uemura, C., Yamamoto, A., and Tohno, S. (1980). Growth of Dislocation-free Undoped InP Crystals. *Jpn. J. Appl. Phys.* **19**, L331–334.

Terashima, K., Nakajima, H., and Fukuda, T. (1982). LEC Growth Technique for Homogeneous Undoped Semi-insulating GaAs Single Crystals with In-situ Melt Purification Process. *Jpn. J. Appl. Phys.* **21**, L452–454.

Waser, J. R. V. (1958). "*Phosphorus and Its Compounds*". Interscience Publishers, Inc., New York.

CHAPTER 4

InP Crystal Growth, Substrate Preparation and Evaluation

O. Oda, K. Katagiri, K. Shinohara, S. Katsura, Y. Takahashi, K. Kainosho, K. Kohiro, and R. Hirano

ELECTRONIC MATERIALS AND COMPONENTS RESEARCH LABORATORIES
NIPPON MINING CO., LTD.
SAITAMA, JAPAN

		List of Symbols.	94
I.		Introduction	94
II.		Polycrystal Synthesis	97
	1.	Various Polycrystal Synthesis	97
	2.	Horizontal Bridgman Technique	100
	3.	Synthesis by Solute Diffusion Technique	106
III.		Single Crystal Growth	111
	4.	Twinning Prevention	111
	5.	Reduction of Dislocation Density	115
	6.	Tin-Doped Single Crystals	123
	7.	Sulphur-Doped Single Crystals	125
	8.	Zinc-Doped Single Crystals	129
	9.	Semi-Insulating Single Crystals	130
	10.	Large-Diameter and Long-Size Single Crystals	136
IV.		Evaluation	138
	11.	Purity	138
	12.	Photoluminescence	140
	13.	Cathodoluminescence	143
	14.	Defects and Their Origins	146
V.		Substrate Preparation	148
	15.	Wafer Processing	148
	16.	Cutting	148
	17.	Lapping	151
	18.	Polishing	152
	19.	Cleaning	155
VI.		Evaluation	159
	20.	Damaged Layer	159
	21.	Surface Waviness	163
	22.	Wafer Inspection	166
VII.		Summary	167
		Acknowledgments	168
		References	169

Copyright © 1990 by Academic Press, Inc.
All rights of reproduction in any form is reserved.
ISBN 0-12-752131-3

List of Symbols

C/C	carrier concentration
T_{in}	temperature of the indium melt zone of a HB furnace
T_{max}	temperature of the synthesis zone of a HB furnace
T_p	temperature of the phosphorus vapor control zone of a HB furnace
g	solidified fraction
R_r	relative rotation number for LEC crystal growth
D_p	diffusion coefficient of phosphorus in indium melt
dT	axial temperature gradient
α	thermal conductivity of crystal
h	cooling constant of Newton's law
R	radius of crystal
τ_{CRSS}	critical resolved shear stress of doped crystal
τ°_{CRSS}	critical resolved shear stress of undoped crystal
dT/dz	axial temperature gradient in crystal or in LEC furnace
R_c	critical radius of crystal in which crystal is dislocation free
n	electron concentration
p	hole concentration
N_a	shallow acceptor concentration
N_{aa}	deep acceptor concentration
N_d	shallow donor concentration
N_{dd}	deep donor concentration
HB	horizontal Bridgman
SSD	synthesis by the solute diffusion
EPD	etch pit density
OEIC	optoelectronic integrated circuit
MISFET	metal-insulator-semiconductor field effect transistor
LD	laser diode
LED	light-emitting diode
APD	avalanche photodiode
LEC	liquid encapsulated Czochralski
SSMS	spark source mass spectroscopy
CL	cathodoluminescence
PL	photoluminescence
FWHM	full width of half maxima
LEED	low energy electron diffraction
AES	Auger electron spectroscopy
XPS	x-ray photoemission spectroscopy

I. Introduction

InP is a III–V compound semiconductor, which is a very promising material for increasing applications in optoelectronic communications, optoelectronic integrated circuits (OEICs), high-speed devices such as metal-insulator-semiconductor field effect transistors (MISFETs), and solar cells.

4. InP CRYSTAL GROWTH

InP was developed later than GaAs and GaP because of the difficulty in obtaining twin-free single crystals. However, since InP has some superior physical properties compared with other III–V compounds, the demand and application are being extended year by year. This extension was apparent after 1981 when optical fiber communication was made practical by the use of InP substrates.

InP is a direct-transition-type compound semiconductor, which has physical properties as shown in Table I. Due to these fundamental physical properties, InP has various characteristics as described next.

(1) The transmission loss of a quartz fiber used for optoelectronic communication has a minimum at the wavelengths of 1.3 μm and 1.55 μm. InP can be used in lattice matching with InGaAsP mixed crystals, the band gap of which can be fitted to the optimum wavelength (Onabe, 1982).

(2) The electron drift velocity of InP is larger than GaAs at higher electric fields ($\sim 10^4$ V/cm) (Nielsen, 1972), and InP is very promising for high-speed FETs with a high cut-off frequency.

(4) The thermal conductivity (Jordan, 1985) is larger than GaAs, so InP is more advantageous for integrated circuits that need larger power dissipation.

(5) The localized states density is smaller than that of GaAs, and it is easy to form n-type inversion layers so that InP is very promising for high-speed MISFETs (Hasegawa and Sawada, 1981).

(6) Since InP has a bandgap of 1.35 eV, it gives a high theoretical conversion efficiency for solar cells (Wysocki and Rappaport, 1960). InP is

TABLE I

PHYSICAL PROPERTIES OF InP

Crystal structure	zinc blend
Lattice constant (Å)	5.869
Density (g/cm^3)	4.8
Melting point (°C)	1062
Dissociation pressure (atm) at M. P.	27
Linear expansion coefficient (10^{-6}/deg)	4.5
Thermal conductivity (W/cm · deg)	0.70
Band gap (eV) at room temperature	1.35
Temperature dependence of the band gap (10^{-4} eV/deg)	-2.9
Optical transition	Direct
Specific dielectric constant	12.5
Intrinsic carrier concentration at R.T. (cm^{-3})	2×10^7
Electron mobility at R.T. (cm^2/V · s)	4500
Hole mobility (cm^2/V · s)	150
Intrinsic resistivity (Ω·cm)	8×10^7

TABLE II

VARIOUS InP AND THEIR APPLICATIONS

Dopant	Carrier concentration (cm^{-3})	Mobility (cm^2/V·s)	Resistivity (Ω·cm)	EPD (cm^{-2})	Application
Non	<10^{16}	3500 ~ 4500	≥0.15	<1 × 10^5	Source material or cover sheets for epitaxitial growth
Sn	0.5 ~ 6 × 10^{18}	1200 ~ 2500	0.5 ~ 6 × 10^{-3}	<5 × 10^4	Laser diodes, Light-emitting diodes
S	≥4 × 10^{18}	800 ~ 1600	≤1.8 × 10^{-3}	<5 × 10^3	Pin-photodiodes
				<500, DF area >6.5 cm^2	Avalanche photodiodes
Zn	≥3 × 10^{18}	50 ~ 60	<3 × 10^{-2}	<5 × 10^3	High-power laser diodes
	2 ~ 5 × 10^{16}	110 ~ 130	1 ~ 10	<5 × 10^4	Solar cells
Fe	—	—	≥10^7	<5 × 10^4	Highfrequency FETs, OEICs

more radiation resistant than GaAs and Si (Yamaguchi et al., 1984a, 1984b), so that it is promising as solar cells for application in space satellites (Weinberg and Brinker, 1986; Okazaki et al., 1988).

(7) It is easier to obtain low dislocation density crystals with low-level impurity doping (Seki et al., 1976; 1978).

(8) Semi-insulating crystals can be easily obtained by doping with iron, which forms a deep acceptor (Iseler, 1979).

Because of these characteristics, low resistivity InP is now used industrially for laser diodes (LDs), light-emitting diodes (LEDs), avalanche photodiodes (APDs) and PIN photodiodes. Fabrication of MISFETs and OEICs has also been extensively studied by using Fe-doped semi-insulating InP substrates. A promising application is radiation-resistant solar cells for the space industry. Table II shows variously doped InP and its application.

The investigation of InP polycrystal synthesis and single crystal growth has already been reviewed by several authors (Sekinobe and Morioka, 1983; Araki, 1984; Saito et al., 1985; Sasaki et al., 1985; Akai et al., 1985; Morioka et al., 1986; Oda and Katagiri, 1987). In this section, the industrial processing of InP is discussed along with the technical difficulties confronted in the research and development process and newly developed technology in our laboratory. Interesting technical developments by other researchers are also presented.

II. Polycrystal Synthesis

1. Various Polycrystal Synthesis

The melting point of InP is 1062°C, lower than that of GaAs, but the dissociation pressure of phosphorus is rather high at the melting point (25–27.5 atm) (Boomgaard and Schol, 1957). Because of this higher dissociation pressure, it is difficult to synthesize indium and phosphorus directly in the high-pressure crystal puller, as it is normally done in the case of GaAs. Therefore, high purity indium and high purity phosphorus are synthesized in a high-pressure chamber to InP polycrystal raw material and then melted in a high-pressure puller to grow single crystals.

In order to synthesize InP polycrystalline material, high-pressure horizontal Bridgman technique, high-pressure gradient freezing technique, synthesis by the solute diffusion technique (SSD technique), and direct synthesis in a high-pressure puller have been studied, as shown in Table III. To select the best polycrystal synthesis technique for industrial production, the following three requirements should be satisfied.

TABLE III
Polycrystal Synthesis Results Reported by Various Researchers

Technique			Carrier Concentration 77 K cm^{-2}	Mobility 77 K cm^2/V·s	Synthesis Temperature °C	Growth Rate mm/h	Synthesis Quantity g	Authors	Year
High Pressure Horizontal Bridgman	Two Zones		$2.0 \sim 13 \times 10^{14}$	$64000 \sim 138500$	1015	1.5	~ 400	J. A. Adamski	1982
	Three Zones		1.8×10^{16}	48235	1003	12	$150 \sim 400$	J. A. Adamski	1983
			3.2×10^{16}	37000	$1050 \sim 1200$	3	250	A. Yamamoto et al.	1981
			$1 \sim 3 \times 10^{14}$	$40000 \sim 60000$	1000	$3 \sim 6$	2000	D. Kusumi et al.	1985
High Pressure Gradient Freezing			$2 \sim 5 \times 10^{15}$	—	1100	$2 \sim 5$	110	W. A. Bonner	1981
			1.5×10^{15}	15000	1060	~ 1hr	1000	W. P. Allred et al.	1981
			$1 \sim 4 \times 10^{14}$	$20000 \sim 40000$	1100	$2 \sim 5$	100	W. A. Bonner et al.	1983
			$5 \sim 10 \times 10^{15}$	$15000 \sim 25000$	1090	—	1100	R. Coquille et al.	1983
			$2.7 \sim 8 \times 10^{15}$	41000	1100	$10 \sim 20$	1600	J. E. Wardill et al.	1983
			$2 \sim 6 \times 10^{15}$	63000	1062	5	2000	S. Yoshida et al.	1987

Method	Type	(300K)		Temp	Time		Author	Year
Synthesis by the Solute Diffusion	Conventional	$5 \sim 10 \times 10^{15}$ (300K)	3700 (300K)	~ 800	$1.8 \sim 2$	—	J. A. Marshall and K. Gillessen	1978
		3.1×10^{15}	40000	$900 \sim 975$	$1.1 \sim 4.9$	—	T. Sugii et al.	1979
		6.4×10^{14} (300K)	79100	910	—	—	E. Kubota and K. Sugii	1981
		$6 \sim 10 \times 10^{14}$	$77000 \sim 110000$	900	—	$10 \sim 30$	M. Ohoka et al.	1985
		$1.11 \sim 11 \times 10^{15}$	$14000 \sim 50000$	850	—	~ 20	W. Siegel et al.	1986
		3.6×10^{14}	126000	$800 \sim 1000$	5	$100 \sim 200$	E. Kubota et al.	1987
	Growth Rate Controlled	—	55800	$800 \sim 1000$	10	$100 \sim 200$	E. Kubota and K. Sugii	1984
Direct Synthesis	P Vapor Injection	—	—	1150	—	600	J. P. Farges	1982
		2.6×10^{15}	30330	1100	rapid	380	S. B. Hyder and C. J. Holloway Jr.	1983
		$3.2 \sim 7.5 \times 10^{14}$ (300K)	$3100 \sim 3700$ (300K)	—	—	3100	Y. Sasaki et al.	1985
		$4 \sim 10 \times 10^{15}$	—	—	$20 \sim 30$ (min)	12000	D. J. Dowling et al.	1986
		2.2×10^{15} (300K)	4190 (300K)	—	—	2200	S. B. Hyder and G. A. Antypas	1986
	Liquid Phosphorus Encapsulated Czochralski	5×10^{15} (300K)	3500 (300K)	$1000 \sim 1100$	rapid	~ 100	T. Inada et al.	1987

(1) Purity of polycrystal should be as high as possible. Industrially speaking, purity, measured as carrier concentration, must be at least less than 10^{16} cm^{-3}, preferably, less than 5×10^{15} cm^{-3}.
(2) Indium inclusions, that is unreacted indium, have to be as few as possible, preferably they should not be observed by optical microscopes.
(3) The batch quantity should be as large as possible, with high synthesis rate from the viewpoint of the production cost.

Considering these requirements, the horizontal Bridgman technique is now industrially accepted as the InP polycrystal synthesis method since the synthesis rate is reasonably fast, and it is possible to grow InP polycrystal with reasonable purity and with few indium inclusions.

The horizontal gradient freezing technique is a very high growth-rate synthesis method. Since this method requires a temperature higher than the melting point, the purity is normally degraded due to contamination by silicon from the quartz ampoule. Work on further purification has currently been reported (Yoshida et al., 1987). Synthesis by the solute diffusion (SSD) is a technique by which it is possible to obtain ultra-pure InP polycrystal because the growth can be performed at lower temperatures and the contamination by silicon from the quartz ampoule is less. The SSD method, however, is not applicable in industry since the synthesis rate is very slow, and the batch quantity is not sufficient, being 10–100 g. The SSD method is, therefore, investigated only at the lanboratory level and is far from commercial production at present. The SSD method, however, might become useful in producing ultra-high purity InP polycrystals to be used as raw materials for liquid-phase epitaxial growth of mixed crystals. The direct synthesis is a rather promising technique since it offers a possibility of producing purer InP crystals with a very rapid synthesis rate, and at a decreased production cost. For the direct synthesis, three methods have been investigated. These include the (1) phosphorus vapor injection technique, (2) the direct synthesis of indium and phosphorus in a very high-pressure chamber, and (3) the liquid phosphorus encapsulation method. These direct synthesis techniques have recently been extensively studied as shown in Table III. Also see Chapter 3 by Inada and Fukuda in this volume.

2. Horizontal Bridgman Technique

Polycrystalline InP is grown industrially by using a high-pressure synthesizer as shown in Fig. 1. The furnace has three zones, indium melting zone, synthesis zone, and phosphorus vapor control zone. High-purity indium in a quartz boat or a pBN boat, and the high-purity phosphorus ingot are sealed in a quartz ampoule and heated in the synthesizer. Each zone is heated up, for example, to the temperatures indicated in the figure. During heating, the

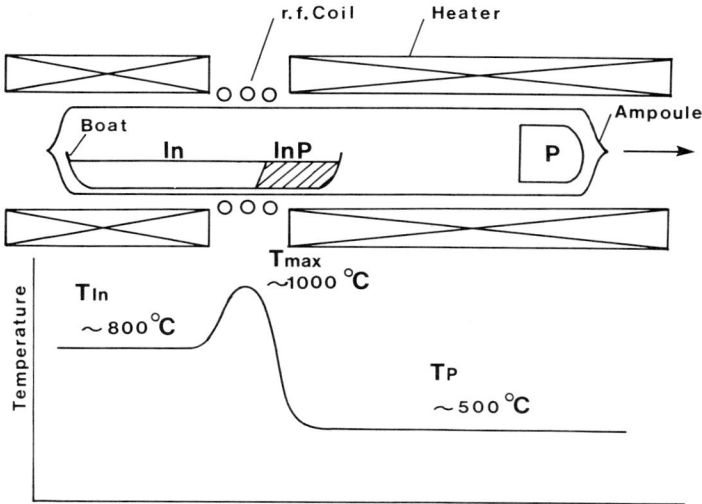

FIG. 1. Horizontal Bridgman furnace for the synthesis of InP polycrystals. Here, T_{In} is the temperature of the indium melt zone, T_{max}, that of the synthesis zone and T_P, that of the phosphorus vapor control zone.

vapor pressure of phosphorus is raised higher than the atmospheric pressure. Inert gas pressure such as N_2 or Ar is, therefore, introduced during the temperature increase in the synthesizer in such a way that the high vapor pressure of phosphorus in the ampoule can be balanced to prevent the ampoule rupturing.

In the horizontal Bridgman technique, to obtain InP polycrystals of reasonable purity, possessing few indium inclusions, and having a reasonable synthesis rate, the following parameters must be optimized:

(1) indium melt temperature
(2) phosphorus vapor pressure
(3) ampoule transfer rate
(4) synthesis zone temperature
(5) synthesis batch quantity
(6) boat material

If the indium melt temperature is higher, the solubility of phosphorus becomes larger. This makes the occurrence of indium inclusions less probable. On the other hand, due to the temperature increase, the purity of the synthesized polycrystalline material is degraded since indium reacts with the boat, and silicon is corporated into the indium melt.

The contamination by silicon takes place in two ways. One is the direct reaction of indium with the quartz boat. This reaction is represented as follows.

$$4\text{In}(\text{melt}) + 3\text{SiO}_2(\text{boat}) \Leftrightarrow 3\text{Si}(\text{in In melt}) + 2\text{In}_2\text{O}_3. \tag{1}$$

This reaction can be prevented by adding In_2O_3 (Pak et al., 1981). The addition of In_2O_3 would, however, increase the oxygen content in indium melt.

The other way that silicon contamination takes place is via the indium vapor. Even though the vapor pressure is low, it reacts with the ampoule wall and generates SiO gas. This SiO gas reacts with the indium melt, and silicon is incorporated. These reactions are represented as follows:

$$4\text{In}(g) + 6\text{SiO}_2 \Leftrightarrow 6\text{SiO}(g) + 2\text{In}_2\text{O}_3 \tag{2}$$

$$3\text{SiO}(g) + 2\text{In}(\text{melt}) \Leftrightarrow 3\text{Si}(\text{in In melt}) + \text{In}_2\text{O}_3. \tag{3}$$

Thermodynamic considerations of the SiO gas and the indium melt have been made by Janson (1983).

Because of this silicon contamination, it follows that the main impurity in InP polycrystals is silicon. Figure 2 shows the relationship between carrier concentration and Si content as obtained from spark source mass spectroscopy (SSMS) analysis of our polycrystalline InP synthesized under various conditions. Even though there is some data scattering, the linearity seen in the figure means that silicon is the main impurity in the InP polycrystals. Yamamoto et al. (1981) also reported that silicon is the main impurity by studying the segregation of doped silicon in InP polycrystals. Dean et al. (1984) concluded that the main donor was sulphur or native defects.

When the phosphorus vapor pressure is higher, InP polycrystal with less indium inclusions can be grown, but the possibility of ampoule rupture becomes greater, so the phosphorus vapor pressure must be less than 20 atm.

The ampoule transfer rate must be optimized at a rate at which the InP polycrystal grows with a minimum density of indium inclusions. During the polycrystal growth, the indium melt is saturated with phosphorus, and InP is precipitated from this saturated indium melt at the temperature of the liquidus line. If the ampoule transfer rate is faster than the precipitation rate, indium melt is unincorporated in the growth interface, and it forms indium inclusions. In Fig. 3., the relationship between the ampoule transfer rate and the indium inclusion occurrence is presented.

The synthesis-zone temperature affects the purity and the growth rate. When this temperature is increased, the saturated composition of phosphorus in the indium melt at the growth interface and the temperature of the liquidus line are increased so that the growth rate is increased, and indium inclusions

4. InP CRYSTAL GROWTH

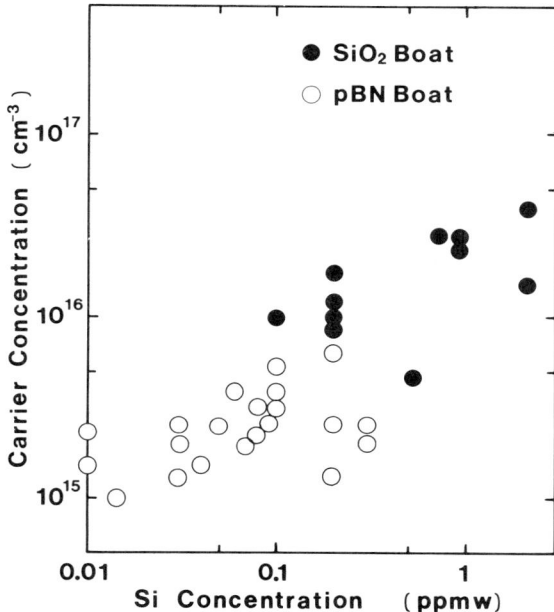

FIG. 2. Relationship between Si concentration and carrier concentration of InP polycrystals synthesized by using quartz boats and pBN boats by various synthesis conditions.

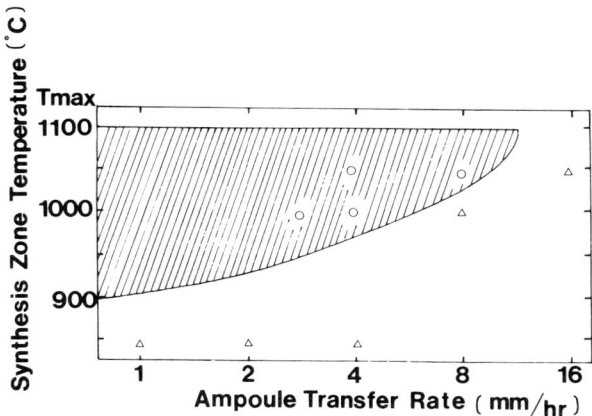

FIG. 3. Indium inclusion occurrence and the polycrystal synthesis conditions. In polycrystals synthesized in the hatched region, indium inclusions were hardly observed.

are decreased. In Fig. 3, the effect of the synthesis-zone temperature is shown with that of the ampoule transfer rate on the indium inclusion occurrence. The increase of the synthesis-zone temperature is, however, undesirable from the viewpoint of the purity as mentioned previously because of the silicon contamination. The results are shown in Fig. 4. The increase of the synthesis-zone temperature clearly increases the carrier concentration of the InP polycrystals.

By increasing the batch quantity, the purity can be slightly improved, as seen in Fig. 5. The purity improvement may be because the relative silicon contamination in the indium melt is decreased when the quantity is increased in the same volume of the quartz ampoule.

By optimizing the previously mentioned synthesis parameters, InP polycrystals with no indium inclusions can now be industrially produced as shown in Fig. 6. A typical 4 kg batch of polycrystal grown by the horizontal Bridgman technique is shown in Fig. 7.

Figure 8 shows the carrier concentration and mobility of InP polycrystals. As seen in the figure, when the quartz boat is used, the carrier concentration becomes more than 1×10^{16} cm^{-3}. However, the purity can be greatly improved when pBN boats are used; and the synthesis conditions are well optimized. The figure shows that, even by the HB technique, it is possible to

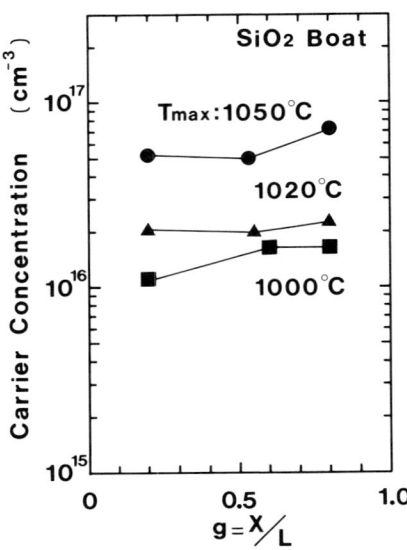

FIG. 4. Effect of synthesis zone temperature (T_{max}) on the carrier concentration of InP polycrystals synthesized in quartz boats. Here, $g = X/L$ means the solidified fraction.

4. InP CRYSTAL GROWTH

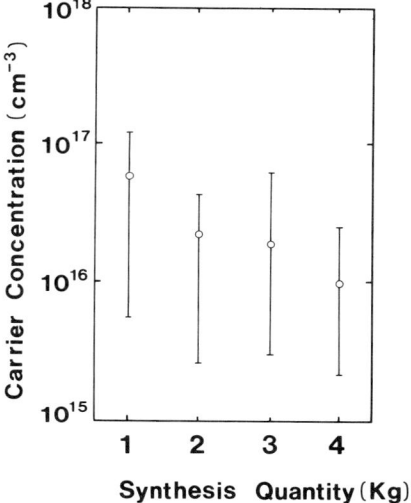

FIG. 5. Carrier concentrations of InP polycrystals of various single batch quantities. Synthesis was performed in quartz ampoules by the HB technique. Error bar means the maximum and minimum carrier concentrations of at least five lots.

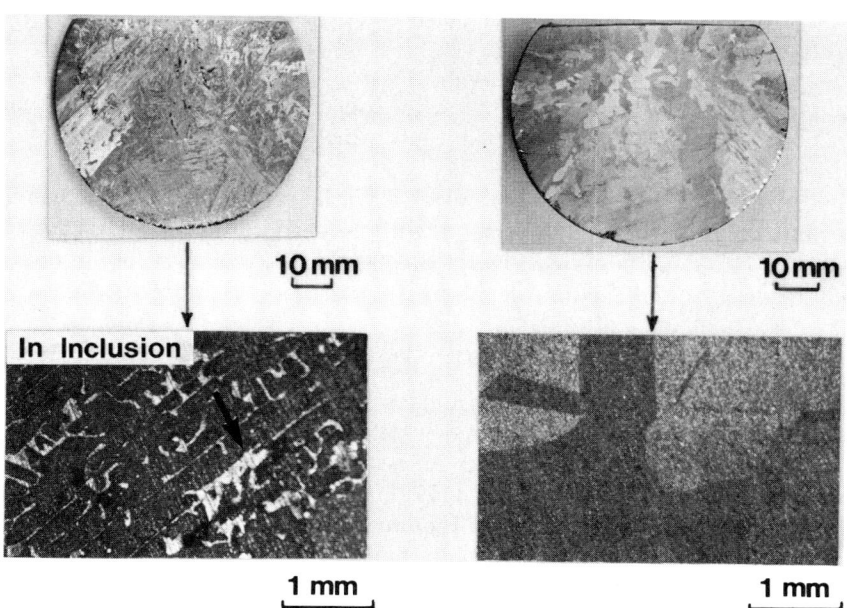

FIG. 6. Indium inclusions in InP polycrystals. (a) InP polycrystal cross section in which indium inclusions are observed as white parts. (b) InP polycrystal cross section in which no indium inclusions are observed.

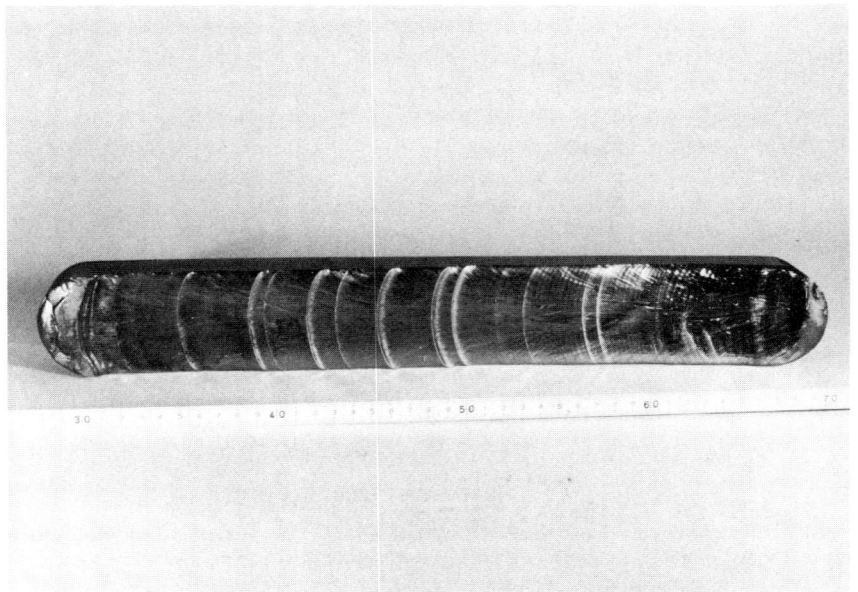

Fig. 7. 4 kg HB InP polycrystal photograph.

grow high-purity polycrystals, the carrier concentration of which is less than 2×10^{15} cm^{-3} and the mobility of which is between 40,000 and 70,000 cm^2/V-sec (77° K).

Figure 9 shows the carrier concentration variation from lot to lot. It can be seen that high-purity InP polycrystals with the carrier concentration range of $1 - 3 \times 10^{15}$ cm^{-3} are grown very consistently. The behavior of impurities will be discussed in Section 11.

3. Synthesis by Solute Diffusion Technique

The synthesis by the solute diffusion technique (SSD) (Kaneko et al., 1973) has been applied for the growth of GaP (Gillessen and Marshall, 1976a; 1976b) and for InP (Marshall and Gillessen, 1978). To produce high-purity InP polycrystals, Kubota and Sugii (1981) have tried to apply this technique. The principle is shown in Fig. 10. Indium and phosphorus are sealed under vacuum in a quartz ampoule, and phosphorus is heated at a certain temperature in order to obtain a phosphorus vapor pressure of less than one atmosphere. The phosphorus vapor is dissolved in the indium melt to the saturation composition. Since the temperature at the bottom of the crucible is lower than that of the indium melt surface, dissolved phosphorus is diffused

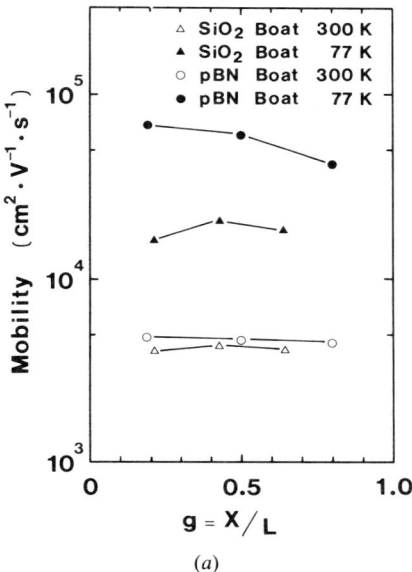

(a)

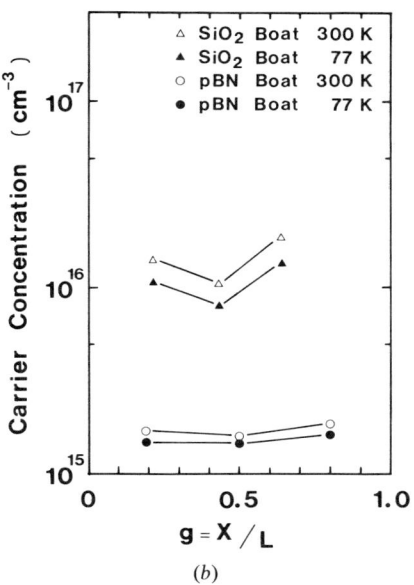

(b)

FIG. 8. Carrier concentrations and mobilities of InP polycrystals synthesized by using quartz boats and pBN boats. Here, $g = X/L$ means the solidified fraction.

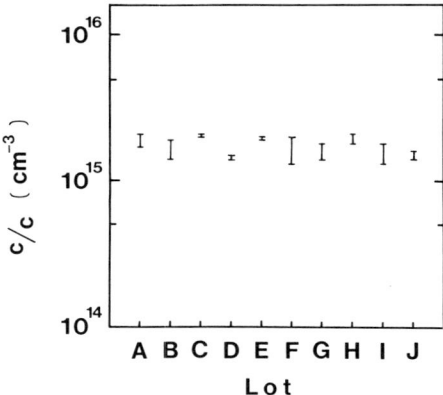

FIG. 9. Carrier concentration variation of InP polycrystals from lot to lot. Synthesis was performed by using pBN boats. The measurement was carried out at room temperature, and the error bar means the maximum and minimum values of six samples taken from head to tail of each synthesis lot.

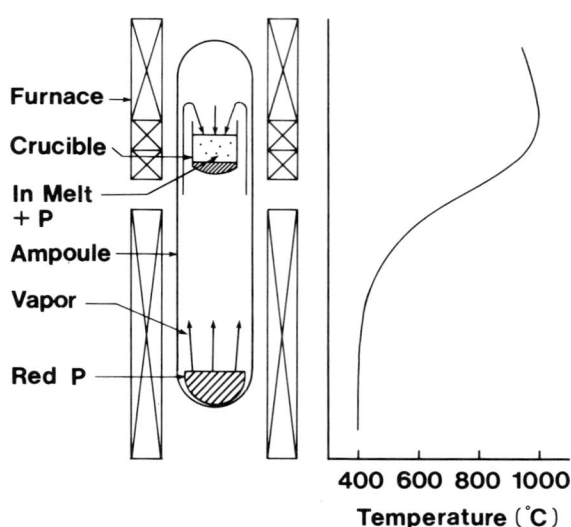

FIG. 10. Principle of the synthesis by the solute diffusion (SSD) technique. Phosphorus vaporized from the bottom of the ampoule is dissolved in the indium melt and InP precipitates at the bottom of the crucible.

from the melt surface toward the bottom of the crucible. When the composition of phosphorus is increased at the bottom of the crucible, and it exceeds the saturation composition, InP polycrystals are grown from the bottom of the crucible. Figure 11 shows a cross section of our SSD polycrystal.

The SSD technique has an advantage in that the growth is at lower temperatures, thus enhancing the purity. Figure 12 shows the carrier concentration and mobility data for SSD polycrystals grown in our laboratory. In the figure, the data on HB polycrystals are also shown for comparison. It can be seen that the SSD technique produces InP polycrystals with higher purity.

The SSD technique has, however, a disadvantage in that the growth rate of the SSD method is very slow compared with that of the HB system. The reason for this is because the phosphorus diffusion in the indium melt is the predominant mechanism of the SSD growth (Kaneko et al., 1973), and the diffusion coefficient is very small. We have studied the growth rate for the SSD technique by using a large diameter (70 mmϕ) crucible with a batch quantity of about 700 g and by interrupting the growth (Kainosho et al., 1985). The growth rate was very small as shown in Fig. 13. The diffusion coefficient of phosphorus in the indium melt at 900°C, D_p, calculated from the data was $2 - 9 \times 10^{-5}$ cm^2/sec.

FIG. 11. Cross section of a SSD polycrystal (diameter; 70 mm). (Kainosho et al., 1985).

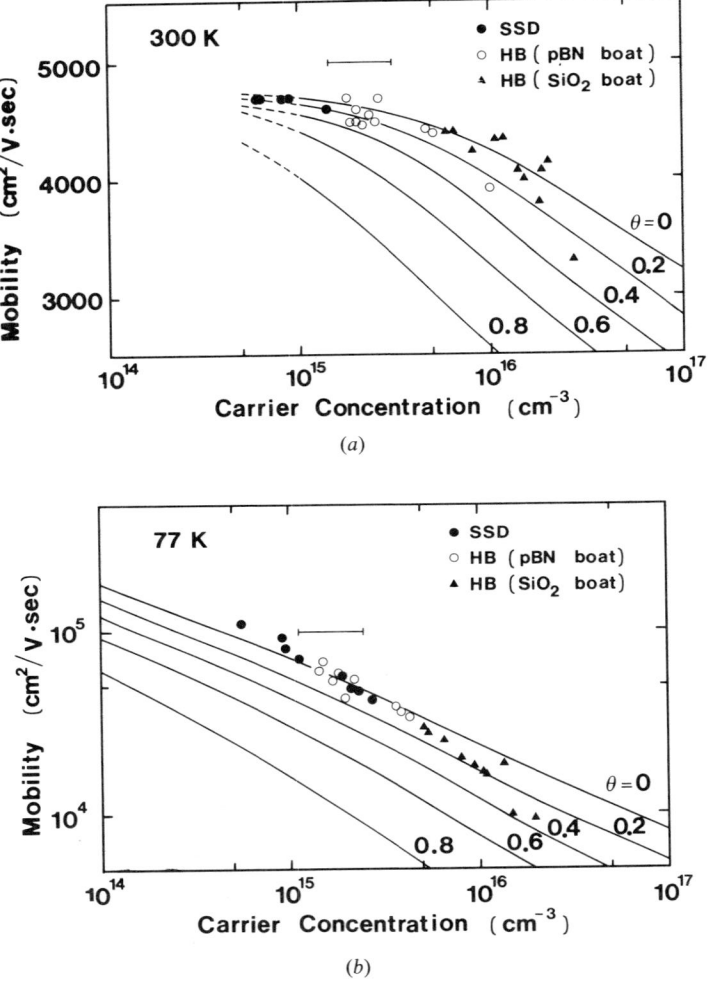

FIG. 12. Electrical properties of SSD polycrystals. The data of HB polycrystals are also shown for comparison. The solid lines are the theoretical calculation (Walukiewicz et al., 1980). Here, θ means the compensation ratio (acceptor concentration N_a/donor concentration N_d).

Kubota and Sugii (1984) have tried to improve the previously mentioned disadvantage of the SSD technique. In their procedure, which they call growth-rate controlled SSD(GRCSSD), they have increased the axial temperature gradient, dT, between the surface of the indium melt and the solid/liquid interface. When dT is increased, the net flux of phosphorus in the indium melt is increased because the flux is proportional to $D_p \cdot dT$. For large

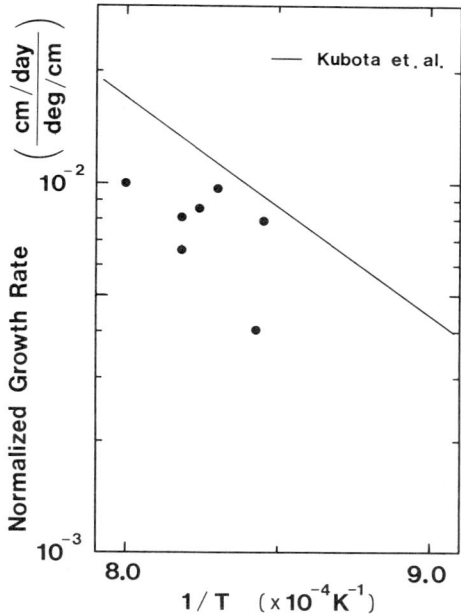

FIG. 13. Normalized growth rate and growth temperature. Normalized growth rate is the growth rate divided by the axial temperature gradient. SSD synthesis was performed with a batch of 700 g. The data are compared with the data (the solid line) by Kubota et al., (1984).

scaling of the SSD technique, it is, however, necessary to achieve the large temperature gradient for a large diameter crucible and for this reason, a precise temperature-controlled SSD furnace must be developed. When this is realized, the SSD technique will become a valuable technique to industry.

III. Single Crystal Growth

4. Twinning Prevention

The details of the LEC technique are explained by Farges in this volume. In the LEC technique, B_2O_3 glass is used as an encapsulant to prevent the loss of phosphorus from the melt as shown in Fig. 14. Even though the LEC technique is advantageous for growing large-diameter single crystals with high single crystal yield, twinning is a large problem in the case of InP. As shown in Table IV, InP has a very small stacking fault energy so that twin boundaries are easily formed during the crystal growth. It is especially so when $\langle 100 \rangle$ orientated single crystals are grown.

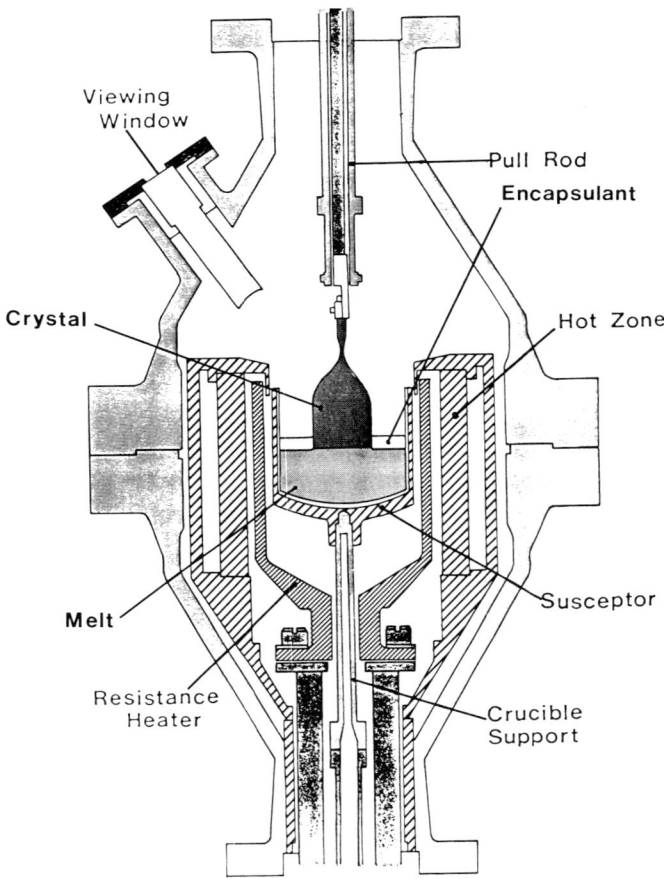

FIG. 14. Typical cross section view of the LEC technique.

TABLE IV

STACKING FAULT ENERGY OF VARIOUS COMPOUND SEMICONDUCTORS (Gottshalk et al., 1978).
(meV/atom)

GaSb	GaAs	InSb	GaP	InAs	InP
53	47	43	33	30	17

There are many factors that could be responsible for the twinning, and these factors must be optimized before one can grow single crystals without any twinning.

B_2O_3 purity. It is well known that the water content in encapsulant B_2O_3 varies from lot to lot. B_2O_3 glass with specified water content is commercially available. From these glasses, one needs to select the water content optimal to his high pressure puller. Blom and Zwichler (1973), Bachmann et al., (1975), and Bonner (1981) recommend use of B_2O_3 encapsulants with lower water content for the twinning prevention.

Rotation conditions (crystal rotation and crucible rotation). The melt convection phenomenon is theoretically explained by Donaghey (1980) as a function of crystal rotation and crucible rotation. By controlling the crystal rotation and crucible rotation, the solid/liquid interface shape can be changed. Figure 15 shows our experimental results. Here, the solid/liquid interface shape has been determined by striation observation. By decreasing the relative rotation number, i.e. the difference between crucible rotation and seed rotation numbers, the solid/liquid interface shape can be made flatter. From the viewpoint of twinning prevention, our experience is that the convex shape toward the bottom of the crucible is preferable. Optimization is very important, however, since crystal bumping phenomenon takes place if the interface is too convex. Recently, Derby and Brown (1986a; 1986b) and Sasaki et al., (1987) have reported theoretical considerations on the solid/liquid interface shapes from the viewpoint of the thermal balance calculation.

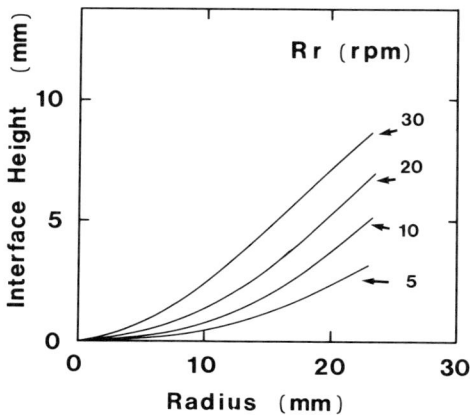

FIG. 15. Solid/liquid interface height and the relative rotation number (R_r). The relative rotation number means the difference between the crucible rotation number and the seed rotation number.

Temperature distribution. For growing twin-free single crystals, a larger axial temperature gradient is desired. The temperature gradient can be controlled in various ways: by changing the crucible heater structure (Shimada et al., 1984), by changing the inert gas pressure (Shinoymama and Uemura, 1985; Shinoyama, 1985), or by changing the amount of B_2O_3 (Kohda et al., 1985). Figure 16(a) shows the case when the axial temperature gradient is changed by using a thermal baffle (Katagiri et al., 1986) for growing three inch InP. It is preferable to increase the temperature gradient for preventing twinning. However, the increase is undesirable for decreasing dislocation density (EPD) as explained in Section 5. The axial temperature distribution is a parameter which should also be taken into account. The radial temperature distribution on the surface of the InP melt affects the solid-liquid interface shape. When it is small, the interface shape becomes flatter, and when it becomes larger, the interface shape becomes convex toward the bottom of the crucible. The radial temperature distribution can also be controlled by changing various factors such as rotation conditions, inert gas pressure, B_2O_3 amount, and the application of a magnetic field.

Temperature fluctuation. The temperature fluctuation is also a factor that affects the occurrence of twinning. Temperature fluctuations are influenced by various parameters such as the atmospheric gas pressure, the structure of the hot zone, and the B_2O_3 thickness. A magnetic field is also effective in decreasing the temperature fluctuation (Osaka and Hoshikawa, 1984). Figure 16 (b) shows a case of how the temperature fluctuation can be decreased by using a thermal baffle for growing three inch diameter InP. It is very difficult to minimize the temperature fluctuation while meeting the other requirements, but this must be achieved for the prevention of twinning.

Crystal shape. The crystal shape, especially the shoulder shape, is a factor that affects twinning. Normally when twinning takes place, it begins from the shoulder part as indicated in Fig. 17. Since the normal of twin planes has a $\langle 111 \rangle$ orientation, twin planes are of two different kinds, the In plane (A plane) toward the growth direction or the P plane (B plane) toward the growth direction. Angle relationships between twin planes and pulling directions are shown in Fig. 17. Bonner (1981) and Shinoyama (1985) recommended growing the shoulder with an angle of less than 19 degrees toward the growth direction when $\langle 111 \rangle$ single crystals are grown. If the crystals are grown with such shallow angles, however, the yield of crystals with the desired diameter becomes lower since the shoulder length becomes long. One can, however, grow single crystals without twins if the starting growth angle is less than 19 degrees, and if the angle is increased gradually as seen in Fig. 18.

Melt composition stoichiometry. The melt composition also seems to be important for preventing twinning. If the intial melt composition deviates from stoichiometry during the crystal growth, the deviation becomes larger and larger as the growth proceeds (see in Fig. 19). In the case of InP, normally the starting material stoichiometry deviates toward the indium-rich side. If this deviation is too large, the melt composition is substantially changed toward the indium-rich side, and twinning and/or polycrystallization takes place easily because of supercooling. This is one of the main reasons why indium inclusions in raw material InP polycrystals should be avoided.

Growth rate. Decreasing the growth rate is a way to prevent twinning, but this is not recommended since it also decreases the number of crystals that can be grown each month.

Optimization of the previous factors is necessary for establishing the technology to grow twin free $\langle 100 \rangle$ or $\langle 111 \rangle$ single crystals reproducibly. Because of the tendency of InP to twin, any factor causes twinning to take place more easily. Figure 18 shows our single crystals grown by the LEC technique industrially.

5. Reduction of Dislocation Density

In the case of the LEC technique for growing single crystals, it is absolutely necessary to have a certain axial temperature gradient during the growth. This necessarily induces thermal stresses in the growing crystals, and these thermal stresses are believed to be a main reason why dislocations are induced in the crystal. Many theoretical considerations and experimental results for this subject have been published by several researchers (Jordan *et al.*, 1980; 1981; 1984; 1985; 1986; Duseaux and Jacob, 1982; Brasen and Bonner, 1983; Szabo, 1985; Kobayashi and Iwaki, 1985).

A simple theoretical equation for estimating the EPD generation can be deduced as follows:

$$\left.\frac{dT}{dz}\right|_{max} \leq 1.8 \times 10^{-5} \frac{\tau_{CRSS}}{\alpha h^{1/2} R^{3/2}} \left(1 - \frac{1}{2} hR\right) \quad (4)$$

Here, α is thermal conductivity of crystal, h is cooling constant of Newton's law, R is the radius of crystal, τ_{CRSS} is a critical resolved shear stress, and dT/dz is the axial temperature gradient in crystal. For the deduction of this equation, we have supposed that (1) crystal is semi-infinitely long, (2) Newton's cooling law is predominant, and (3) solid-liquid interface is flat.

Figure 20 shows the calculation results of Eq. (4). The unknown constant in Eq. (4) was fitted to the experimental results obtained by Shinoyama (1980). In Fig. 20, the empirical equation $(dT/dz) Rc^2 = 124(\text{deg cm})$, deduced by Shinoyama is also shown as a dotted line. Here, Rc is a radius of crystal in which crystal is dislocation free. It may be discerned that, by assuming that the critical resolved shear stress is 0.8 MPa (at 700°C, from the

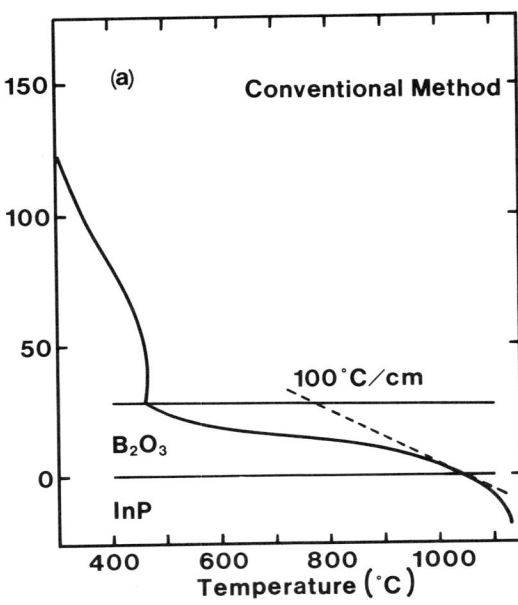

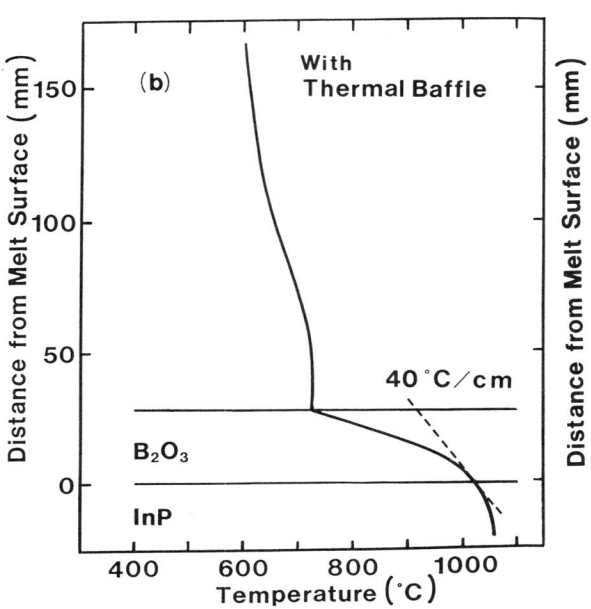

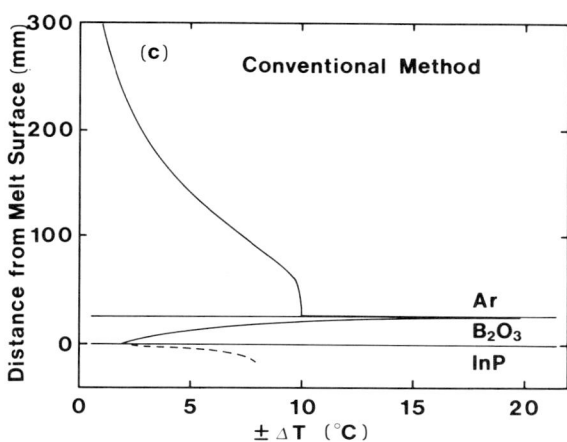

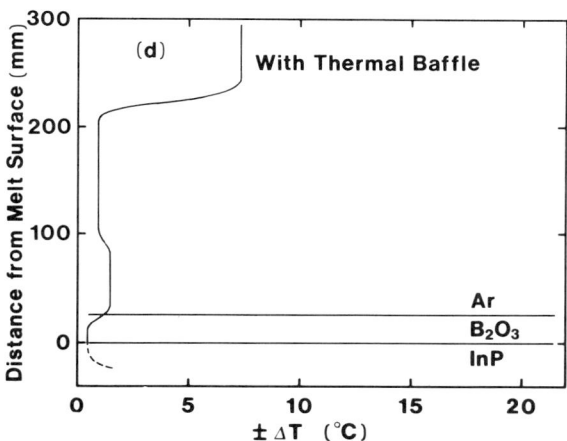

FIG. 16. Axial temperature gradients (a, b) (Katagiri et al., 1985) and the temperature fluctuations (c, d) in the case of conventional method and in the case when a thermal baffle was used for growing three-inch diameter InP.

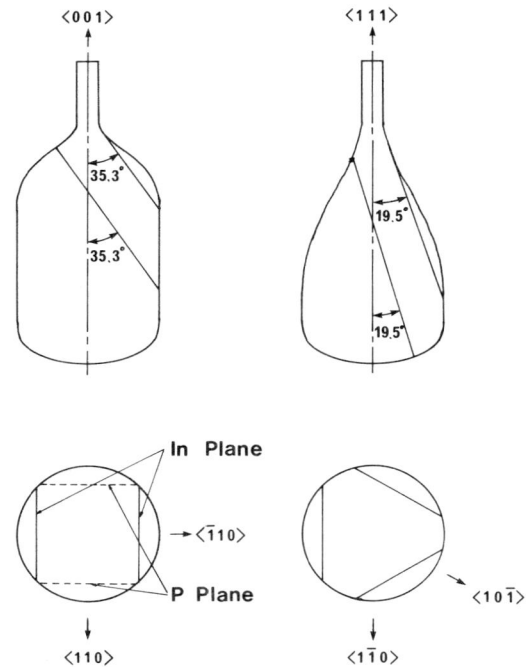

FIG. 17. Twinning planes relative to the crystal pulling directions.

data of Brasen and Bonner, 1983) and only by fitting the cooling constant, h, of Eq. (4) to Shinoyama's data, the curve of Eq. (4) becomes very similar to Shinoyama's empirical equation. In Fig. 20, our experimental results of sulphur-doped single crystals are also plotted.

From Fig. 20, it may be discerned that with a normal temperature gradient (40°C/cm), it is possible to grow dislocation-free crystals when the crystal diameter is small. Actually, Shinoyama et al., (1980) have reported the growth of dislocation-free single crystals with the diameter of 20 mm. When the crystal diameter becomes larger, however, the temperature gradient to grow dislocation-free single crystals becomes quasi-exponentially small. For example, for obtaining two inch diameter dislocation-free crystals, dT/dz must be lower than 5°C/cm, and for three inch diameter dislocation-free crystals, dT/dz must be lower than $2 \sim 3$°C/cm. In reality, when dT/dz is lower than about 20°C/cm, because of the seed crystal decomposition or because of twinning occurrence, crystal growth itself becomes impossible.

The vertical Bridgman or horizontal Bridgman techniques are, therefore, promising techniques for obtaining low-dislocation-density single crystals,

4. InP CRYSTAL GROWTH

FIG. 18. InP single crystals. Inside from the left to the right: three-inch diameter (100); length 160 mm; two-inch diameter (111); length 100 mm; two-inch diameter (100); length 80 mm. Foreground: two-inch diameter (100); length 210 mm.

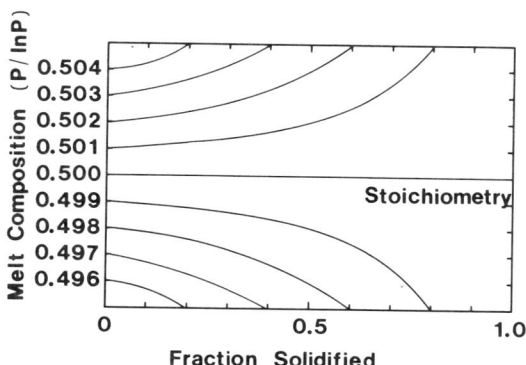

FIG. 19. Melt composition variation during crystal growth. The calculation was carried out by supposing that there is no phosphorus loss during crystal growth. If there is a significant loss of phosphorus, all lines are deviated toward the indium rich side.

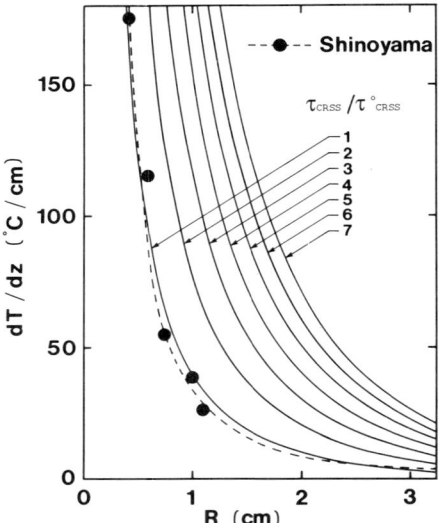

FIG. 20. Axial temperature gradient (dT/dz) and the critical crystal radius (R), in which dislocation-free single crystals are obtained, calculated according to Eq. (14). The calculation was made by changing τ_{CRSS}. τ_{CRSS}° is the resolved critical shear stress of undoped InP single crystal. Black points are the data by Sinoyama (1981), and the dotted line is the empirical law of Shinoyama.

since by these techniques, the temperature gradient can be reduced up to about 5°C/cm. The other way to decrease the EPD is to decrease the resolved critical shear stress in Eq. (4). This was first found by Seki et al., (1976) in the case of Zn, S, or Te doping in InP. After this discovery, various dopants have been studied for decreasing the EPD of various compound crystals as shown in Table V.

This effect of dopants in decreasing the EPD is referred to as "impurity hardening." Even though various explanations for this effect have been proposed, the theoretical work is still continuing. Seki et al., (1976) tried to explain this impurity hardening from the binding energies of the matrix elements and dopant elements. Ehrenreich and Hirth (1985) claim that this is because of dislocation pinning by impurities as in the case of metal hardening. The deviation from stoichiometry due to impurity doping is also considered to be the reason of the EPD reduction. Jordan and Parsey (1986) have theoretically examined the thermal stresses by changing the CRSS value in order to explain the impurity hardening effect.

Critical resolved shear stress measurements of sulphur-doped InP have also been reported by Brasen and Bonner (1983) and Gall et al., (1986). Even though the definitive explanation for this impurity hardening phenomenon is

TABLE Va
Impurity Hardening Studies Made By Various Researchers (Single Doping)

Dopant	Diameter (mm)	Length (mm)	Direction	Doping Level (cm^{-3})	EPD (cm^{-2})	Mobility 300 K (cm^2/V·s)	Authors	Year
Zn	20 ~ 30	—	111	1.4×10^{18}	500 ~ 900	—	Y. Seki et al.	1976
S				5.4×10^{18}	0	—		
				2.0×10^{18}	$(2 \sim 3) \times 10^2$	—		
				4.7×10^{18}	0	—		
Te				3.0×10^{18}	800 ~ 1500	—		
				1.0×10^{19}	300 ~ 400	—		
Ge	~ 25	~ 10	111	$\geq 1 \times 10^{18}$	D.F.	700 ~ 900	G. T. Brown et al.	1981
S	~ 20	~ 40	111	$\geq 2 \times 10^{18}$	D.F.	—	P. J. Roksnoer and M. M. B. Van Rijbroek-van den Boom	1984
Zn	~ 20	~ 40	111	$\geq 1 \times 10^{18}$	D.F.	—		
Ga	20 ~ 25	50 ~ 70	111	$(5 \sim 8) \times 10^{18}$	D.F.	—	S. Tohno and A. Katui	1986
Sb				$(2 \sim 3) \times 10^{18}$	D.F.	—		
As				$(4 \sim 5) \times 10^{18}$	D.F.	—		
Ga	20	60	111	3×10^{20}	D.F.	—	A. Katsui et al.	1986
S	40	—	001	$\geq 6 \times 10^{18}$	D.F.	—	J. P. Farges and C. Schiller	1987
	37	—	001	$(2 \sim 3) \times 10^{18}$	D.F.	—		
	37 ~ 42	—	111	$(2 \sim 3) \times 10^{18}$	D.F.	—		

TABLE Vb

IMPURITY HARDENING STUDIES MADE BY VARIOUS RESEARCHERS (DOUBLE DOPING)

Dopant	Diameter (mm)	Length (mm)	Direction	Doping Level (cm^{-3})	EPD (cm^{-2})	Mobility 300 K (cm^2/V·s)	Authors	Year
S,Zn	30	80	111	$3 \times 10^{18} \sim 3 \times 10^{19}$	$10^2 \sim 10^3$	—	A. A. Ballman et al.	1983
Ga,Sb	20 ~ 25	60 ~ 80	111	5.0×10^{19}(Ga) 1.5×10^{18}(Sb)	D.F.	2900 ~ 3300	S. Tohno et al.	1984
Ge,S	—	—	111	$4 \times 10^{18} \sim 1.7 \times 10^{19}$	D.F.	—	B. Cockayne et al.	1986
Ga,As	20 ~ 25	50 ~ 70	111	$\geq 4.0 \times 10^{19}$	D.F.	—	S. Tohno and A. Katsui	1986
Ga,Sb				$\geq 1.0 \times 10^{19}$	D.F.	—		
Ga,As,Sb				$\geq 2.5 \times 10^{19}$	D.F.	—		
Ga,Sb	25	—	111	$\geq 8 \times 10^{18}$	D.F.	3700	A. Katsui and S. Tohno	1986
Fe,Ga	50	—	001 111	1.2×10^{19}(Ga)	D.F.	—	R. Coquille et al.	1987

not yet given, impurity hardening has become a technique accepted in industry for obtaining low dislocation density crystals. In Fig. 20, the curves calculated from Eq. (4) are shown by changing the value of τ_{CRSS}. From the figure, it can be seen that if dT/dz is 40°C/cm, the attainable temperature gradient in the actual LEC growth as shown in Fig. 16, two-inch diameter dislocation-free crystals can be grown if the τ_{CRSS} is seven times larger than that of undoped τ_{CRSS}. In this way, if the τ_{CRSS} can be increased up to several times more than that of undoped InP, one can grow dislocation-free single crystals of the diameter predicted from Fig. 20. The actual results will be shown in Sections 7 and 8.

For reducing the EPD of Fe-doped semi-insulating InP, isoelectronic inpurity doping has been studied by several authors, as shown in Table V. Since isoelectronic dopants such as Ga, As, and Sb do not affect the electrical properties, they are convenient for EPD reduction in semi-insulating InP. In order to reduce the EPD effectively by these dopants, however, they have to be doped with one order of magnitude higher doping level than Zn and S. Because of the different segregation coefficients for each isoelectronic element, in order to decrease the EPD from top to tail of the grown crystal, two or three isoelectronic elements have been used simultaneously.

6. Tin-Doped Single Crystals

Tin-doped single crystals are mainly used for laser diodes. Since tin does not have any effect on reducing the dislocation density, the EPD level of tin-doped crystals grown industrially is in the range of 1×10^4 to 5×10^4 cm^{-2}. For reducing the EPD, as mentioned in Section 5, the decrease of the axial temperature gradient is very effective (Katagiri et al., 1985). If the solid-liquid interface shape is made flatter, this also has an effect on reducing the EPD. By changing rotation conditions, the solid-liquid interface shape can be controlled as shown in Fig. 15. Application of a magnetic field also provides a way to control the solid-liquid interface shape. Reducing the axial temperature gradient and making the solid-liquid interface flatter are effective for the reduction of the EPD, but these conditions are contradictory to the prevention of twinning. It is therefore necessary to find a compromise.

Figure 21 shows an example of EPD distributions from top to tail of a tin-doped crystal. It can be seen that the EPD distribution is not always a strong W-shape as is often seen in the case of LEC-grown GaAs and as is predicted by the theoretical thermal stress calculation. Another point that should be noted is that in the case of InP single crystals, the feature of the EPD distribution is quite different from the case of GaAs (Chen and Holmes, 1983). Even though the EPD level is nearly the same as LEC GaAs, strong lineages and cell structures, which are inevitably observed in GaAs, are not

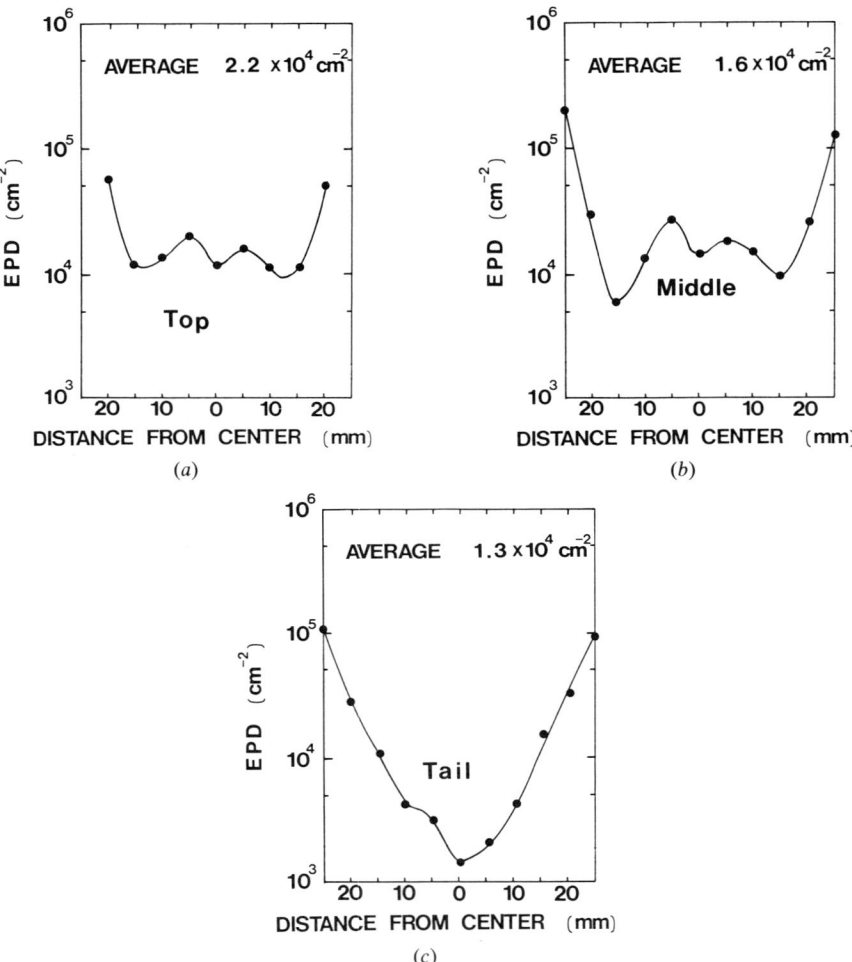

FIG. 21. EPD distribution of tin-doped two-inch diameter (100) InP. Each black point is the average value measured toward $\langle 011 \rangle$, $\langle 010 \rangle$, and $\langle 01\bar{1} \rangle$ directions.

observed in InP. In InP, all etch pits are distributed in isolation with each other. This phenomenon is not yet clearly explained but there is a possibility that it is concerned with the ionicity of the crystal. In the case of GaAs, the ionicity is very weak, and As atoms can be situated in Ga sites as antisite defects. In the case of InP, it is hardly possible that P atoms are situated in In sites. Because of these essential physical differences, dislocation generation-multiplication mechanisms would not be similar between InP and GaAs.

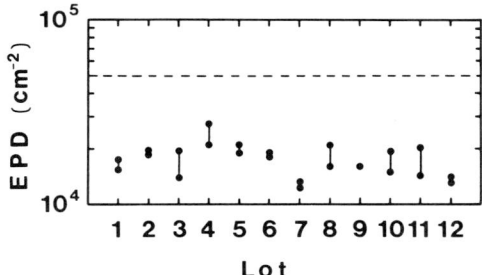

FIG. 22. Average EPD data variation from lot to lot of tin-doped two-inch diameter InP. Three wafers from top, middle, and tail parts are measured for each ingot lot. For each wafer, the EPD was measured for 13 points including $\langle 011 \rangle$, $\langle 010 \rangle$, and $\langle 01\bar{1} \rangle$ directions. The dashed line shows a commercially specified EPD.

Figure 22 gives the EPD level variation from lot to lot. It can be seen that tin-doped crystals with an EPD level less than 3×10^4 cm^{-2} are routinely grown.

Figure 23 shows the EPD distribution of tin-doped three-inch diameter InP. By decreasing the axial temperature gradient, the EPD can be decreased to an average EPD level of less than 5×10^4 cm^{-2}. The decrease of the axial temperature gradient is thus effective in decreasing the EPD.

7. Sulphur-Doped Single Crystals

Sulphur-doped crystals are mainly used for photodetectors. In the case of photodetectors, dislocations act as recombination centers, and they degrade the detection efficiency of photodectors. As already explained, sulphur has a large effect on reducing the EPD. Figure 24 shows how the EPD is decreased as a function of the carrier concentration for various diameter crystals and for different pulling directions. Normally, when the carrier concentration exceeds a certain critical value, the EPD is decreased, and when the carrier concentration exceeds a certain other higher value, fully dislocation-free areas begin to be obtained. These critical values are different for various diameters, lengths, and pulling directions as seen in Fig. 24.

The relationship between the average EPD and the carrier concentration can, however, be influenced by crystal growth conditions such as the solid-liquid interface, the axial temperature gradient, and the pulling rate. Figure 25, for instance, shows how the relationship is changed by changing the crucible rotation number.

To obtain whole area dislocation-free crystals, the sulphur concentration must be made larger, as seen from Fig. 20, to increase the τ_{CRSS}. If the doping concentration is increased, however, the grown crystals are easily twinned so

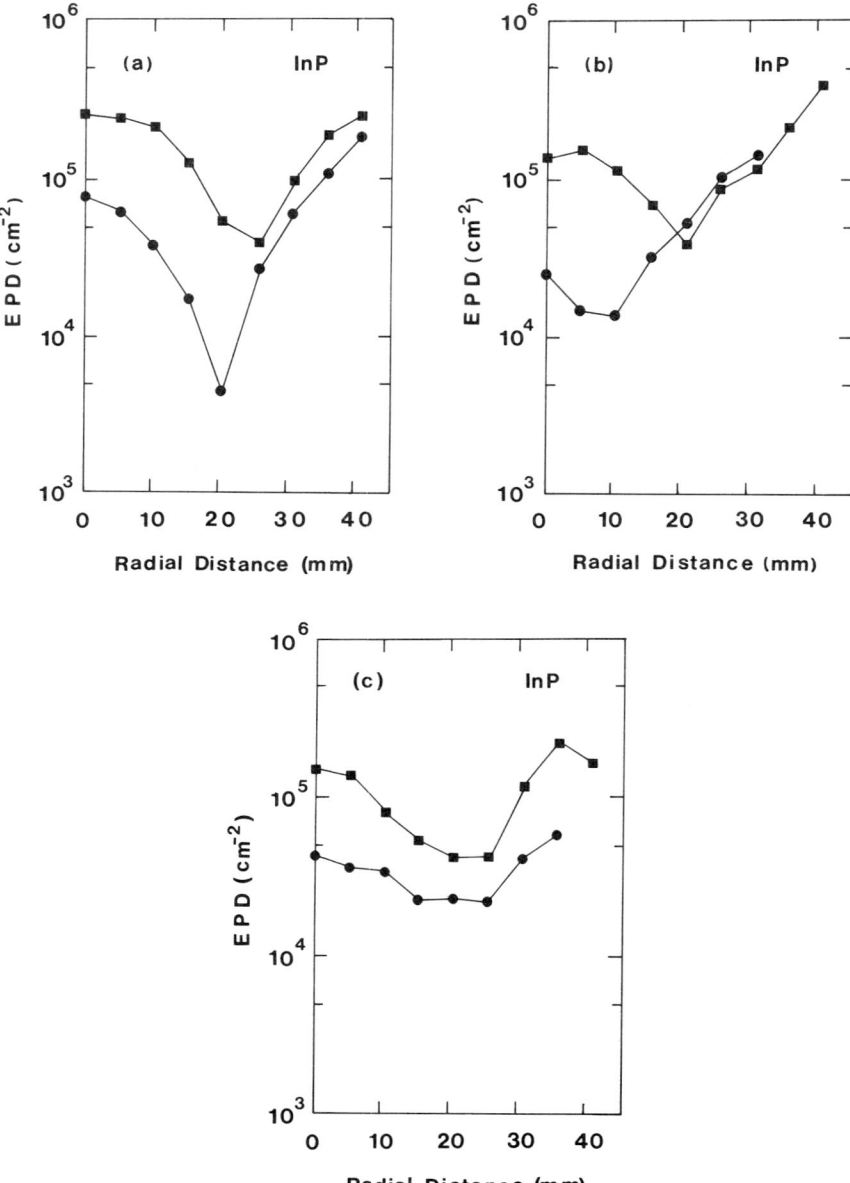

FIG. 23. EPD distribution of three-inch diameter tin-doped InP. Two crystals are compared. One is that grown by a conventional method (■), and the other is that grown with a thermal baffle (●) (Katagiri et al., 1985, IOP Publishing Limited).

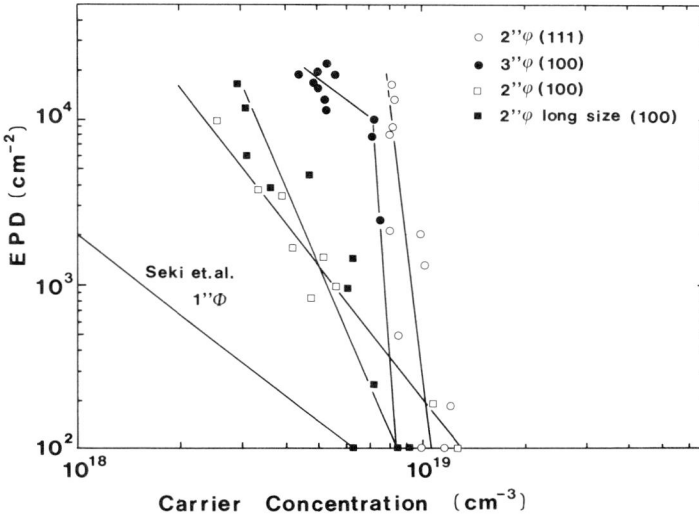

FIG. 24. EPD variation of sulphur-doped InP single crystals as a function of carrier concentration.

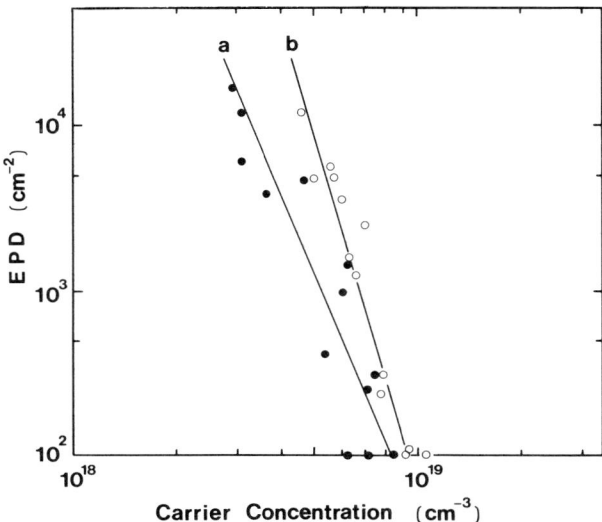

FIG. 25. Crucible rotation effect on the EPD reduction of two-inch diameter long-size sulphur-doped InP single crystals. (a) is the case of 20 rpm and (b) is the case of 30 rpm.

there is a maximum limit for doping. In the case of $\langle 100 \rangle$ direction crystals, this doping limit is 5×10^{18} cm^{-3} at the solidified fraction of 0.2. In the case of $\langle 111 \rangle$ direction crystals, this doping limit is higher, 3×10^{19} cm^{-3}, so that in $\langle 111 \rangle$ direction crystals, it is easier to obtain a larger dislocation-free area.

Figure 26 shows how the dislocation-free area can be increased as a function of carrier concentration. It can be seen that three-inch diameter crystals are more advantageous than smaller diameter crystals to obtain larger dislocation-free areas at the same carrier concentration. Figure 27 shows an example of dislocation-free areas of a three-inch diameter wafer compared with a two-inch diameter wafer. Figure 28 shows how the dislocation-free area is changed from top to tail of a two-inch diameter, 200 mm long sulphur-doped single crystal. As can be seen in the figure, even at the top part, a dislocation-free area of about 6.5 cm^2 is obtained, and the area is increased toward the tail part. Figure 29 shows $\langle 111 \rangle$ wafers with two different sulphur concentrations after etching for revealing EPD. It can be seen that by increasing the carrier concentration, a completely dislocation-free wafer over the 60 m diameter whole area can be obtained without any dislocation slip lines even at the periphery of the wafer.

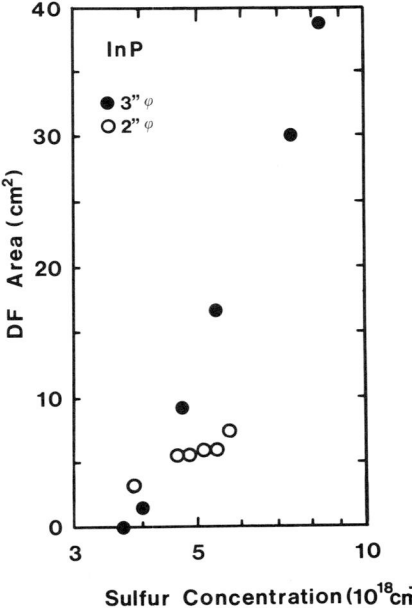

FIG. 26. Dislocation-free area and carrier concentration of sulphur-doped InP single crystals.

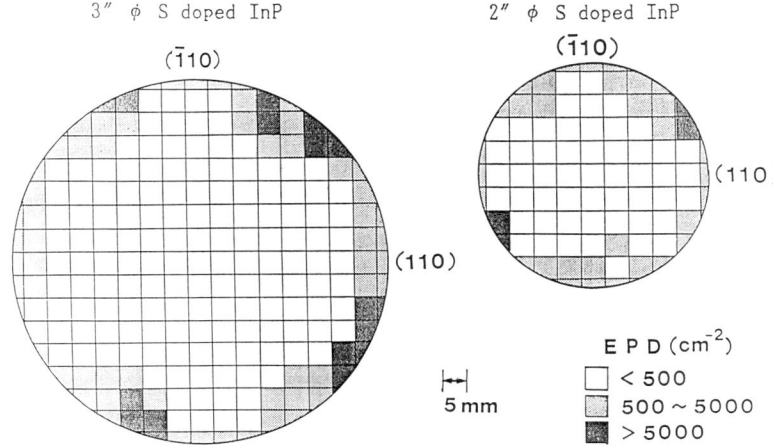

FIG. 27. Dislocation-free area of three-inch diameter and two-inch diameter sulphur-doped InP wafers.

8. ZINC-DOPED SINGLE CRYSTALS

Zinc also has an effect in reducing the EPD, and low EPD zinc-doped crystals are used for high-power lasers, where low EPD is very important for improving the lifetime of lasers. Figure 30 shows how the EPD is reduced as a function of zinc concentration. Dislocation-free areas can be obtained when the zinc concentration exceeds 5×10^{18} cm^{-3}, nearly the same concentration as in the case of sulphur doping.

With zinc doping it is difficult to obtain a higher concentration of zinc. If metallic zinc is added with InP polycrystal into the crucible and heated, the necessary concentration is not obtained because of the high vapor pressure of zinc, and it is difficult to control the zinc concentration. We have, therefore, developed a technique to synthesize zinc-doped InP polycrystals prior to the crystal growth.

Figure 31 shows a typical EPD distribution of zinc-doped single crystals. It can be seen that dislocation-free areas of more than 6.5 cm^2 even from the top part of the crystal and completely dislocation-free wafers can be obtained from the tail part. Figure 32 shows the relationship between mobility and carrier concentration in the case of zinc-doped InP single crystals. By decreasing the carrier concentration, a mobility up to 150 cm^2/V-s can be obtained. Low-zinc-concentration InP single crystals are useful for solar cell application as shown by Yamaguchi et al., (1984a; 1984b).

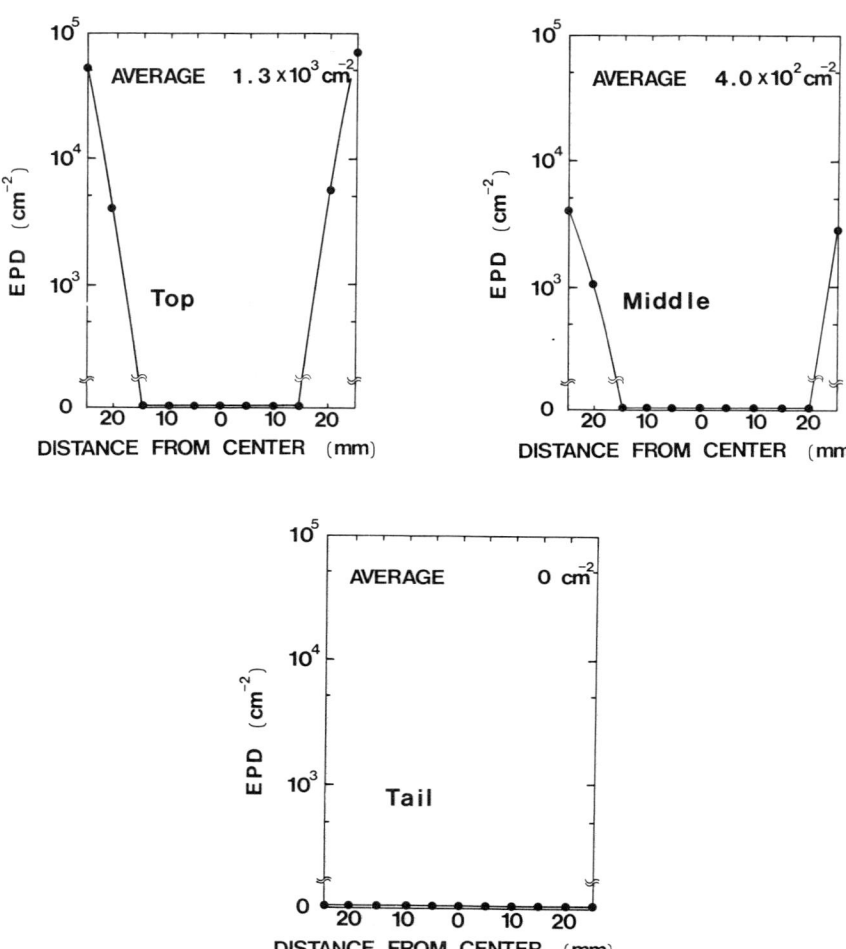

FIG. 28. EPD distribution of a long-size two-inch diameter InP. For each wafer, the EPD was measured for 13 points including $\langle 011 \rangle$, $\langle 010 \rangle$, and $\langle 011 \rangle$ directions.

9. Semi-Insulating Single Crystals

Semi-insulating InP is required for MISFETs and OEICs, because substrates for these applications need to be of very high resistivity in order to isolate each device.

Semi-insulating single crystals can be grown by doping with impurities such as Fe, Cr, and Co, which form deep acceptors, or by doping with Ti, which forms deep donors. Since the band gap of InP is 1.35 eV in theory, InP

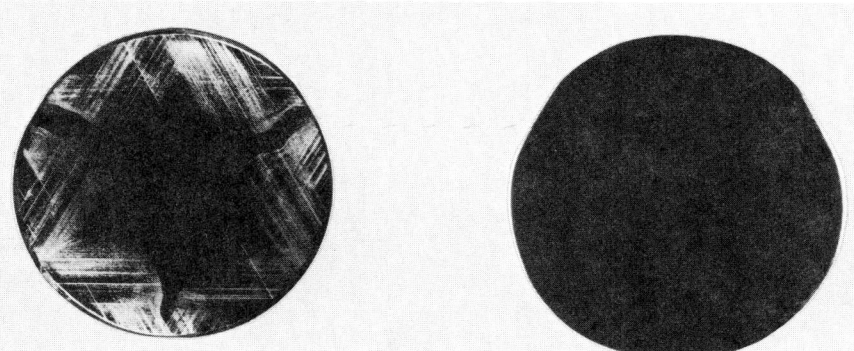

FIG. 29. Etched wafers of two-inch diameter (111) InP single crystals with different doping levels. (a) carrier concentration: 1.2×10^{19} cm^{-3}, (b) carrier concentration: 2.0×10^{19} cm^{-3}.

becomes semi-insulating if the purity of InP is sufficently high. From a calculation, intrinsic InP will have a resistivity of 8×10^7 $\Omega \cdot$cm and a carrier concentration of 2×10^7 cm^{-3}. However, for obtaining this intrinsic semi-insulating InP, the purity must be 15 nines and such an ultra-high purity cannot be obtained in reality. Even for silicon, the best purity is 12 nines. Semi-insulating InP is, therefore, realized only by doping with the previously

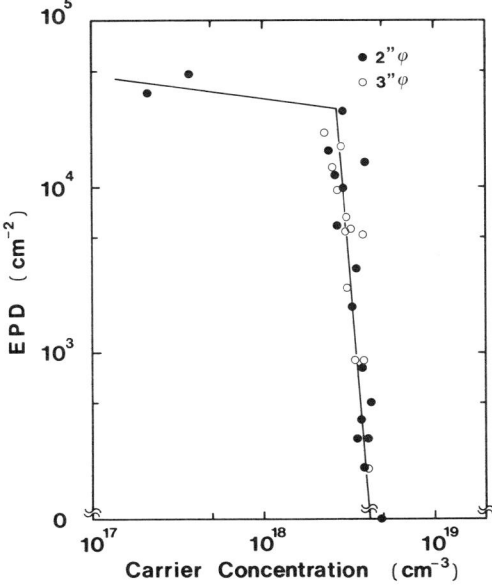

FIG. 30. EPD variation of zinc-doped InP single crystals as a function of carrier concentration.

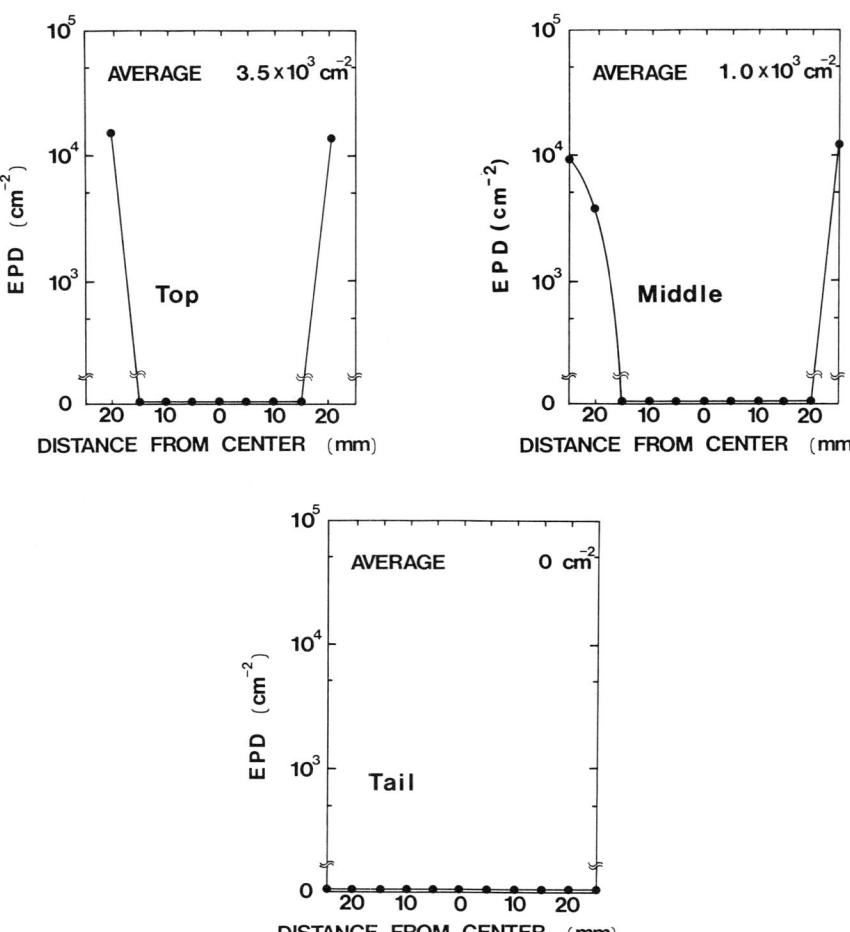

FIG. 31. EPD distribution of two-inch diameter zinc-doped InP single crystals. For each wafer, the EPD was measured for 13 points including $\langle 011 \rangle$, $\langle 010 \rangle$, and $\langle 011 \rangle$ directions.

mentioned impurities, which form deep levels. In the case of deep acceptor impurities, they can compensate shallow donor impurities such as silicon or sulphur in InP. In the case of Ti doping, the shallow acceptor concentration must be higher than that of the shallow donors for obtaining semi-insulating InP since Ti is a deep donor. Shallow acceptors such as Zn, Cd, or Be must be intentionally added in the case of Ti doping (Iseler and Ahren, 1986; Brandt et al., 1986). Figure 33 shows calculated examples of Shockley diagrams (Martin and Jacob, 1983) of Fe-doped and Ti-doped semi-insulating InP.

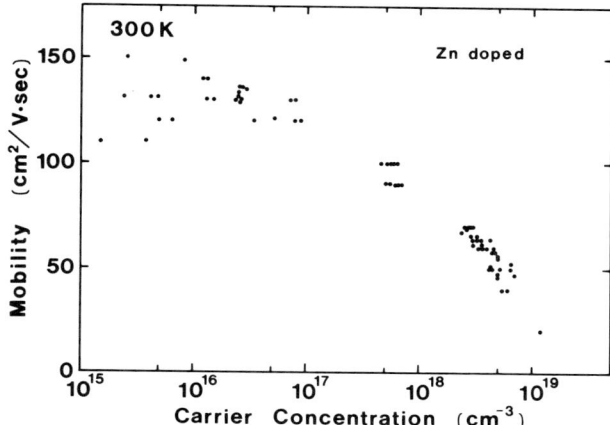

FIG. 32. Mobility and carrier concentration of zinc-doped p-type InP single crystals.

Among these dopants by which semi-insulating InP is realized, only Fe is industrially used. Rumsby et al., (1980), Iseler (1979), Kubota et al., (1981), Morioka et al., (1981), Cockayne et al., (1981), and Zeiss et al., (1983) have all studied extensively Fe-doped InP single crystals. In the case of Co and Cr (Harris et al., 1983; Straughan et al., 1974), because of the precipitation of these impurities, they are not adequate as dopants for semi-insulating crystals. In the case of Ti doping, it is still in the research stage and up to now it has yet to be industrially applied.

Figure 34 shows the relationship between the Fe content and the resistivity from our crystal data. As can be seen, semi-insulating ($>10^7$ $\Omega\cdot$cm) becomes possible when the Fe content is more than 0.2 ppmw (1×10^{16} cm^{-3}). Since the segregation coefficient is as shown in Table VI, for growing semi-insulating ingots from top to tail, more than 0.03 wt.% Fe has to be added in InP raw material. The thermal stability of the grown semi-insulating InP is evaluated by heat treatment at 700°C for 15 min. As can be seen in Table VII, grown crystals show satisfactory thermal stability.

For further improvement of semi-insulating InP, we need to purify the InP in order to decrease the Fe content as low as possible. Since Fe forms deep acceptors in electronic devices, these acceptors will act as deep traps with slow detrapping times, and for fabricating high-speed electronic devices, these impurities need to be as few as possible. Our recent study showed that the Fe content can be decreased by using highly purified HB InP polycrystals as raw materials (Kainosho et al., 1989a).

For fabricating FETs with consistent device performance, semi-insulating wafers need to have homogeneous electrical properties over the wafer, from

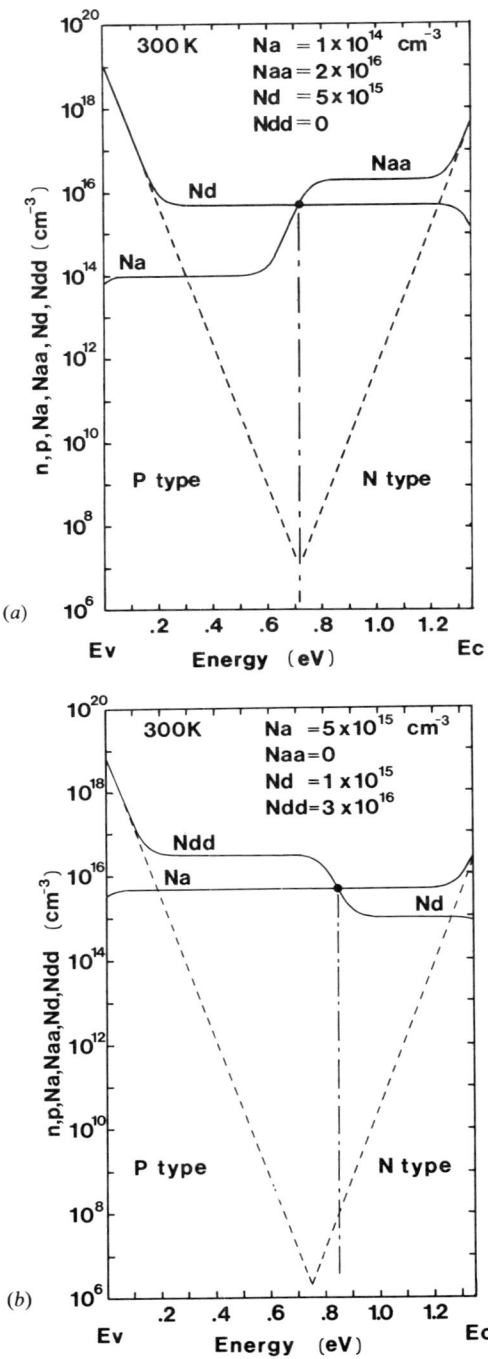

FIG. 33. Shockley diagrams of (a) Fe-doped and (b) Ti-doped semi-insulating InP.

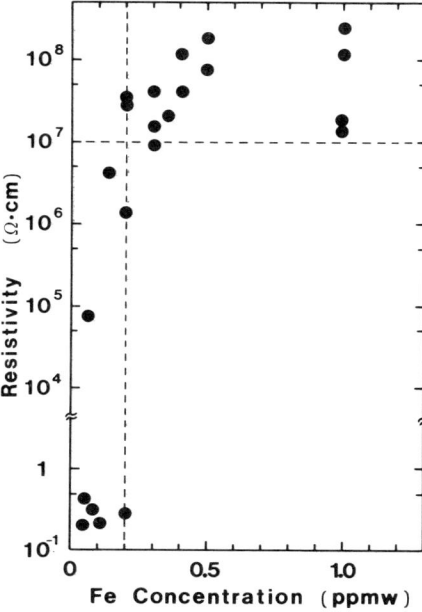

FIG. 34. Resistivity and iron concentration.

TABLE VI

SEGREGATION COEFFICIENTS OF IRON IN InP

1.6×10^{-3}	Lee et al. (1977)
2×10^{-4}	Iseler (1979)
2.5×10^{-4}	Cockayne et al. (1981a)
$1 \sim 4 \times 10^{-4}$	Cockayne et al. (1981b)
$2.9 \sim 4.7 \times 10^{-4}$	Kusumi et al. (1985)

TABLE VII

THERMAL STABILITY OF IRON-DOPED SEMI-INSULATING InP

Fe content (ppmw) by SSMS	Resistivity at 300K ($\Omega \cdot$cm)	
	as-grown	after thermal treatment
0.2	3×10^7	2×10^7
0.4	1×10^8	7×10^7
0.5	8×10^7	2×10^7
0.5	2×10^8	4×10^7
1.0	1×10^8	1×10^8

wafer to wafer, and from lot to lot. As-grown InP has rather homogeneous electrical properties compared with semi-insulating GaAs (Kim et al., 1986), but its homogeneity was not as good as that of ingot-annealed GaAs. Our recent research, however, clarified that the homogeneity of InP could be improved better than that of ingot-annealed GaAs (Kainosho et al., 1989b). In order for Fe-doped InP to be applied industrially as substrates for MISFETs, purification and homogenization of electrical properties are very important.

10. LARGE-DIAMETER AND LONG-SIZE SINGLE CRYSTALS

From the history of silicon single crystal development, to justify growing larger diameter and longer sized crystals, it was very important that the crystal be used extensively in its device application field. This would be the same for InP. In the case of GaAs, it is now possible to grow four-inch diameter GaAs and/or three-inch diameter GaAs with a length longer than 100 mm. In the case of InP, since its application is still limited only to the optoelectronic devices used for longer wavelengths, the consumption in the world was estimated to be approximately 42,000 two-inch diameter wafers per year. This situation did not cause a decrease in price and a consumption increase is inversely related to price. However, InP also has a large potential probability that it will be used for high-speed MISFETs (Itoh, et al., 1989), which are useful for high-speed telecommunication, and solar cells (Yamaguchi et al., 1984a; 1984b; Okazaki et al., 1988), which are very promising for satellite application.

It should be noted that the annual indium production is 100 tons, which is about 2% of silicon usage and twice current gallium usage. Even though the annual indium production is small, only 1% of this total is needed to supply the InP single crystal production of 42,000 two-inch diameter wafers per year. Indium production can be easily increased by a factor of two. If all of indium is used for producing InP wafers, 14 million two-inch diameter wafers per year could be produced. This amount is 340 times more than the present usage. If the scraps of InP are recycled, this amount is further increased. Even though there is a limitation for indium production, the present usage is still far from this limitation.

In the case of InP, growing large-diameter single crystals is quite difficult compared with GaAs. Normally, two-inch diameter InP crystals are grown by using four-inch diameter crucibles with the material charge in the range of 800–1000 g in a rather small high-pressure puller. For growing three-inch diameter InP or long crystals, however, it is necessary to use at least a six-inch diameter crucible for a material charge of about $3 \sim 4$ kg and a large high-pressure puller as shown in Fig. 35. As mentioned previously, in the case

4. InP CRYSTAL GROWTH

FIG. 35. Large high-pressure LEC puller.

of InP crystal growth, the system pressure must be more than 35 atm, typically 40 atm. This pressure is 2 ~ 6 times higher than the case of GaAs crystal growth. This high pressure in the large chamber of the crystal puller then induces a strong gas convection during crystal growth. This strong gas convection is the main reason why it is much more difficult to grow single crystals from a six-inch diameter crucible than from a four-inch diameter crucible.

In order to solve this convection problem, it is very important to optimize hot zone structures in the high-pressure puller. Figure 16 gave an example of the axial temperature gradients and the temperature fluctuations after hot-zone structure improvement. It can be seen that in a high-pressure puller with the improved hot-zone structure, the temperature fluctuations became very small at the B_2O_3 and melt interface. We were able to grow industrial three-inch diameter InP with the length of 160 mm and two-inch diameter InP with the length of 210 mm after making various improvements, including that of the previously mentioned hot-zone structure. These crystals were shown in Fig. 18.

IV. Evaluation

11. Purity

Because the purity of InP affects not only its electrical properties and its crystallinity but also the quality of electrical devices made by using InP substrates, growing higher purity InP is one of the most important factors in InP crystal growth. The purity of In and InP is normally examined by using solid state mass spectroscopy, and the purity of phosphorus is examined by using atomic absorption spectroscopy.

Table VIII shows the typical purity of 6N In and 7N In purified in NMC. As can be seen in the table, most impurities are undetectable. In 7N In, the silicon concentration is significantly lower than it is in 6N In.

Phosphorus is purified by nitric acid leaching and vacuum distillation. Table IX shows the purity of phosphorus available from several suppliers. It can be seen that in phosphorus, the main impurity is arsenic. Since arsenic behavior during purification is similar to sulphur, it is important to decrease the arsenic concentration. In the case of sulphur concentration, since the detection limit is high, useful information is not obtained. However, sulphur is supposed to be one of the main donor impurities (Dean et al., 1984; Coquille and Guillou, 1986). If the silicon contamination is avoided, sulphur would become the main donor impurity. It should be noted that the vapor pressure of sulphur is nearly the same as phosphorus, and sulphur cannot be

TABLE VIII

SSMS ANALYSIS DATA OF 6N INDIUM AND 7N INDIUM
(ppmw)

Element	Detection limit	6N			7N		
		Lot 1	Lot 2	Lot 3	Lot 1	Lot 2	Lot 3
B	0.0003	0.004	0.0006	0.001	ND	0.0004	ND
Mg	0.001	ND	ND	ND	ND	ND	ND
Al	0.001	ND	0.008	0.003	0.001	ND	0.002
Si	0.003	0.01	0.009	0.01	0.003	ND	ND
S	0.04	ND	ND	ND	ND	ND	ND
K	0.002	0.06	0.03	0.02	0.008	0.005	0.01
Ca	0.002	0.005	0.01	0.003	ND	ND	0.003
V	0.004	ND	ND	ND	ND	ND	ND
Cr	0.004	0.009	ND	ND	ND	ND	ND
Mn	0.004	ND	ND	ND	ND	ND	ND
Fe	0.005	0.02	0.009	0.008	ND	0.006	0.005
Co	0.005	ND	ND	ND	ND	ND	ND
Ni	0.02	ND	ND	ND	ND	ND	ND
Cu	0.008	ND	ND	ND	ND	ND	ND
Zn	0.02	ND	ND	ND	ND	ND	ND
As	0.007	ND	ND	ND	ND	ND	ND
Br	0.02	ND	ND	ND	ND	ND	ND
Se	0.02	ND	ND	ND	ND	ND	ND
Ag	0.02	ND	ND	ND	ND	ND	ND
Cd	0.01*	ND	ND	ND	ND	ND	ND
Sn	0.06	ND	ND	ND	ND	ND	ND
Sb	0.03	ND	ND	ND	ND	ND	ND
Tl	0.03*	ND	ND	ND	ND	ND	ND
Pb	0.05*	ND	ND	ND	ND	ND	ND
Bi	0.02*	ND	ND	ND	ND	ND	ND

*Analysis by atomic absorption spectroscopy because of the lower detection limit than SSMS

decreased during the HB synthesis. Reducing the sulphur concentration in phosphorus is an important goal.

As explained previously, the main donor impurity in synthesized InP polycrystal is silicon, and it can be decreased by using pBN boats during the synthesis. Table X shows the purity of InP polycrystals and undoped InP single crystals. By using pBN boats and crucibles, the purity can be improved compared with quartz boats and crucibles. It can be seen that after crystal growth, several impurities are extracted. The extraction of silicon during crystal growth is prominent. This is due to segregation of silicon into the melt.

TABLE IX

PURITY OF PHOSPHORUS AVAILABLE FROM SEVERAL SUPPLIERS
(ppmw)

Supplier	Grade	Lot	Elements						
			Cu	Pb	Fe	Cd	Zn	S*	As
	(Detection limit)		0.01	0.01	0.1	0.01	0.1	0.2	0.01
A	6N	1	ND	ND	ND	ND	ND	ND	1.8
		2	ND	ND	ND	ND	ND	ND	0.28
		3	ND	ND	ND	ND	ND	ND	0.51
B	6N	1	ND	0.04	0.28	ND	ND	ND	7.4
		2	0.01	0.04	0.30	ND	ND	0.3	5.7
		3	0.01	0.03	0.21	ND	ND	ND	5.6
C	6N	1	ND	ND	ND	ND	ND	ND	0.25
		2	ND	ND	ND	ND	ND	ND	0.72
	6N +	1	ND	ND	ND	ND	ND	ND	ND
		2	ND	ND	ND	ND	ND	ND	ND

Analysis by flameless atomic absorption spectroscopy,* absorptiometry

12. PHOTOLUMINESCENCE

Photoluminescence is an effective technique for the evaluation of the purity of InP. Figure 36 shows the photoluminescence spectra of InP polycrystals. As can be seen in the figure, when the carrier concentration is decreased, fine structures at the edge emission band (~ 1.42 eV) are revealed. This feature

TABLE X

PURITY OF InP POLYCRYSTALS AND UNDOPED InP SINGLE CRYSTALS
(ppmw)

Element	Detection limit	Polycrystal by pBN boat	Single crystal by pBN crucible	Polycrystal by quartz boat	Single crystal by quartz crucible
Si	0.003	0.02	0.007	0.2	0.06
S	0.1	ND	ND	ND	ND
Zn	0.05	ND	ND	0.06	Trace
Mg	0.003	ND	ND	Trace	Trace
As	0.02	0.07	0.04	0.07	0.08
Ni	0.04	ND	ND	ND	ND
B	0.0008	ND	0.003	ND	0.04
Cr	0.01	Trace	0.03	Trace	ND
Mn	0.01	0.02	0.02	ND	ND
Fe	0.01	0.02	0.04	0.04	ND

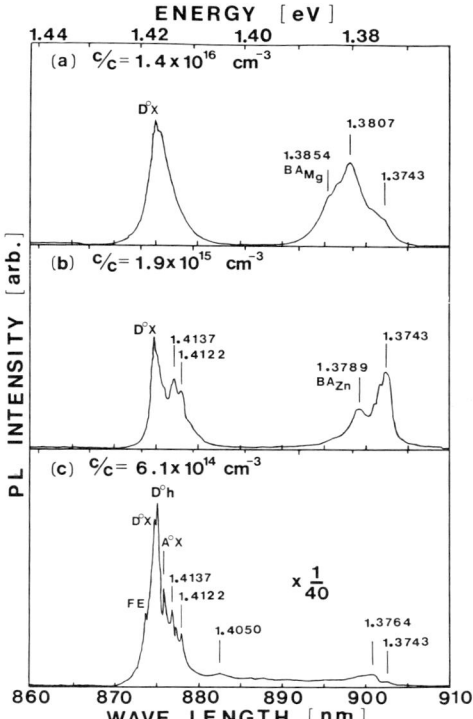

FIG. 36. Photoluminescence spectra of HB and SSD polycrystals. (a) HB polycrystal grown by a quartz boat, (b) HB polycrystal grown by a pBN boat, (c) SSD polycrystal. c/c: carrier concentration.

has also been reported by Sakagami (1986). In SSD polycrystals, $D°X$, $D°h$, and $A°X$ peaks are clearly revealed. Even the free exciton peak is also observed. In the case of $1.37 \sim 1.38$ eV peaks due to acceptor impurities, by decreasing the carrier concentration peaks could be well resolved as seen in Fig. 36 (b). However, $1.37 \sim 1.38$ eV peaks were quite small compared with the edge emission peak in the case SSD polycrystals. It should also be noted that the edge emission peak intensity becomes very strong in the case of SSD polycrystals.

Figure 37 shows the relationship between the photoluminescence intensity and the EPD in the case of undoped, Sn-doped, and S-doped InP single crystals (Inoue and Takashi, 1987). As can be seen in the figure, the 1.4 eV band edge emission peak height, the 1.3 eV peak height, and the peak height at the 1.06 eV deep level are all correlated with the EPD. This means that dislocations will act as nonradiative centers.

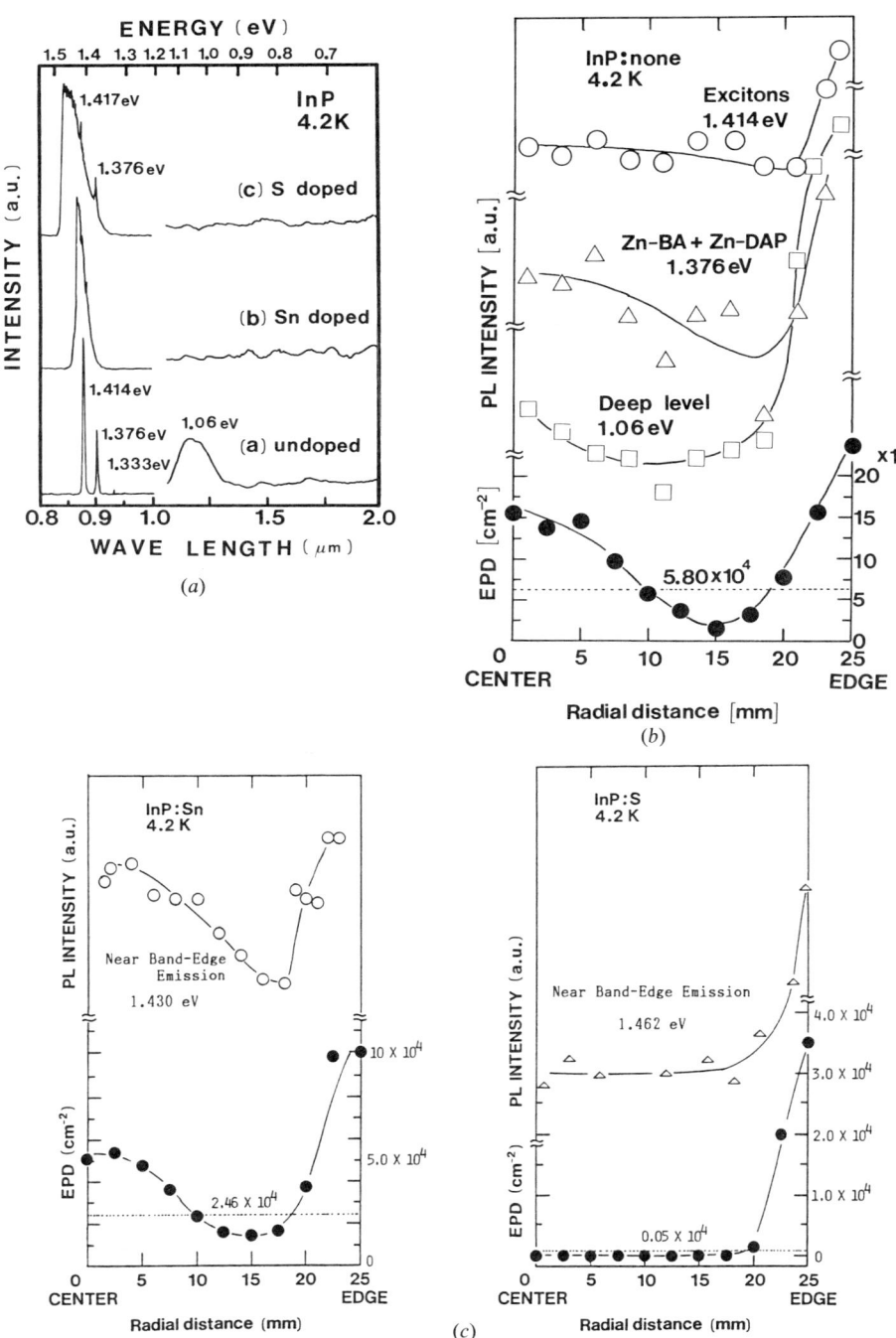

FIG. 37. Photoluminescence spectra and the intensity distribution of two-inch diameter undoped, tin-doped, and sulphur-doped InP single crystals. (Inoue and Takahashi, 1987).

13. CATHODOLUMINESCENCE

Cathodoluminescence is also a technique that is effective in evaluating crystal quality, especially microscopic inhomogeneity. It is well known that in the case of undoped GaAs grown by LEC technique, CL image shows a strong inhomogeneity along dislocation cells as shown in Fig. 38(a). Figures 38(b)–(d) show the CL images of undoped and doped InP single crystals. It can be seen that in the case of InP, the CL images are normally very homogeneous except for small, isolated black points, which are due to dislocations. This feature of strong homogeneity of CL images is also an advantage of InP single crystals. The homogeneity would be due to dislocation structrures. As already mentioned, in the case of InP single crystals, dislocations are isolated from each other. This homogeneous dislocation structure can also be observed in CdTe single crystals but not in LEC GaAs. In LEC GaAs, dislocations easily make cell structures, and these cell structures are very detrimental for fabricating FET devices with high yield. In this sense, since dislocations in InP do not constitute cell structures as they do in GaAs, InP would be a more appropriate crystal for FET device fabrication.

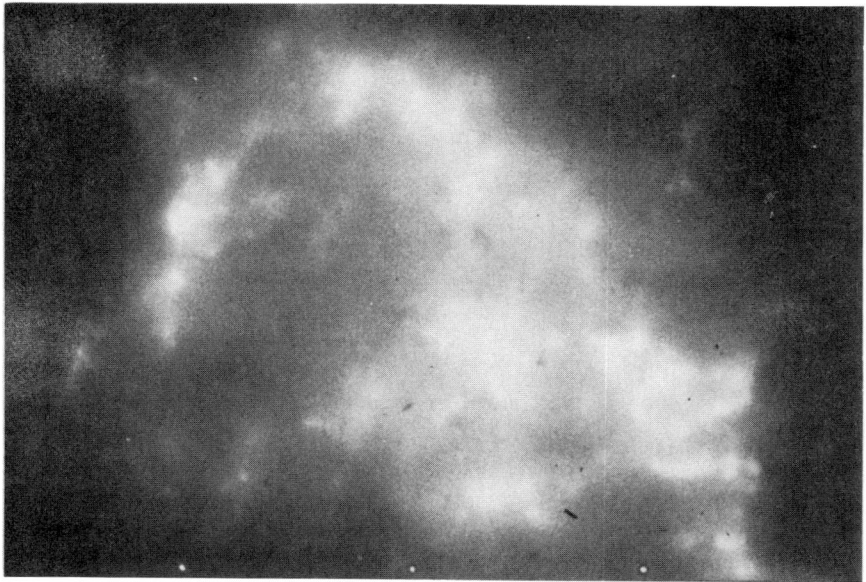

(a)

FIG. 38. Cathodoluminescence photographs of (a) undoped GaAs, (b) undoped InP, (c) Sn-doped InP, (d) Zn-doped InP, and (e) S-doped InP.

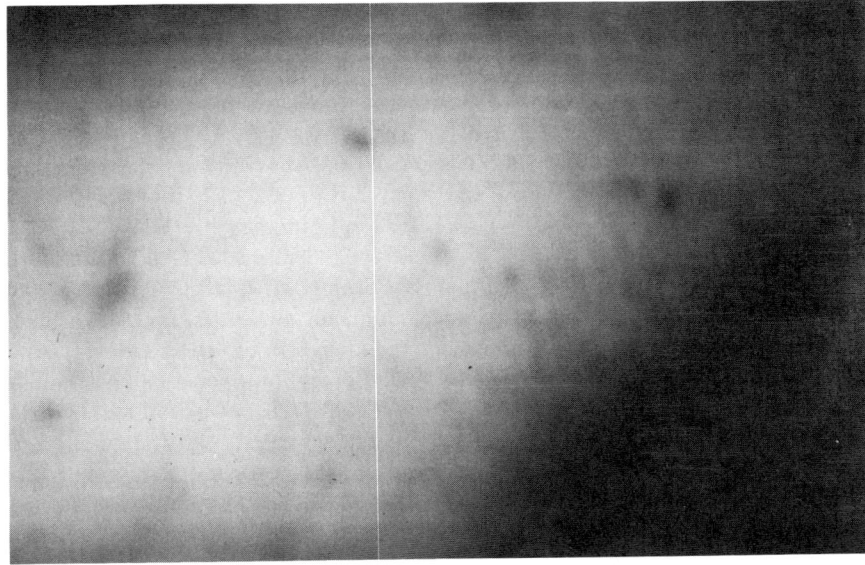

(b)

(c)

FIG. 38. (*Continued*)

4. InP CRYSTAL GROWTH

(d)

(e)

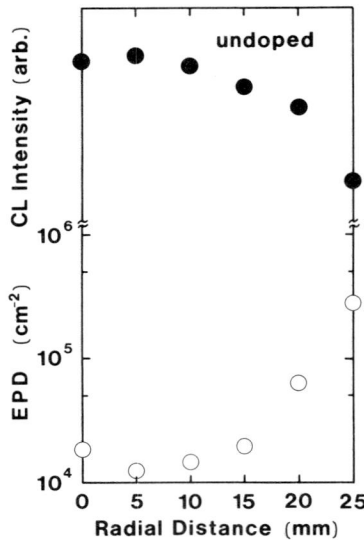

FIG. 39. Cathodoluminescence intensity distribution of undoped InP single crystal.

Figure 39 shows CL intensity distribution over an undoped InP wafer. As can be seen in the figure, CL intensity has a strong correlation with the EPD. This means that dislocations act as nonradiative centers as in the case of photoluminescence.

14. DEFECTS AND THEIR ORIGINS

InP single crystals have various defects other than dislocations. Typical defects are grappes or clusters as shown in Fig. 40. The origin of these defects has been discussed by several authors (Augustus and Stirland, 1982; Brown et al., 1984; Franzosi et al., 1984). These defects, however, can be completely eliminated by paying attention to the purification process. Grappes are many etch pits gathered like grapes, and they were formed in impure InP. Grappes are defects that had been observed sometimes in InP single crystals grown in the beginning of InP crystal development when the purity of In, phosphorus, InP polycrystals and B_2O_3 was not carefully examined.

Clusters are defects that have a kind of center in the etch pit as shown in Fig. 40. These clusters can only be observed by careful observation using a high magnification microscope. Table XI shows the data of cluster densities before and after the improvement of the process. Attention had been given to the improvement not only of the purity of B_2O_3 but also of the purity of InP and the crystal growth process. InP single crystals free from these defects were

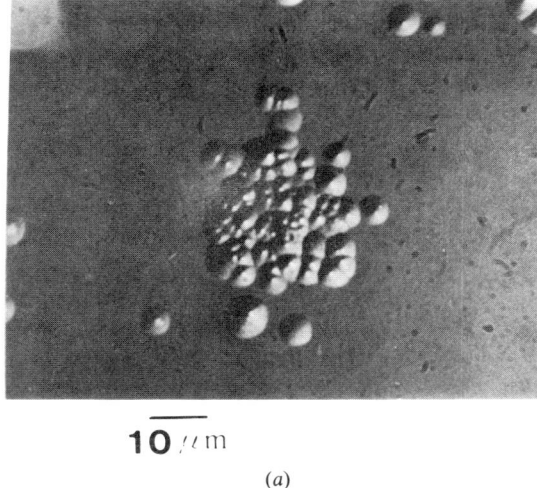

10 μm

(a)

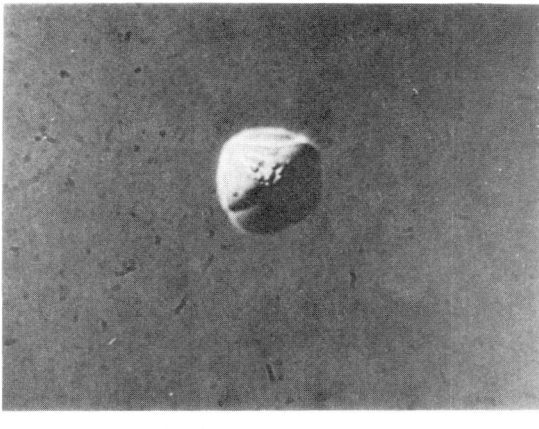

10 μm

(b)

Fig. 40. Microphotographs of defects revealed by Huber etchant: (a) grappe and (b) cluster.

TABLE XI

CLUSTER DENSITIES OF VARIOUS InP SINGLE CRYSTALS
BEFORE AND AFTER IMPROVEMENT
(cm^{-2})

	Before improvement	After improvement
undoped	30 ~ 400	0 ~ 5
S doped	10 ~ 120	0 ~ 5
Zn doped	0 ~ 5	0
Sn doped	100 ~ 700	0 ~ 5
Fe doped	0 ~ 25	0 ~ 5

of high purity with respect to Si, Mg, and Ni. We, therefore, believe that these defects are due to impurities segregated in InP even though their forms are not yet exactly identified. A possibility is the oxides.

V. Substrate Preparation

15. WAFER PROCESSING

Typical InP production process is shown in Fig. 41. Top and bottom parts of InP single crystal boules are cut with the surface orientation within less than 0.3 degrees from $\langle 100 \rangle$ or $\langle 111 \rangle$ directions. The cut ingot is then cylindrically ground, and then flats are ground for marking the orientation and index flats. For determining the surface orientation and the orientation flat, an x-ray cut-surface measurement instrument (the accuracy of which is within 0.01 degree), is applied. The cylindrically ground ingot is then cut to wafers, the thickness of which is between 500 and 600 μm depending on the specified thickness of the polished wafers. The cut wafers are then bevelled, making the final orientation and index flats as shown in Fig. 42. The bevelled wafers are then lapped, polished, cleaned, dried, and then inspected according to specified inspection criteria before the delivery.

16. CUTTING

For cutting or slicing compound semiconductor materials, three different techniques are available, ID (internal diameter) saw, multi-wire saw, and multi-blade saw. Most of these techniques were mainly developed for silicon slicing, and they have been reviewed by Iscoff (1982).

The ID saw is a machine by which the material is cut or sliced by a stainless steel wheel with a large center hole and a diamond-coated rim. During

4. InP CRYSTAL GROWTH

SCHEMATIC FLOW OF InP WAFER PRODUCTION

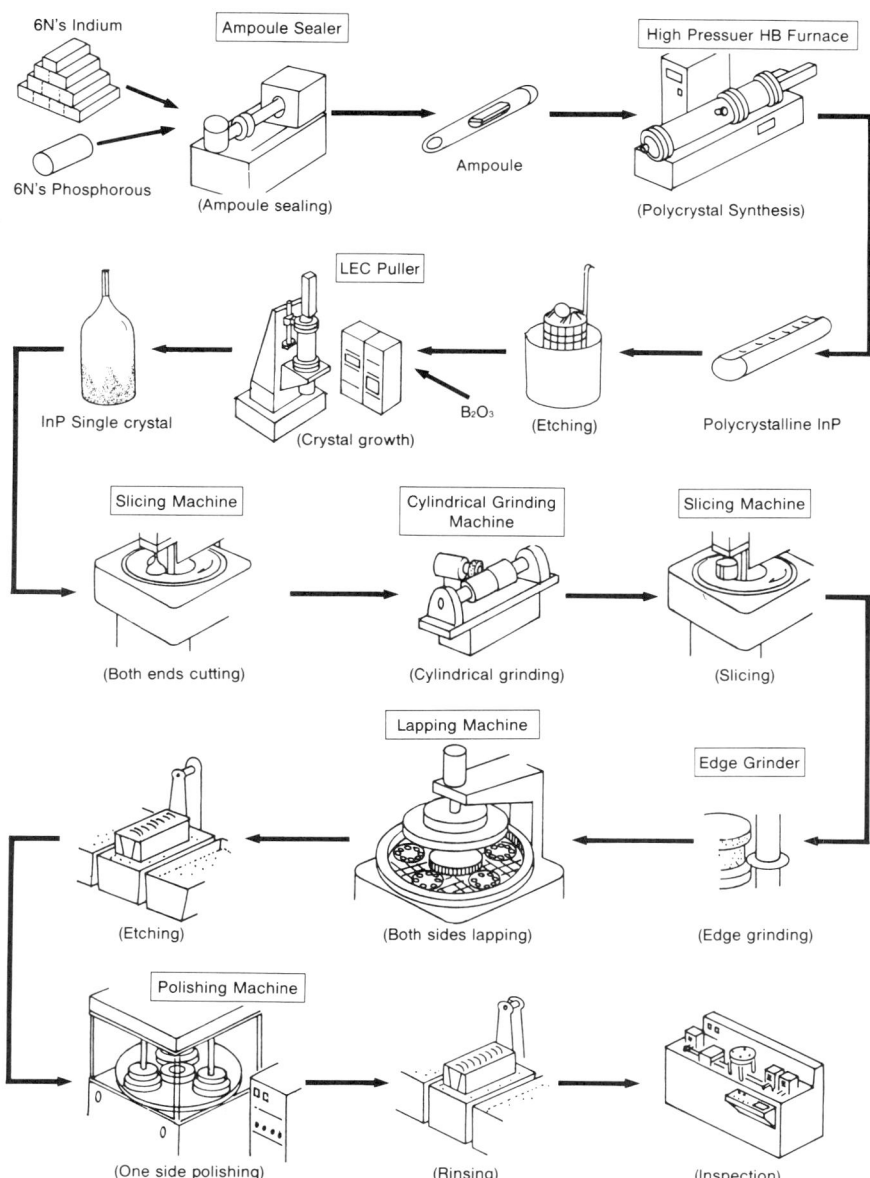

FIG. 41. Typical production process for InP wafers.

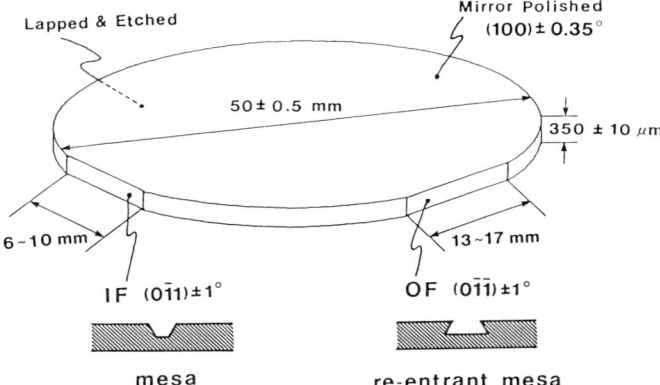

FIG. 42. Wafer shape and orientation and index flats.

cutting, the blade is cooled by spraying oil coolant or water with a surface active agent. The technology of the ID saw is discussed by several authors (Meek and Huffstutler, 1969; Forman and Rhines, 1972; Kachajian, 1972; Grandia and Hill, 1978; Chen, 1982). Since InP is very brittle compared with silicon and gallium arsenide, when the ID saw is applied, it is very important to optimize various operation parameters such as wheel tensile strength, diamond coating cross-section shape, blade tensioning, blade dressing, cutting feed rate, coolant temperature, coolant spraying angle, and blade lifetime control. If these parameters are not well selected and controlled, most InP wafers are easily broken during slicing or after slicing. Even if they are well sliced, severe work damage, such as saw marks, can be left on the wafer. The damage cannot be removed even if it is lapped and polished. The ID saw thus needs a careful optimization of operating conditions, since InP is more fragile than silicon or GaAs because of its brittleness. The ID saw has, however, an advantage in that it is easy to obtain a precise orientation because the axis of the rod that supports the ingot can be precisely adjusted by measuring the sliced wafer orientation.

The wire saw and multi-blade saw are machines by which the material can be sliced by using abrasives such as SiC or Al_2O_3 powders suspended in oil (Jensen, 1973; Iscoff, 1982). Since the material is sliced by lapping with the abrasive, which is rubbed on the material with piano wires or stainless bands, the thickness of the damaged layer is expected to be less than that produced by the ID saw cutting process. For cutting InP whose brittleness is greater than silicon or GaAs, these machines are advantageous for obtaining less-damaged wafers. However, their shortcomings are that to obtain precise wafer orientation is more difficult than it is with the ID saw, and that when there is misoperation, all wafers from one ingot are spoiled, since they are simultaneously sliced. The other shortcoming is that these machines need a high degree of skill to be used for production with a resonable yield of wafers.

Figures 43 and 44 show the wafer orientation and the bow variation of several InP ingots. As can be seen, by optimizing the cutting conditions, the wafer can be cut with the orientation within 0.2 degrees, and the bow can be made less than 15 μm.

17. LAPPING

Sliced wafers are bevelled and etched before the lapping process. Bevelling is for preventing edge breaking in the subsequent process. The bevelled wafers are then lapped to control the thickness, the parallelism, and the flatness, and also to remove damaged layers, such as saw marks, which are induced by the cutting process. Al_2O_3, SiC, or ZrO_2 abrasives are used for lapping InP.

For lapping, there are single-side lapping and double-side lapping techniques. In single-side lapping, sliced wafers are attached to stainless steel plates by using wax. On the plates, bolts with diamond tips are set for determining the lapping thickness by adjusting the bolts before lapping. Single-side lapping is an easy way to obtain reproducible wafer thicknesses with good accuracy. Since one needs to stick each wafer to a plate twice to lap both sides, however, it is very time consuming. Second, in the case of single-side lapping the parallelism and flatness are largely affected not only by the accuracy of the plate but also by the uniformity of the wax used for mounting the wafers.

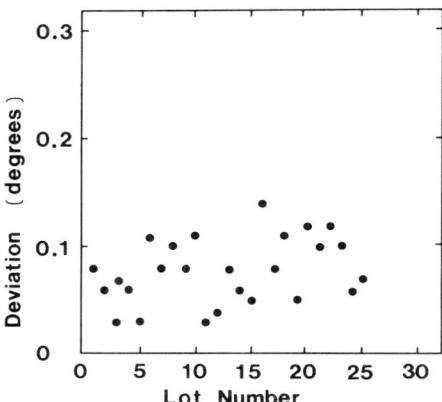

FIG. 43. Wafer surface orientation variation from lot to lot.

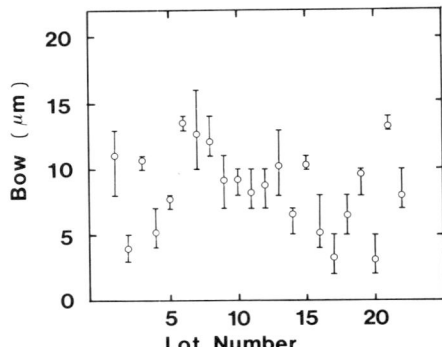

FIG. 44. Bow variation of all wafers sliced from several InP ingots.

In the case of double-side lapping, even though the technique is very advantageous from the viewpoint of the lapped wafer quality compared with single-side lapping the technique is very complicated, and one needs to optimize many parameters such as abrasive particle size, lapping pressure, plate rotating speed, lapping time, and stopping technique. When any of these parameters is not correct, most wafers are broken during lapping or in the subsequent process, because of the brittleness of InP. Double-side lapping is, however, very attractive since it gives good quality lapped wafers with good productivity.

Figure 45 gives the photograph of a double-side lapping machine, and Table XII gives typical processing accuracy data for lapped InP wafers. It is known that double-side lapping is a good technique to obtain excellent parallelism and flatness.

18. POLISHING

By polishing lapped wafers, one can obtain damage-free, mirror-like wafer surfaces and on these high-quality polished wafers, one can make epitaxial layers of various mixed compounds for the purpose of optoelectronic device fabrication. To polish for high quality is therefore one of the most important processes for wafer fabrication. Polishing technology of compound semiconductors has generally been reviewed by Jensen (1973), and some experimental results have been reported on GaAs. However, hardly any published reports that are concerned with InP polishing itself can be found. Only Tuck (1975), Iiyama (1982), and Tuppen and Cohen (1987) have reported on the polishing of InP by Br_2-methanol solutions.

Before polishing, it is necessary to remove by an etching process, any abrasive particles and damaged layers left on the surface of lapped wafers.

FIG. 45. Photograph of a double-side lapping machine.

The etchant selection is very important because various etchants have different effects on wafer surface quality such as change in parallelism and flatness. If the etchant has preferential etching characteristics, after etching, parallelism and flatness can be degraded, and the situation cannot be reimproved in the subsequent process. Table XIII shows the experimental results of various etchant characteristics on parallelism.

Polishing normally consists of two processes: the first one is for removing the damaged layer introduced by lapping process, and the second one is for obtaining the final high-quality surfaces. For the first polishing process, single-side or double-side polishing is used according to the specification.

TABLE XII

MACHINING ACCURACY OF AS-SLICED AND AS-LAPPED WAFERS (μm/2″ wafer).

	Parallelism	Flatness
As-sliced	$3 \sim 14\ \mu$m	$4 \sim 12\ \mu$m
One-side lapping	$3 \sim 5\ \mu$m	$3 \sim 5\ \mu$m
Both-side lapping	$1 \sim 3\ \mu$m	$1 \sim 3\ \mu$m

TABLE XIII

Characteristics of Various Etchants

Etchant (mixed by volume ratio)	Etching temperature (°C)	Etching depth after 3 min. (μm)	Parallelism after 3 min. etching (μm)
Br_2 1: Methanol 99	25 ± 5	12.3 ± 1.3	7.8 ± 2.3
Br_2 0.05: Methanol 100	25 ± 5	1.3 ± 0.4	2.8 ± 1.1
HBr 1: HNO_3 1:H_2O 20	25 ± 5	~ 0	1.7 ± 0.9
2N $K_2Cr_2O_7$ 3: H_2SO_4 1: HCl 2	25 ± 5	4.3 ± 0.5	3.3 ± 1.5
HBr 1: H_3PO_4 1	25 ± 5	14.7 ± 2.6	2.0 ± 0.5
H_2SO_4 1: H_2O_2 1: H_2O 1	60 ± 5	5.7 ± 3.7	3.5 ± 0.5

The parallelism of as-lapped wafers before etching was all in the range of 0 ~ 2 μm.

As in the case of lapping, double-side polishing is better than single-side polishing from the viewpoint of the machining accuracy. When double-side polishing is applied, however, the disadvantage is that one cannot distinguish the right and the wrong sides of the wafer because both sides become mirror polished. This is especially significant since, in the case of InP, most wafers are used as optoelectronic devices, and wafers are normally cleaved into rectangular pieces before their use.

For the polished surface quality inspection, there are various items to be inspected and for obtaining good quality, the best polishing solution should be selected for each polishing process. These polishing conditions are generally proprietary know-how for each wafer manufacturer.

Mechanochemical reagents such as colloidal silica-containing solution, which are normally used for silicon polishing, are not recommended since these mechanochemical reagents can easily introduce visible and/or latent scratches in InP. As chemical polishing reagents, there are two kinds, solutions with strong chemical reaction and solutions with weak chemical reaction. Typical solutions with strong chemical reaction are Br_2-methanol solutions. By using these strong acid reagents, one can polish off the wafer surface very rapidly without new damaged layer formation, but since they are too reactive, it sacrifices wafer quality such as parallelism, flatness, or waviness. Typical solutions with weak chemical reactions are NaClO-based solutions and weak-acid-based solutions. These chemically weak polishing reagents are advantageous for obtaining good parallelism, flatness, or waviness. Their shortcomings are their low polishing speed and the possibility of forming visible or latent scratches, because these chemically weak reagents have more mechanical characteristics.

In the real polishing process, the best combination of polishing reagents is selected. By changing other polishing conditions such as pad material selection, polishing pressure, plate rotation conditions, polishing time, reagent feeding rate, and reagent feeding position, the quality of polished wafers is varied. When the chemical factors are too strong, that is, the use of a chemically strong reagent, soft pad material with low polishing pressure and long polishing time, it appears that pits and hazes take place and parallelism, flatness, and waviness become worse. When the mechanical factors are too strong, that is, the use of a chemically weak reagent, hard pad material with high polishing pressure, then visible and/or latent scratches take place and damaged layers cannot be removed but good parallelism, flatness, and waviness can be achieved. These polishing parameters should, therefore, be well optimized to satisfy all of the qualities necessary for wafers. Table XIV shows typical polished wafer qualities industrially realized. The polishing procedure is not yet perfected in the case of InP and further research is necessary to develop still better technology.

19. Cleaning

After polishing, the wafer surface has to be cleaned to remove organics or impurities adhered or absorbed on the surface. In the case of silicon, because of its long history of cleaning-technology development, the highly sophisticated cleaning procedure is already established while in the case of InP, the study is not yet complete. For establishing a good cleaning procedure, it is necessary to utilize effective evaluation techniques. In the case of silicon, it was ellipsometry by which surface-oxide layer thickness could be measured very precisely. In the case of InP, Aspnes and Studna (1982a; 1982b) have studied the effect of various etchants on the cleanliness by using spectroscopic ellipsometry. Singh et al., (1982) have studied the cleaning effect of various cleaning solutions by measuring the C and O Auger peak heights. Guivarc'h et al., (1984) have examined the cleaning effect of various etchants by XPS.

In order to clean the surface, one needs to use an organic reagent to remove organic residue, then alkaline solution and/or acid solution for removing the

TABLE XIV

MACHINING ACCURACY OF POLISHED TWO-INCH WAFERS

Orientation accuracy	$< 0.2°$
Thickness accuracy	$\pm 10 \ \mu m$
Parallelism	$< 10 \ \mu m$
Bow	$< 15 \ \mu m$
Surface waviness	$< 20 \ \text{Å}$
Edge rounding	none

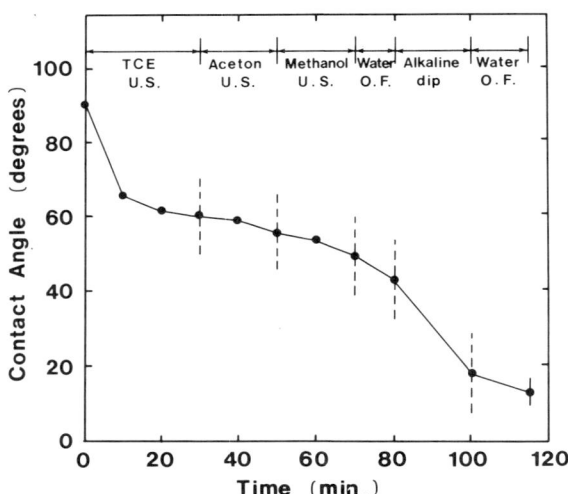

FIG. 46. Water drop contact angle evolution as a function of cleaning process. Here, U.S. means ultrasonic cleaning, and O.F. means overflow cleaning.

residual impurities. We have used the water-drop contact angle measurement technique to improve the cleaning procedure since, by this technique, the measurement could be promptly performed immediately after cleaning. Figure 46 shows an example of contact angle evolution as a function of the cleaning process. As can be seen, the organic solution cleaning has the effect of decreasing the contact angle up to a certain level of about 50 degrees. For decreasing it further, however, another solution is necessary. By using an alkaline solution, the contact angle can be decreased down to the level of 15 degrees. However, the surface is still hydrophobic, and this means the surface is not yet well cleaned. For decreasing the contact angle further, various cleaning solutions have been examined. Table XV shows the effect of various

TABLE XV

WATER DROP CONTACT ANGLES AFTER CLEANING BY VARIOUS SOLUTIONS

Cleaning solution	Temperature	Time (min.)	Contact angle (degrees)
Organics	—	—	30 ~ 40
conc. H_2SO_4	R.T.	15	19
H_2SO_4 4: H_2O_a 1: H_2O 1	R.T.	15	7
	80	15	< 2
HF 1: H_2O 1	R.T.	15	< 2
HF 4: H_2O_2 1: H_2O 1	R.T.	15	13
conc. H_3PO_4	R.T.	15	10

TABLE XVI

AES AND XPS MEASUREMENT RESULTS OF InP WAFERS CLEANED BY VARIOUS SOLUTIONS

Cleaning solution	Cleaning condition		AES				XPS			
	temperature	time (min.)	P/In	C/In	O/In	S/In	P/In	C/In	O/In	S/In
organics	R.T.	—	0.15	0.28	1.01	—	0.154	0.123	0.203	—
Br_2 0.05: Methanol	R.T.	15	—	—	—	—	0.136	0.036	0.140	—
conc. H_2SO_4	R.T.	10	0.25	0.12	0.59	0.12	0.154	0.063	0.081	0.002
conc. HNO_3	R.T.	15	—	—	—	—	0.130	0.050	0.198	—
H_3PO_4 based solution	R.T.	4	0.52	0.05	0.25	—	0.127	0.021	0.066	—

For AES measurements; P(120 eV), C(266 eV), In(411 eV), O(507 eV) and S(152 eV)
For XPS measurements; P(132.9 eV), C(284.6 eV), In(443.6 eV), O(531.6 eV) and S(164.05 eV)

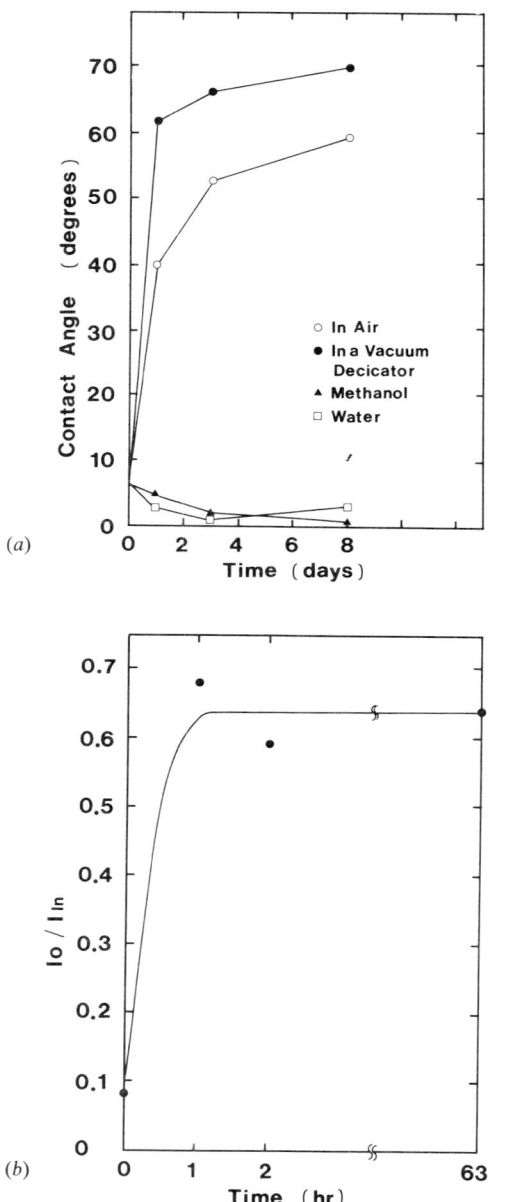

FIG. 47. Water drop contact angle evolution of InP wafers maintained in various media (a) and AES intensity evolution of a InP wafer in air at room temperature (b). For the AES measurement, a polished InP wafer was ion sputtered before beginning the measurement.

cleaning solutions on decreasing the contact angle. It can be seen that various solutions have a different effect on the contact angle. Table XVI shows the AES and XPS measurement results for cleaned and dried wafers. It can be shown that in the case of sulfuric acid cleaning, sulphur is detected on the surface.

After polishing and drying, polished wafers should be stored under the proper conditions. Figure 47 shows the contact angle evolution when wafers are kept in various media. As can be seen in the figure, when the wafers are dried and kept in air or in a vacuum desiccator, the contact angle is increased rapidly. This is because of room temperature oxidation. AES measurement also showed that the oxygen peak height increases rapidly when the wafer is exposed to air after ion etching. On the other hand, when the wafer is kept in solutions such as methanol or even water, there is almost no increase of the contact angle. This means that these solutions have the effect of at least preventing further oxidation of InP.

VI. Evaluation

20. Damaged Layer

During wafer processing, damaged layers are induced in the vicinity of wafer surfaces and by the final polishing these damaged layers should be eliminated. In order to decide the thickness to be removed in each case, one needs to evaluate the damaged thickness for each wafer process.

For evaluating the damaged layer thickness, we have used the double crystal x-ray diffraction technique (Takahashi et al., 1987). The results are shown in Fig. 48. The full widths of half maxima (FWHM) of x-ray rocking curves are measured by etching the wafer step by step. As can be seen in the figure, the damaged layer of as-sliced wafers is about ~ 9 μm. For each wafer, after deep etching, the FWHM became $11 \sim 13$ arc-sec. We can, therefore, consider that this level is the value without detectable damaged layers. In the case of as-polished wafers, without any etching, the FWHM value gave $11 \sim 13$ arc-sec, the value of bulk material.

We have also applied surface x-ray topography to evaluate the damaged layer (Takahashi et al., 1987). Figure 49 shows the photographs of as-lapped wafers. The surface x-ray topographs have been taken by etching the as-lapped wafer step by step. It can be seen that even if the surface is etched off more than the thickness of damaged layers evaluated by the double x-ray diffraction technique, one can still observe micro crack defects. This means that the double crystal x-ray diffraction evaluation is not sufficient for a complete damaged layer evaluation. In the case of the double crystal x-ray diffraction, one can only evaluate the damaged layer thickness when the

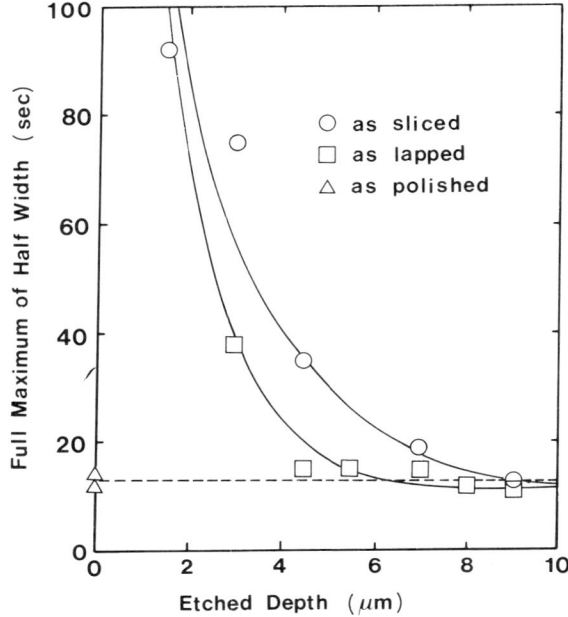

FIG. 48. Double crystal x-ray rocking curve widths as a function of etching depth of as-sliced, as-lapped, and as-polished wafers. (Takahashi *et al.*, 1987, reprinted by permission of the publisher, The Electrochemical Society, Inc).

lattice constant has been changed in the x-ray reflected area, and the diffracted x-ray wavelength can be affected by a reasonable amount. On the other hand, by surface x-ray topography, one can evaluate micro defects only where the lattice constant is changed and that gives an abrupt change of the x-ray diffraction contrast. Normally speaking, when the wafer is cut or lapped, the damage consists of two layers. In the upper layer, because of many defects and dislocations, the lattice is imperfect over the whole area, while in the lower layer, the lattice is nearly perfect in most of the area except for micro cracks or defects. As shown in the Fig. 49, we have evaluated the upper layer and the lower layer thickness by using the double crystal x-ray diffraction and the surface x-ray topography. In the case of polished wafers, no damaged layer was detected by either technique.

Even though the damaged layer is not detected by the x-ray diffraction technique, it does not mean there is no damaged layer on the polished surface since the x-ray penetration depth is about $0.5 \sim 1$ μm. If there is a damaged layer of several hundred Å, it cannot be detected by x-ray diffraction techniques.

(a)

(b)

FIG. 49. Surface x-ray topographs of as-lapped wafers and as-polished wafers. In the lower figure, the estimated damaged layer is shown.

(c)

(d)

FIG. 49. (*Continued*)

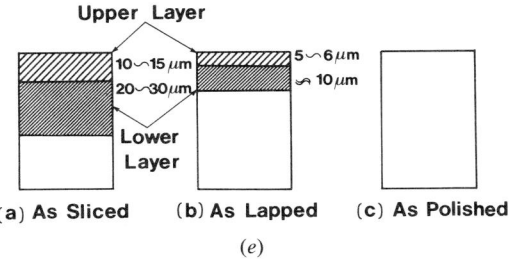

FIG. 49. (*Continued*)

We therefore used low-energy electron diffraction (LEED) in order to evaluate very thin damaged layers. The penetration depth of low-energy electrons (100–300 eV) is about 5–20 Å (Seah and Dench, 1979), so if several hundred Å damaged layers are left on the surface, one cannot observe any diffraction spots. Figure 50 shows an example of the LEED pattern and an AES spectrum of an as-polished InP wafer, set in ultra-high vacuum equipment and heat-treated at 500°C for 60 min. Here, it should be noted that the surface was not decomposed by the heat treatment. This was confirmed after the LEED/AES measurement by a Nomarski microscope. We could observe diffraction spots of (1×1) structure, as seen in Fig. 50, probably due to oxygen adsorption. Even though some oxygen was left on the surface, the fact that the diffracted spots were observed means that atoms beneath the adsorbed oxygen are arranged periodically and that there were hardly any damaged layers. This, however, does not exclude the possibility that a thin damaged layer had existed before heat treatment, but even if it had existed, the thickness would have been sufficiently thin for the deviated atoms to easily be reconstructed during the heat treatment.

21. SURFACE WAVINESS

Surface waviness is a very important quality for InP wafers. This is especially so when epitaxial layers are grown by the metal-organic chemical-vapor deposition (MOCVD) technique because the waviness of the substrate affects the surface morphology of the grown epitaxial layer.

Figure 51 shows micrographs of polished wafers and the surface waviness measurement data. As explained previously, the surface waviness can be largely improved by optimizing the polishing etchant and the polishing conditions. As can be seen in the figure, on the badly polished wafer, one can easily see that the surface is wavy. As seen in Fig. 51 (b), the surface waviness is about 80 Å in this case. If the surface waviness is more than 100 Å, the wavy surface can be observed directly with the naked eye, but when it is less than

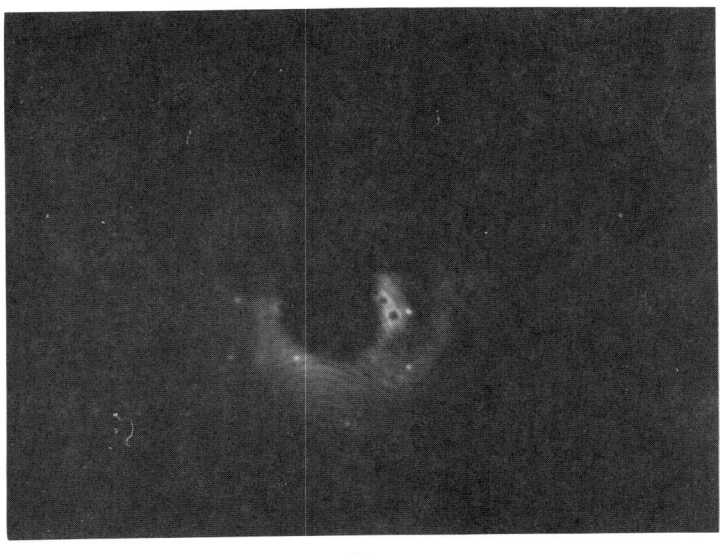

(a)

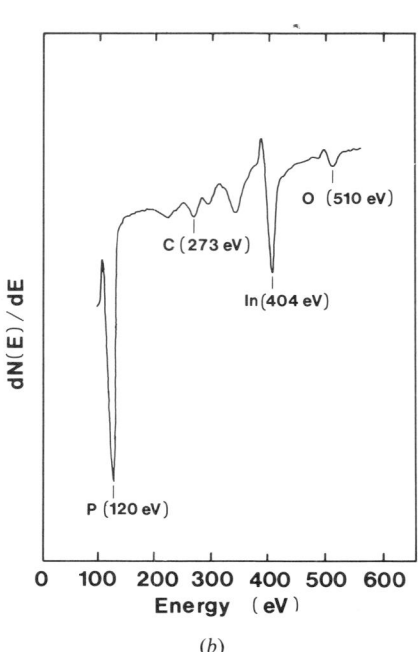

(b)

FIG. 50. LEED pattern and Auger electron spectrum of an InP wafer after heat treatment at 500°C for 60 min. Leed pattern was obtained using an electron energy of 76 eV.

(a)

(b)

FIG. 51. Nomarski photographs (left) and surface waviness (right) measured by a Talystep. (a) and (c) are badly polished wafers and (b) and (d) are polished wafers of good quality.

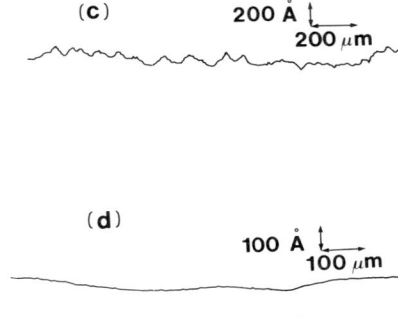

Fig. 51. (Continued)

100 Å, the Nomarski microscope is a good instrument for observing surface waviness. Figures 51 (b) and (d) show the Nomarski micrograph and the surface waviness measured by a Talystep of a good polished wafer. With the microscope, there is nothing to be observed. When the surface waviness is sufficently good, it is impossible to focus on the surface directly. For focusing, one has to observe the edge of the wafer surface. The surface waviness also shows that the measured value is less than the detection limit, 5 Å of the Talystep. The same quality surface as seen in Fig. 51(b) is reported by Gormley et al., (1981) when the wafer is polished by the hydroplaner polishing technique using Br_2-methanol solution.

22. Wafer Inspection

Polished wafers are inspected for various imperfections. In each case, the wafers must meet specifications. The most important specifications relate to surface waviness, latent scratches, and surface irregularities after etching. Surface waviness and latent scratches are due to the polishing conditions. When the polishing conditions are optimized, surface waviness is always less than the standard specified value. Even though scratches are not observed in as-polished wafers, when wafers are etched by Br_2-methanol or HF + FBr solutions, if some damaged layers exist, they are easily revealed. These are latent scratches, which are formed when the polishing condition is too mechanical. These invisible shallow scratches are made visible as deep scratches by selective etching.

The etching technique is also a very sensitive method to determine if the surface cleaning process was adequate. Even if wafers are judged as having good qualities from the previously mentioned quality inspection, if the surface cleaning process is not adequate, one can observe surface irregularities by Br_2-methanol etching. These surface irregularities include very shallow scratches, pits, trapezoid mountainlike defects, brushedlike defects, and so on. This irregular morphology can be eliminated when an adequate cleaning process is selected. It is quite difficult to determine the exact origin of these surface irregularities, but our speculation is that it is because of inhomogeneous oxidation in air after drying. When the surface is not well cleaned, leaving organics such as wax, metal impurities and/or water drop traces, these make irregular surface oxidation possible. If the surface oxide is etched off by Br_2-methanol, irregular surface morphology is revealed since the oxidation rate is different when the surface cleanliness is inhomogeneous over the wafer. Our speculation supports the fact that these surface irregularities can be revealed when the wafer is left for a long time in air and that these irregularities can be eliminated by improving the cleaning process.

The major optoelectronic device fabrication methods are changing from the LPE to the MOCVD technique. With MOCVD, the surface state of the substrate affects the epitaxial layer morphology greatly. Thus, the wafer surface quality inspection is becoming more and more important.

VII. Summary

InP is a III–V compound semiconductor that has been developed subsequent to GaAs and GaP and is now being utilized for various optoelectronic devices. It will be used not only for high frequency FETs and OEICs but also for solar cells which are useful in space.

To meet both present and future requirements, InP substrates need further quality improvement and development. First, the purity should be further improved, not only in raw materials such as indium and phosphorus, but also InP polycrystals and InP single crystals. Further purification of raw materials is continuously studied by their suppliers. For improving the purity of InP polycrystals, we need to develop new techniques that would replace HB techniques where silicon contamination is more or less inevitable. Direct synthesis technique is one of the promising methods and is now studied by various researchers. This purification work will be helpful in developing high-quality single crystals with less inclusionlike defects and Fe-doped single crystals with more stable electrical properties.

Second, EPD reduction efforts should be continued. Even though the LEC technique has its limitation for EPD reduction because of its high axial temperature gradient, the optimization of the crystal growth conditions would have a beneficial effect on the EPD reduction. In fact, vapor pressure controlled LEC techniques have been studied by Tada *et al.*, (1987), Kawase *et al.*, (1988), and Kohiro *et al.*, (1989). These techniques are promising for EPD reduction because of the realization of the low axial temperature gradient. Impurity hardening is also a very effective technique and was found to be industrially applicable for S-doped, Zn-doped dislocation-free single-crystal growth. Double impurity doping techniques are also interesting and have been studied by several researchers. The discovery of new dopants that would be effective for EPD reduction is desired. As for the crystal growth techniques that have a possibility of reducing the EPD, the vertical or horizontal Bridgeman or gradient freezing techniques should be studied in greater depth. Monberg *et al.* (1987) reported their success in the InP single crystal growth with the vertical gradient freezing technique. Their work will stimulate further investigations by these techniques.

Third, machining accuracy in the production of polished wafers should be improved in order that the level will approach that of silicon wafers. The surface cleanliness must also be improved according to fundamental studies on the surface phenomena. The extent of these surface studies still do not approach those of silicon and GaAs. Surface studies should be promoted further in order to establish the technology for high-quality InP substrates.

Development of large-diameter and long single-crystal growth techniques should now be accelerated in order to stimulate further application of InP. Even though InP found its niche in the application for optoelectronic devices, for further consumption increase and for an expansion of its field of application, a reduction in the cost of InP substrates is a very important factor. For instance, in order for InP to be extensively used for solar cells in space, large-diameter and long single-crystal growth is necessary. Large-diameter and long single growth also gives the possibility of dislocation-free area expansion in the case of impurity-hardening effects.

InP is a material that has potential for a large increase in its consumption by the semiconductor industry. For this to happen, however, the previously mentioned quality development and cost-reduction efforts must be carried out with worldwide collaboration.

Acknowledgments

The authors are very grateful to Drs. T. Ogawa, I. Tsuboya, I. Kyono, and H. Araki for their continuous encouragement, advice, and support throughout our studies on InP.

We also wish to thank everyone who has been involved in InP research and development. They are as follows: Drs. H. Araki, H. Nakata, H. Kurita, K. Suga, Y. Taniguchi, Messrs. A. Takagi, K. Yamazaki, K. Iwanami, D. Kusumi, M. Ohoka, and U. Sawafuji for their intial development of single-crystal growth; Messrs. A. Kano, K. Iwanami, D. Kusumi, M. Ohoka, and H. Hirooka for the development of the HB polycrystal synthesis; Messrs. M. Ohoka and H. Tsuchiya for the preliminary study of the SSD method; Messrs. T. Fukui and S. Kanbara for their preliminary fundamental study on the wafer processing work; Dr. J. Takahashi and Mr. T. Inoue for the photoluminescence measurement; Messrs. N. Maruyama and H. Kurita for the measurement by a LEED/AES system; Messrs. I. Shimooka, T. Osada, K. Suzuki, and T. Kanazawa for their work on InP single crystal production and industrial wafer production.

The authors are particularly grateful to Messrs. T. Obata and Y. Kobayashi for their continuous work on the analysis method establishment and the actual analysis of many samples. We finally thank Miss Y. Matsuda for electrical property measurement, data accumulation, and preparation of all figures and photographs in this section.

REFERENCES

Adamski, J. A. (1982). High Purity Polycrystalline InP. *J. Crystal Growth* **60**, 141–143.
Adamski, J. A. (1983). Synthesis of Indium Phosphide. *J. Crystal Growth* **64**, 1–9.
Akai, S., Tada, K., Morioka, M., Tatsumi, M., Kotani, T., and Shimizu, A. (1985). Large Diameter InP Single Crystals (in Japanese). "25th Proceedings of the Semiconductor Speciality Seminar" p. 95–124. (Yamagata, Japan).
Allred, W. P., Burns, J. W., and Hunter, W. L. (1981). Synthesis of Indium Phosphide in a Pressure Balanced System. *J. Crystal Growth* **54**, 4144.
Araki, H. (1984). Indium Phosphide as Applied in Optoelectronic Communication Devices. *JEE* January, 38–42.
Aspnes, D. E., and Studna, A. A. (1982a). Chemical Etching and Cleaning Procedures for Si, Ge and Some III-V Compound Semiconductors. *Appl. Phys. Lett.* **39**, 316–318.
Aspnes, D. E., and Studna, A. A. (1982b). Summary Abstract: Preparation of High-Quality Surfaces on Semiconductors by Selective Chemical Etching. *J. Vac. Sci. Technol.* **20**, 488–489.
Augustus, P. D., and Stirland, D. J. (1982). A Study of Inclusions in Indium Phosphide. *J. Electrochem. Soc.* **129**, 614–621.
Bachmann, K. J., Buehler, E., Shay, J. L., and Strand, R. A. (1975). Liquid Encapsulated Czochralski Pulling of InP Crystals. *J. Electron. Mater.* 4, 389–406.
Ballman, A. A., Glass, A. M., Nahory, R. E., and Brown, H. (1983). Double Doped Low Etch Pit Density InP with Reduced Optical Absorption. *J. Crystal Growth* **62**, 198–202.
Blom, G. M., and Zwicher, W. K. (1973). Growth of GaP Single Crystals by Liquid Encapsulated Czochralski Pulling. *Acta Electron.* **16**, 315–322.
Bonner, W. A. (1981). InP Synthesis and LEC Growth of Twin-free Crystals. *J. Crystal Growth* **54**, 21–31.
Boomgaard, J., and Schol, K. (1957). The P-T-x Phase Diagrams of the Systems In-As, Ga-As and In-P. *Philips Res. Rep.* **12**, 127–140.
Brandt, C. D., Hennel, A. M., Pawlowicz, L. M., Wu, Y. T., Bryskiewicz, T., Lagowski, J., and Gatos, H. C. (1986). New Semi-insulating InP: Titanium Midgap Donors. *Appl. Phys. Lett.* **48**, 1162–1164.
Brasen, D., and Bonner, W. A., (1983). Effect of Temperature and Sulfur Doping on the Plastic Deformation of InP Single Crystals. *Mat. Sci. Eng.* **61**, 167–172.
Brown, G. T., Cockayne, B., Elliott, C. R., Regnault, J. C., Stirland, D. J., and Augustus, P. D. (1984). A Detailed Microscopic Examination of Dislocation Clusters in LEC InP. *J. Crystal Growth* **67**, 495–506.
Brown, G. T., Cockayne B., and MacEwan, W. R. (1981). The Growth of Dislocation-free Ge-doped InP. *J. Crystal Growth* **51**, 369–372.
Chen, C. P. (1982). Minimum Silicon Wafer Thickness for ID Wafering. *J. Electrochem. Soc.* **129**, 2835–2837.
Chen, R. T., and Holmes, D. E. (1983). Dislocation Studies in 3-inch Diameter Liquid Encapsulated Czochralski GaAs. *J. Crystal Growth* **61**, 111–124.

Cockayne, B., Bailey, T., and MacEwan, W. R. (1986). Dislocation-free LEC Growth of InP Doped with Ge and S. *J. Crystal Growth* **76**, 507–510.

Cockayne, B., MacEwan, W. R., Rown, G. T., and Wilgoss, W. H. E. (1981a). A Systematic Study of the Electrical Properties of Fe-doped InP Single Crystals. *J. Materials Science* **16**, 554–557.

Cockayne, B., Brown, G. T., and MacEwan, W. R. (1981b). The Growth and Perfection of Single Indium Phosphide Produced by the LEC Technique. *J. Crystal Growth* **54**, 9–15.

Coquille, R., Guillou, Y. L. (1986). Preparation de Substrat d'InP. *Vides Couches Minces* **41**, 237–238.

Coquille, R., Toudic, Y., Gauneau, M., Grandpierre, G., and Paris, J. C. (1983). Synthesis, Crystal Growth and Characterization of InP. *J. Crystal Growth* **64**, 23–31.

Coquille, R., Toudic, Y., Haji, L., Gauneau, M., Moisan, G., and Lecrosnier, D. (1987). Growth of Low-Dislocation Semi-insulating InP (Fe, Ga). *J. Crystal Growth* **83**, 167–173.

Dean, P. J., Skolnick, M. S., Cockayne, B., MacEwan, W. R., and Iseler, G. W. (1984). Residual Donors in LEC Indium Phosphide. *J. Crystal Growth* **67**, 486–494.

Derby, J. J., and Brown, R. A. (1986a). Thermal-capillary Analysis of Czochralski and Liquid Encapsulated Czochralski Crystal Growth. I. Simulation. *J. Crystal Growth* **74**, 605–624.

Derby, J. J., and Brown, R. A. (1986b). Thermal-capillary Analysis of Czochralski and Liquid Encapsulated Czochralski Crystal Growth. II. Processing Strategies. *J. Crystal Growth* **75**, 227–240.

Donaghey, L. F. (1980). Hydrodynamics of Crystal Growth Processes. "Crystal Growth", B. R. Pamplin ed. p. 65–103. Pergamon Press.

Dowling, D. J., Wardill, J. E., Brunton, R. A., Crouch, D. A. E., and Thompson, A. J. (1986). The Large Scale Synthesis and Growth of Polycrystalline InP. 1986 *NATO Sponsored InP Workshop*.

Duseaux, M., and Jacob, G. (1982). Formation of Dislocations during Liquid Encapsulated Czochralski Growth of GaAs Single Crystals. *Appl. Phys. Lett.* **40**, 790–793.

Ehrenreich, H., and Hirth, J. P. (1985). Mechanism for Dislocation Density Reduction in GaAs Crystals by Indium Addition. *Appl. Phys. Lett.* **46**, 668–670.

Farges, J. P. (1982). A Method for the "in-situ" Synthesis and Growth of Indium Phosphide in a Czochralski Puller. *J. Crystal Growth* **59**, 665–668.

Farges, J. P. and Schiller, C. (1987). Growth of Large Diameter Dislocation-Free InP Ingots, *J. Crystal Growth* **83**, 159–166.

Forman, S. E., and Rhines, W. J. (1972). Vibration Characteristics of Crystal Slicing ID Saw Blades. *J. Electrochem. Soc.* **119**, 686–690.

Franzosi, P., Salviati, G., Cocito, M., Taiariol, F., and Ghezzi, C. (1984). Inclusion-like Defects in Czochralski Grown InP Single Crystals. *J. Crystal Growth* **69**, 388–398.

Gall, P., Kallel, A., Lauret, N., Peyrade, J. P., Brousseau, B., and Mazzashi, J. (1986). An Experimental Evidence of the Destruction of Complex Defects by the Introduction of Dislocations in InP. *Phys. Stat. Sol. (a)* **97**, K25–K27.

Gillessen, K., and Marshall, A. J. (1976). Growth of GaP Single Crystals by the Synthesis, Solute Diffusion Method. *J. Crystal Growth* **32**, 216–220.

Gillessen, K., and Marshall, A. J. (1977). Optimization of Growth Conditions in the Synthesis, Solute Diffusion Growth of GaP. *J. Crystal Growth* **33**, 356.

Gormley, J. V., Manfra, M. J., and Calawa, A. R. (1981). Hydroplane Polishing of Semiconductor Crystals. *Rev. Sci. Instrum.* **52**, 1256–1259.

Gottshalk, H., Patzer, G., and Alexander, H. (1978). Stacking Fault Energy and Ionicity of Cubic III-V Compounds. *Phys. Stat. Sol. (a)* **45**, 207–217.

Guivarc'h, A., L'haridon, H., and Pelous, G. (1984). Chemical Cleaning of InP Surfaces: Oxide Composition and Electrical Properties. *J. Appl. Phys.* **55**, 1139–1148.

Grandia, J., and Hill, J. C. (1978). Improved Slicing and Orientation Technique for I.D. Sawing. *Solid State Tech.* Feb., 40–42.

Hasegawa, H., and Sawada, T. (1981). III-V Compound Semiconductor MIS Interfaces and their Applications (in Japanese). *Ohyou Butsuri (Appl. Phys.).* **50,** 1289–1303.

Harris, I. R., Smith, N. A., Cockayne, B., and MacEwan, W. R. (1983). Precipitate Identification in Co-doped InP. *J. Crystal Growth* **64,** 115–120.

Hyder, S. B., and Holloway, C. J., Jr. (1983). In-situ Synthesis and Growth of Indium Phosphide. *J. Electron. Mat.* **12,** 575–585.

Hyder, S. B., and Antypas, G. A. (1986). Synthesis and "in-situ" LEC Growth of (100) Oriented InP Single Crystals. 1986 *NATO Sponsored InP Workshop.*

Liyama, S. (1982). Mirror Polishing of InP Substrates (in Japanese). *Seikigakkai Ronbunshu* **57,** 572–573.

Inada, T., Fujii, T., Eguchi, M., and Fukuda, T. (1987). Technique for the Direct Synthesis and Growth of Indium Phosphide by the Liquid Phosphorus Encapsulated Czochralski Method. *Appl. Phys. Lett.* **50,** 86–88.

Inoue, T., and Takahashi, J. (1987). Photoluminescence Study of Undoped, Sn-Doped and S-doped InP Single crystals. *Jpn. J. Appl. Phys.* **26,** L249–252.

Iscoff, R. (1982). Crystal Slicing Equipment Directions. *Semicon. International.* Feb., 51–71.

Iseler, G. W. (1979). Properties of InP Doped with Fe, Cr or Co. *Inst. Phys Conf. Ser.* 45, 144–153.

Iseler, G. W., and Ahern, B. S. (1986). Titanium-doped Semi-insulating InP Grown by the Liquid Encapsulated Czochralski Method. *Appl. Phys. Lett.* **48,** 1656–1657.

Itoh, T., Asano, K., Kasahara, K. Ozawa, T., Ando, Y., and Ohata, K. (1989). MBE Grown AlGaAs/InP MIS System and Its Hetero-MIS Gate InP FETs. "The Proceedings of the First International Conference on InP and Related Materials for Advanced Electronics and Optical Devices" (Oklahoma, U.S.A., 1989). to be published.

Janson, M. (1983). Optimum Conditions for Baking of Indium. *J. Electrochem. Soc.* **130,** 960–962.

Jensen, E. W. (1973). Polishing Compound Semiconductors. *Solid State Tech.* Aug., 49–52.

Jordan, A. S. (1980). An Evaluation of the Thermal and Elastic Constants Affecting GaAs Crystal Growth. *J. Crystal Growth* **49,** 631–642.

Jordan, A. S., Caruso, R., and Von Neida, A. R. (1980). A Thermoelastic Analysis of Dislocation Generation in Pulled GaAs Crystals. *Bell System Tech. J.* April, 593–637.

Jordan, A. S., Von Neida, A. R., and Caruso, R. (1984). The Theory and Practice of Dislocation Reduction in GaAs and InP. *J. Crystal Growth* **70,** 555–573.

Jordan, A. S. (1985). Some Thermal and Mechanical Properties of InP Essential to Crystal Growth Modeling. *J. Crystal Growth* **71,** 559–565.

Jordan, A. S., Brown, G. T., Cockayne, B., Brasen, D., and Bonner, W. A. (1985). An Analysis of Dislocation Reduction by Impurity Hardening in the Liquid-encapsulated Czochralski Growth of $\langle 111 \rangle$ InP. *J. Appl. Phys.* **58,** 4383–4389.

Jordan, A. S., Caruso, R., Von Neida, A. R., and Nielsen, J. W. (1981). A Comparative Study of Thermal Stress Induced Dislocation Generation in Pulled GaAs, InP, and Si Crystals. *J. Appl. Phys.* **52,** 3331–3336.

Jordan, A. S., Von Neida, A. R., and Caruso, R. (1986). The Theoretical and Experimental Fundamentals of Decreasing Dislocations in Melt Grown GaAs and InP. *J. Crystal Growth* **76,** 243–262.

Jordan, A. S., and Parsey, J. M., Jr. (1986). The Role of Crystal Diameter and Impurity Hardening on the Threshold for Dislocation Formation in LEC GaAS. *J. Crystal Growth* **79,** 280–286.

Kachajian, G. S. (1972). A System Approach to Semiconductor Slicing to Improve Wafer Quality and Productivity. *Solid State Tech.* Sept., 59–64.

Kainosho, K., Ohoka, M., Kusimi, D., and Oda, O. (1985). Synthesis of InP Polycrystals by the SSD Technique. "*Extended Abstracts (the 33rd Spring Meeting), The Japan Society of Applied Physics and Related Societies*" p. 743.

Kainosho, K., Shimakura, H., Kanazawa, T., Inoue, T., and Oda, O. (1989). Low Fe Doped InP Single Crystals and their Ion Implantation Characteristics, published in "*Proceedings of the 16th International Symposium on GaAs and Related Compounds*" (Karuizawa, Japan, 1989).

Kainosho, K., Shimakura, H., Yamamoto, H., Inoue, T., and Oda, O. (1989). Annealing Conditions for Fe Doped Semi-insulating InP, to be published in "*Proceedings of the First International Conference on Indium Phosphide and Related Materials for Advanced Electronic and Optical Devices*" (Norman, Oklahoma, 1989).

Kaneko, K., Ayabe, M., Dosen, M., Morizane, K., Usui, S., and Watanabe, N. (1973). A New Method of Growing GaP Crystals for Light-emitting Diodes. *Proceedings of the IEEE* **61**, 884–890.

Katagiri, K., Yamazaki, S., Takagi, A., Oda, O., Araki, H. and Tsuboya, I. (1986). LEC Growth of Large Diameter InP Single Crystals Doped with Sn and S. *Inst. Phys. Conf. Ser.* (79) 6772.

Katsui, A., and Thono, S. (1986). Growth and Characterization of Isoelectrically Double-doped Semi-insulating InP (Fe) Single crystals. *J. Crystal Growth* **79**, 287–290.

Katsui, A., Thono, S., Homma, Y., and Tanaka, T. (1986). LEC Growth and Characterization of Ga Doped InP Crystals. *J. Crystal Growth* **74**, 221–224.

Kawase, T., Araki, T., Nakagawa, M., Tatsumi, M., and Tada, K. (1988). Evaluation of Sn-doped InP Single Crystal by Modified LEC Technique. "*Extended Abstracts (the 49th Autumn Meeting), The Japan Society of Applied Physics*," p. 227.

Kim, H. K., Hwang, J. S., Noh, S. K., and Chung, C. H. (1986). Characteristics of a Semi-insulating InP:Fe Wafer. *Jpn. J. Appl. Phys.* **25**, L888–L890.

Kobayashi, N., and Iwaki, T. (1985). A Thermoelastic Analysis of the Thermal Stress Produced in a Semi-infinite Cylindrical Single Crystal during Czochralski Growth. *J. Crystal Growth* **73**, 96–110.

Kohda, H., Yamada, K., Nakanishi, H., Kobayashi, T., Osaka, J. and Hoshikawa, K. (1985). Crystal Growth of Completely Dislocation-free and Striation-free GaAs. *J. Crystal Growth* **71**, 813–816.

Kohiro, K., Mori, M., Yamamoto, H., and Oda, O. (1989). Growth of Sulphur Doped InP Single Crystals by the P Pressure Controlled LEC Method. "*Extended Abstracts (the 36th Spring Meeting)* No. 1, *The Japan Society of Applied Physics and Related Societies*," p. 279.

Kubota, E., Ohmori, Y., and Sugii, K. (1981). High-quality Semi-insulating Indium Phosphide Single Crystals Grown by Novel ADC System. *Inst. Phys. Conf. Ser.* **63**, 31–36.

Kubota, E., and Sugii, K. (1981). Preparation of High-purity InP by the Synthesis, Solute Diffusion Technique. *J. Appl. Phys.* **52**, 2983–2986.

Kubota, E., and Sugii, J. (1984). Growth of InP Single Crystals by Grown-rate Controlled Synthesis, Solute-Diffusion Technique. *J. Crystal Growth* **68**, 639–643.

Kusumi, D., Kainosho, K., Katagiri, K., Oda, O., and Araki, H. (1985). Synthesis of High-purity InP Polycrystals. "*Extended Abstracts (the 46th Autumn Meeting), The Japan Society of Applied Physics*," p. 743.

Lee, R. N., Norr, M. K., Henry, R. L. and Swiggard, E. M. (1977). Precipitates in Fe-doped InP. *Mat. Res. Bull.* **12**, 651–655.

Marshall, A. J., and Gillessen, K. (1978). Growth of InP Crystals by the Synthesis, Solute Diffusion Method. *J. Crystal Growth* **44**, 651–652.

Martin, S., and Jacob, G. (1983). Proprietes Electriques des Lingots d'Arsenuire de Gallium. *Acta Electronica* **25**, 123–132.

Meek, R. L., and Huffstutler, M. C., Jr. (1969). ID-diamond Sawing Damage to Germanium and Silicon. *J. Electrochem. Soc.* **116**, 893–898.

Miyairi, H., Inada, T., Eguchi, M., and Fukuda, T. (1986). Growth and Properties of InP Single Crystals Grown by the Magnetic Field Applied LEC Method. *J. Crystal Growth* **79**, 291–295.
Miyazawa, S., and Koizumi, H. (1982). Eutectic Precipitates in Fe-doped InP. *J. Electrochem. Soc.* **129**, 2335–2338.
Monberg, E. M., Gault, W. A., Dominguez, F., Simchock, F., Chu, S. N. G., and Stiles, C. M. (1988). The Growth and Characterization of Large Size, High Quality, InP Single Crystals. *J. Electrochem. Soc.* **135**, 500–503.
Morioka, M., Kikuchi, K., Kohe, K., and Akai, S. (1981). Effect of Fe Content on the Thermal Stability of LEC-grown Semi-insulating InP. *Inst. Phys. Conf. Ser.* (63), 37–42.
Morioka, M., Tada, K., and Akai, S. (1986). Development of InP Crystal Substrates towards the Low EPD and the Purification (in Japanese). *Nikkei Microdevices* July, 89–107.
Nielsen, L. D. (1972). Microwave Measurement of Electron Drift Velocity in Indium Phosphide for Electric Fields up to 50 kV/cm. *Phys. Lett.* A **38**, 221–222.
Oda, O., and Katagiri, K. (1986). InP (in Japanese). "*Manual for Optoelectronics Materials*" (*Tokyo, Optronics Co.*), p. 492–497.
Ohoka, M., Tsuchiya, H., and Araki, H. (1985). Fabrication of InP Polycrystals by the SSD Technique. "*Extended Abstracts* (*the 46th Autumn Meeting*), *The Japan Society of Applied Physics.*" p. 630.
Okazaki, H., Takamoto, T., Takamura, H., Kamei, T., Ura, M., Yamamoto, A., and Yamaguchi, M. (1988). "*Proceedings of the 20th IEEE Photovoltaic Specialists Conference*" (*Las Vegas, September, 1988*). p. 26–30.
Onabe, K. (1982). Calculation of Miscibility Gap in Quaternary InGaPAs with Strictly Regular Solution Approximation. *Jpn. J. Appl. Phys.* **21**, 797–798.
Osaka, J., and Hoshikawa, K. (1984). Homogeneous GaAs Grown by Vertical Magnetic Field Applied LEC Method. "*Semi-Insulating III-V Materials*" D. C. Look and J. S. Blakemore, eds. p. 126–133.
Pak, K., Nakano, T., and Nishinaga, T. (1981). Doping Effect of Oxygen on Horizontal Bridgeman Grown InP. *Jpn. J. Appl. Phys.* **20**, 1815–1819.
Roksnoer, P. J., and Van Rijbroek-van den Boom, M. M. B. (1984). The Single Crystal Growth and Characterization of Indium Phosphide. *J. Crystal Growth* **66**, 317–326.
Rumsby, D., Ware, R. M., and Whittaker, M. (1980). The Growth and Properties of Large Semi-insulating Crystals of Indium Phosphide. "*Semi-insulating III-V Materials*," G. J. R. Rees ed. p. 59–67. Nottingham.
Saito, Y., Sakagami, K., and Watanabe, K. (1985). Technology of InP Single Crystal Growth (in Japanese). *Semiconductor World* June, 110–116.
Sakagami, K. (1986). High Purity InP Polycrystals (in Japanese). *Nikkei Materials* **4**, 51–54.
Sasaki, Y., Kuma, S., and Nakagawa, J. (1985). Development of InP Wafers for Long Wavelength Optical Devices (in Japanese). *Hitachi Densen* **5**, 33–36.
Sasaki, Y., Shibata, M., Inada, T., and Kuma, S. (1987). Heat Flow Variation through B_2O_3 Layer in the LEC Growth (in Japanese). "*Proceedings of the Seminar of the Electrical Society*" (*Electronic Materials Seminar, June, 1987*). EFM-87-14, p. 51–57.
Seah, M. P., and Dench, W. A. (1979). Quantitative Electron Spectroscopy of Surfaces: A Standard Data Base for Electron Inelastic Mean Free Paths in Solids. *Surface Interface Anal.* **1**, 2–11.
Seki, Y., Matsui, J., and Watanabe, H. (1976). Impurity Effect on the Growth of Dislocation-free InP Single Crystals. *J. Appl. Phys.* **47**, 3374–3376.
Seki, Y., Watanabe, H., and Matsui, J. (1978). Impurity Effect on Grown-in Dislocation Density of InP and GaAs Crystals. *J. Appl. Phys.* **49**, 822–828.
Sekinobe, M., and Morioka, M. (1983). Present Status and Future Trend of the Technology of InP Single Crystal Growth (in Japanese). *Semiconductor World* July, 55–86.

Shimada, T., Obokata, T., and Fukuda, T., (1984). Growth and Resistivity Characteristics of Undoped Semi-insulating GaAs Crystals with Low Dislocation Density. *Jpn. J. Appl. Phys.* **23**, L441–L445.

Shinoyama, S. (1985). Study on High-quality InP Single Crystal Growth. *Doctor thesis (Tohoku University)*.

Shinoyama, S., and Uemura, C. (1985). Growth of Dislocation-free InP Crystals by the LEC Method (in Japanese). *Zairyoo Kagaku (Materials Sciences)* **21**, 273–278.

Shinoyama, S., Uemura, C., Yamamoto, A., and Thono, S. (1980). Growth of Dislocation-free Undoped InP Crystals. *Jpn. J. Appl. Phys.* **19**, L331–L334.

Singh, S., Williams, R. S., Van Uitert, L. G., Schlierr, A., Camlibel, I., and Bonner, W. A. (1982). Analysis of InP Surface Prepared by Various Cleaning Methods. *J. Electrochem. Soc.* **129**, 447–448.

Straughan, B. W., Hurle, D. T. J., Lloyd, K., and Mullin, J. B. (1974). Eutectic Formation in Chromium-doped Indium Phosphide. *J. Crystal Growth* **21**, 117–124.

Sugii, K., Kubota, E., and Iwasaki, H. (1979). Large-sized InP Single crystals by the Synthesis, Solute-Diffusion Technique. *J. Crystal Growth* **46**, 289–292.

Szabo, G. (1985). Thermal Strain during Czochralski Growth. *J. Crystal Growth* **73**, 131–141.

Tada, K., Tatsumi, M., Nakagawa, M., Kawase, T., and Akai, S. (1987). Growth of Low-dislocation Density InP by the Modified CZ Method in the Atmosphere of Phosphorus Vapour Pressure. *Inst. Phys. Conf. Ser.* (91), 439–442.

Takahashi, Y., Fukui, T., and Oda, O. (1987). Evaluation of the Damaged Layers Formed during the Wafer Processing of InP Wafers. *J. Electrochem. Soc.* **134**, 1027–1028.

Thono, S., Kubota, E., Shinoyama, S., Katsui, A., and Uemura, C. (1984). Isoelectronic Double Doping Effect on Dislocation Density of InP Single Crystal. *Jpn. J. Appl. Phys.* **23**, L72–L74.

Thono, S., and Katsui, A. (1986). Single Crystal Growth and Characterization of Isolectronically-doped InP. *J. Crystal Growth* **78**, 249–256.

Tuck, B. (1975). Review: The Chemical Polishing of Semiconductors. *J. Mater. Sci.* **10**, 321–339.

Tuppen, C. G., and Conen, B. H. (1987). Ultra-flat InP Substrates Produced using a Chemo-mechanical Polishing Technique. *J. Crystal Growth* **80**, 459–462.

Yamaguchi, M., Uemura, C., and Yamamoto, A. (1984a). Radiation Damage in InP Single Crystals and Solar Cells. *J. Appl. Phys.* **55**, 1429–1436.

Yamaguchi, M., Uemura, C., Yamamoto, A., and Shibukawa, A. (1984b). Electron Irradiation Damage in Radiation Resistant InP Solar Cells. *Jpn. J. Appl. Phys.* **23**, 302–307.

Yamamoto, A., Shinoyama, S., and Uemura, C. (1981). Silicon Contamination of InP synthesized under High Phosphorus Pressure. *J. Electrochem. Soc.* **128**, 585–589.

Yoshida, K., Suzuki, Y., Nakayama, M., and Kikuta, T. (1987). Crystal Growth of High Purity InP by GF Method. "*Extended Abstracts*" (*The 48th Autumn Meeting*) *No. 1*; *The Jpn. Soc. Appl. Phys.*, p. 190.

Walukiewicz, W., Lagowski, J., Jastrebski, L., Rava, P., Lichtensteiger, M., Gatos, C. H., and Gatos, H. C. (1980). Electron Mobility and Free-carrier Absorption in InP; Determination of the Compensation Ratio. *J. Appl. Phys.* **51**, 2659–2668.

Wardill, J. E., Doling, D. J., Brunton, R. A., Crouch, D. A. E., Stockbridge, J. R., and Thompson, A. J. (1983). The Preparation and Assessment of Indium Phosphide. *J. Crystal Growth* **64**, 15–22.

Weinberg, I., and Brinker, D. J. (1986). Indium Phosphide Solar Cells-Status and Prospects for Use in Space. "*Proc. Intersoc. Energy Convers. Eng. Conf.*" 21st **3**, 1431–1435.

Wysocki, J., and Rappaport, P. (1960). Effect of Temperature on Photovoltaic Solar Energy Conversion. *J. Appl. Phys.* **31**, 571–578.

Zeiss, C. R., Antypas, G. A., and Hopkins, C. (1983). Electron and Iron Concentration in Semi-insulating Indium Phosphide. *J. Crystal Growth* **64**, 217–221.

CHAPTER 5

InP Substrates: Production and Quality Control

Koji Tada, Masami Tatsumi, Mikio Morioka, Takashi Araki, and Tomohiro Kawase

SUMITOMO ELECTRIC INDUSTRIES, LTD.
OSAKA, JAPAN

	LIST OF ACRONYMS.	175
I.	INTRODUCTION	176
II.	APPLICATIONS AND REQUIREMENTS	176
III.	MATERIAL FEATURES	178
IV.	PRODUCTION OF InP CRYSTAL.	180
	1. Synthesis of InP Polycrystal	181
	2. Growth of InP Single Crystal	184
V.	NEW DEVELOPMENTS IN InP CRYSTAL GROWTH	193
	3. Preparation of Polycrystal with High Purity	194
	4. Reduction of Dislocation Density	195
VI.	EVALUATIONS.	204
	5. On-line Evaluations	205
	6. Off-line Evaluations	209
	7. Simulations	220
VII.	QUALITY CONTROL OF CRYSTAL GROWTH.	221
VIII.	WAFER PROCESSING	222
	8. Fundamental Properties of InP Single Crystal	223
	9. Fundamental Specification for InP Substrates	224
	10. Wafer Processing for InP Substrates	225
	11. Wafer Characteristics and Their Evaluations.	235
IX.	SUMMARY	238
	REFERENCES	238

List of Acronyms

AI	artificial intelligence
EPD	etch-pit density
FEC	fully encapsulated Czochralski
GDMS	glow discharge mass spectrometry
GF	gradient freezing
GFA	graphite furnace atomic absorption
HB	horizontal Bridgman
ICP	inductively coupled plasma
ID	inner diameter
LEC	liquid encapsulated Czochralski

m/c   mass/charge
OEIC   optoelectronic integrated circuits
ppma   parts per million of atoms
SIMS   secondary ion mass spectrometry
SSD   synthesis solute diffusion
SSMS   spark source mass spectrometry
TTV   total thickness variation
VGF   vertical gradient freezing

I. Introduction

Indium phosphide single crystals have been used exclusively as conductive substrates for infrared optical devices. They are also under development as semi-insulating substrates usable in higher frequency microwave devices, such as MISFETs and optoelectronic integrated circuits. Although the effects of such microdefects as dislocations, stacking faults, and precipitates on the properties of microwave devices, optoelectronic devices, and future integrated circuits are not yet clear, there is a great demand for the development of higher quality indium phosphide single crystals. For this purpose, the technologies reviewed in this article, especially those concerning both the crystal growth and the processing of this material, are very important for future developments.

This article covers the whole scope of techniques related to indium phosphide single crystal materials and their characterization. In Section 2, a general overview of the applications of InP crystals and the requirements for their quality will be described. In Section 3, the crystal's material features will be briefly discussed. In Sections 4 and 5, the details of crystal growth techniques will be reviewed, including recent developments obtained in our laboratories. In the last three sections, the techniques used in the evaluation, quality control, and wafer processing will be described.

II. Applications and Requirements

There are two main applications of InP single crystals. InP single crystals are used for optical devices, including optical sources (LED, LD) and detectors (PD, APD), which are used in optical communication systems, and for electrical devices, which include high-speed microwave devices, such as MISFET, and optoelectronic integrated circuits (OEIC). For these applications, crystal growers can now supply InP substrates having the standard specifications shown in Table I. Materials for laser diodes, especially n-type (Sn-doped) substrates, are used in large quantities. P-type (Zn-doped) substrates are used in high-power pulse lasers or in solar cells. Low

TABLE I
USAGE AND STANDARD SPECIFICATION OF InP SUBSTRATES

Dopant	Device application	Process	Specification		
			Electrical property	EPD	Size
Sn	LD LED PIN-PD		$n = 1\text{-}4 \times 10^{18}$ cm^{-3}	$\leqq 5 \times 10^4$ cm^{-2}	Rectangular or $2''\phi$
S	LD LED PIN-PD, APD	LPE VPE OMVPE	$n = 4\text{-}20 \times 10^{18}$ cm^{-3}	$\leqq 5 \times 10^3$ cm^{-2} ($\leqq 5 \times 10^2$ cm^{-2} for APD)	Rectangular or $2''\phi$ ($3''\phi$)
Zn	LD Solar Cell[a]	MBE[a] I$^{2\,a}$	$P = 4\text{-}6 \times 10^{18}$ cm^{-3} $P = 2\text{-}10 \times 10^{18}$ cm^{-3}	$\leqq 5 \times 10^3$ cm^{-2} $\leqq 1 \times 10^5$ cm^{-2}	Rectangular or $2''\phi$
Fe	FET[a] OEIC[a]		$\rho \geqq 1 \times 10^7$ $\Omega\cdot$cm	$\leqq 1 \times 10^5$ cm^{-2}	Rectangular or $2''\phi$ ($3''\phi$)
Undoped	Source material	LPE	$n \leqq 1 \times 10^{16}$ cm^{-3}	—	Rectangular or 50–60 mmϕ
Single Poly					

[a]Under development

dislocation density is required for both pin-PDs, and APDs, since dislocations increase the dark current, which affects their sensitivities. As for semi-insulating substrates, Fe-doped InP single crystals are chiefly employed in MISFETs. The substrates must be highly pure and, at $600 \sim 800°C$, must be thermally stable. These substrates are usually supplied in rectangular form as the liquid phase epitaxial technique is employed for fabricating optical devices. Recently, however, round wafers have been required for such new techniques as OMVPE, VPE, and MBE. As device technology advances, it becomes essential for InP crystals to have better qualities than those described in Table I. The essential qualities are low dislocation density (S-doped: slip free; Sn-doped: EPD $< 2 \times 10^4$ cm^{-2}, Fe-doped: EPD $< 1 \times 10^4$ cm^{-2}) and high purity (amount of residual impurities: $\langle 1-5 \times 10^{15}$ cm^{-3}). Other necessary qualities in the wafer surface include minimum damage, extreme cleanliness, and flatness. Crystal growers will soon be able to achieve these qualities by further developing crystal growth and wafer processing techniques.

III. Material Features

InP single crystals exhibit several useful properties (Bimberg et al., 1982) that are superior to those of GaAs single crystals for several optical and electrical applications. One of the features of InP crystals is that InP has a suitable lattice constant that matches perfectly to that of alloy semiconductors such as InGaAsP, which can be applied to optical devices used in optical communication systems (Foyt, 1981). Other features include the high mobility and the high saturated drift velocity of electrons in InP crystal, which are very important properties for constructing high-speed electron devices (Cameron, 1984).

Fundamental properties on InP crystal are shown in Table II, which includes those of GaAs and Si crystals. In order to produce high-quality InP single crystals, it is essential to know the thermodynamic (Bachmann and Buehler, 1975; Richman, 1963), crystallographic (Donnay, 1963), and mechanical (Jordan, 1985) properties.

The complete phase diagram of the InP system has not been obtained yet due to the high pressure of phosphorus equilibrated with the melt, especially in the phosphorus-rich region. The phase diagram (Boomgaard and Schol, 1957) is shown in Fig. 1. The dissociation pressure is about 27.5 atm at the melting point (1062°C), which is much higher than the 0.9 atm for GaAs. Solid InP also exhibits high dissociation pressure of phosphorus, e.g., 0.6 atm at 1,000°C. Therefore, it becomes necessary to prepare an optimum temperature profile and a suitable atmosphere around the crystal during growth in

5. InP SUBSTRATES: PRODUCTION AND QUALITY CONTROL

TABLE II
PROPERTIES OF SEMICONDUCTOR MATERIALS

Property Material	InP	GaAs	Si
Crystal structure	Zinc blend	Zinc Blend	Diamond
Lattice constant (nm)	0.5869	0.5654	0.543
Density (g/cm^3)	4.787	5.316	2.329
Melting point (K)	1335	1511	1683
Vapor pressure at Tm (atm)	27.5	0.9	$3\text{--}4 \times 10^{-4}$
Heat capacity (cal/gK)	0.073	0.086	0.167
Dielectric constant	12.0	11.1	12.0
Thermal expansion coefficient (K^{-1})	4.75×10^{-6}	5.93×10^{-6}	2.33×10^{-6}
Lattice thermal conductivity (W/cmK)	0.7	0.455	1.412
Band structure	direct	direct	indirect
Band gap energy (eV)	1.35	1.40	1.106
Intrinsic resistivity (Ωcm)	8.2×10^7	3.8×10^8	2.3×10^5
Intrinsic carrier concentration (cm^{-3})	1.6×10^7	2.25×10^6	1.5×10^{10}
Electron mobility (cm^2/Vs)	4600	8800	1400
Hole mobility (cm^2/Vs)	150	400	425

(at 300 K)

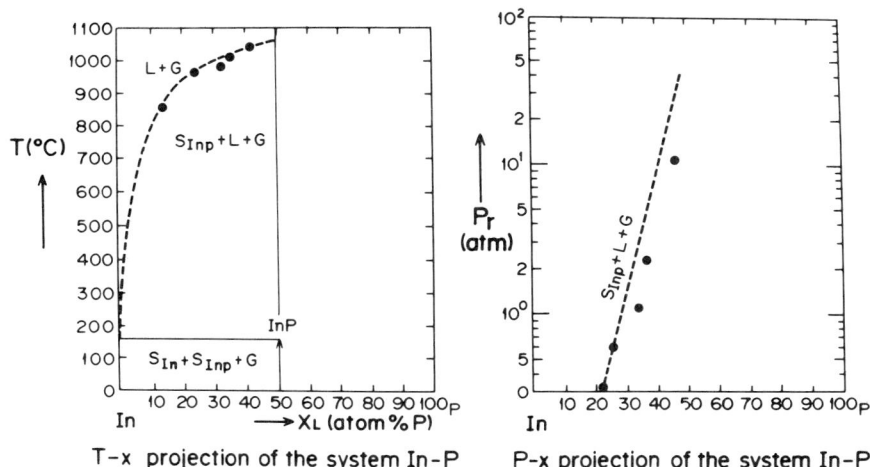

FIG. 1. Phase diagrams of the In–P system.

TABLE III

STACKING FAULT ENERGY AND LOWER RESOLVED YIELD STRESS OF III–V COMPOUNDS

	GaSb	GaAs	InSb	GaP	InAs	InP
Stacking fault energy (meV/atom)[a]	53	47	43	33	30	17
Lower resolved yield stress (N/mm^2)[a]	15.8	1.9	5.0	4.0	0.8	1.8

[a] At $T = 0.65\,T_m$, where T_m is the absolute temperature of the melting point

order to suppress the dissociation. This condition makes the growth of InP single crystals more difficult than that of GaAs.

The other important properties on InP for growing single crystals are stacking fault energy and lower resolved yield stresses (critical resolved shear stresses, shown in Table III). The stacking-fault energy is related to the probability of generation of twinning in InP single crystals. As shown in Table III, the stacking-fault energy in InP is much lower than that in other III–V compounds, so the growth of InP single crystal is more frequently impeded by the occurrence of twinning. The critical shear stress is the minimum amount of stress that will generate dislocations in a single crystal. Although dislocations are easily generated or multiplicated under less stress than the critical shear stress, this value is assumed as a condition of limitation for growing single crystals having low dislocation density. The critical shear stress of InP is nearly equal to that of GaAs. The other properties related to the generation of stress, such as thermal conductivity, thermal expansion coefficient, etc. also exhibit practically the same values for both GaAs and InP crystals. According to the Jordan estimation, the dislocation density in InP should have the same order of magnitude as that in GaAs (Jordan et al., 1981).

IV. Production of InP Crystal

Developments in the manufacturing techniques of a promising material generally result in both higher quality and lower cost of the material. In order to realize these benefits, crystal growers have been making great efforts. In order to achieve lower cost and mass production, it is necessary to produce, using sophisticated procedures, long single crystals with large diameters and well-controlled shapes. In order to obtain better properties having uniform distribution in single crystals, it is necessary to achieve a high level of purity, low dislocation density, and stoichiometric composition in the growth

5. InP SUBSTRATES: PRODUCTION AND QUALITY CONTROL

process. InP single crystals have been successfully grown on the mass production level as a result of many improvements in the previous matters.

In the production of InP single crystals, there are two major processes, namely, the synthesis of polycrystals and the growth of single crystals. Polycrystals of InP can be synthesized by several methods, including horizontal Bridgman (HB) (Yamamoto et al., 1981), gradient freezing (GF) (Adamski, 1983), synthesis solute diffusion (SSD) (Kubota, 1984; Kaneko et al., 1973), and injection of phosphorus into In melt (Farges, 1982). Single crystals were initially grown by liquid encapsulated Czochralski (LEC) (Bonner, 1981; Shinoyama, 1981). Recently, the vertical gradient freezing method (VGF) (Gault et al., 1986) and modified LEC (Tada et al., 1987) have been developed to obtain single crystals having low dislocation density. Table IV shows some of the synthesis and single crystal growth methods. At the present time, the HB and LEC methods are being employed because they are the most suitable methods for production, based on the reasons mentioned previously.

1. SYNTHESIS OF InP POLYCRYSTAL

a. Techniques of Polycrystal Synthesis

(1) HB technique. This method is the dominant production technique used for the synthesis of InP polycrystals, because of its ability to control stoichiometry, as well as its high growth rate. The typical furnace arrangement and temperature profile for the HB technique are illustrated in Fig. 2. Pure indium is held in a boat, which is placed in the high temperature region, and red phosphorus is placed in the low temperature region. InP is synthesized by the reaction of the indium melt with the phosphorus vapor. The synthesis takes place in the narrow region where the indium melt is locally heated to a higher temperature by r.f. heating. This local heating method is effective in reducing the contamination due to the quartz ampoule in comparison with the method of totally heating the indium region. The phosphorus pressure depends on the lowest temperature in the ampoule, which must be the equilibrium pressure of InP at the melting point. The synthesis proceeds as the ampoule moves from the left side to the right side (Fig. 2). A similar method is the gradient freezing (GF) technique in which the portion having a temperature gradient is continuously moved by reducing the heating power of the higher temperature region. This method is seldom employed in the production process since contamination is inevitable due to the higher temperature in the system.

(2) SSD technique. This is a low temperature solution growth technique that has the advantage of providing InP polycrystals with high purity. It is

TABLE IV

CHARACTERIZATION OF TYPICAL SYNTHESIS METHODS AND CRYSTAL GROWTH METHODS.

Method	Construction	Way of Heating	Atmosphere	Temperature of synthesis	Growth rate	Shape of grown crystal	Diameter	Carrier concentration
HB	Sealed quartz ampoule	Resistance heater and r.f. coil	Phosphorus ~ 27 atm	$\sim 1100°C$	~ 10 mm/h	Boat shape	~ 75 mmϕ	5×10^{15} cm^{-3}
GF	Sealed quartz ampoule	Resistance heater	Phosphorus ~ 27 atm	$\sim 1100°C$	~ 10 mm/h	Boat shape	~ 75 mmϕ	5×10^{15} cm^{-3}
SSD	Sealed quartz ampoule	Resistance heater	Phosphorus ~ 7 atm	$\sim 1100°C$	~ 5 mm/day	Disk shape	~ 50 mmϕ	4×10^{14} cm^{-3}
LEC	High pressure chamber	Resistance heater or r.f. coil	Nitrogen or argon $30 \sim 50$ atm	—	~ 15 mm/h	Cylindrical shape	~ 75 mmϕ	$3 \sim 4 \times 10^{15}$ cm^{-3}
modified LEC	High pressure chamber	Resistance heater	Phosphorus and nitrogen $35 \sim 40$ atm	—	~ 6 mm/h	Cylindrical shape	~ 50 mmϕ	$3 \sim 4 \times 10^{15}$ cm^{-3}
VGF	High pressure chamber	Resistance heater	Phosphorus $27 \sim 40$ atm	—	—	Cylindrical shape	~ 75 mmϕ	—

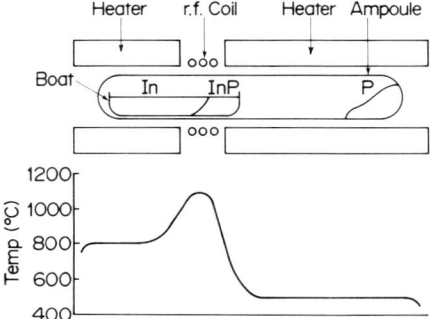

FIG. 2. Schematic drawing of a horizontal Bridgman furnace and its typical temperature profile. The r.f. coil is suitable for locally heating the growth interface.

unsuitable for mass production, however, because of its low growth rate. A schematic diagram of the apparatus and its temperature profile are shown in Fig. 3. Indium held in a crucible and red phosphorus are located in the upper and lower portions, respectively, of the quartz ampoule. Phosphorus vapor pressure is determined by the lowest temperature in the ampoule. Phosphorus vapor reacts with indium on the surface of the liquid indium to result in synthesized InP. This solid InP dissolves into the indium melt and then diffuses into the liquid. Then, InP is precipitated at the bottom of the crucible, where the temperature is the lowest. The ampoule is drawn down at a rate equal to that of the crystal growth, due to keeping a constant temperature profile at the solid-liquid interface, which results in the growth of polycrystals with uniform composition.

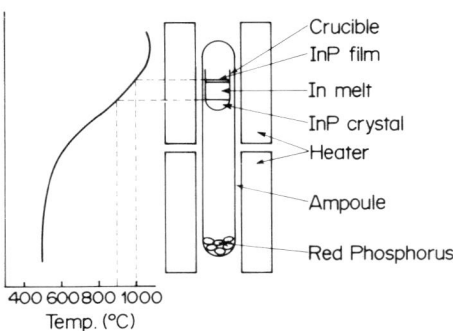

FIG. 3. Schematic drawing of a SSD furnace and its typical temperature profile.

b. Production Processes of the HB Method for Polycrystal Synthesis

Table V shows a typical production process flow of the synthesis of InP polycrystals by the HB technique. In this process, improvements have been attempted on the growth rate and level of purity. Attention must be paid to the purity of both the source materials and the reaction system. The quartz ampoule and the quartz boat, which are main sources of contamination, are etched in the aqua regia and baked in a high vacuum chamber (e.g., 1×10^{-5} torr, 1200°C). The appropriate amount of red phosphorus to be used has been experimentally determined under various conditions. Red phosphorus is usually weighed and loaded into a quartz ampoule in an atmosphere of inert gas, since it is apt to ignite. After loading indium metal into the boat, the ampoule is evacuated to less than 5×10^{-6} torr. It is then sealed at the side opposite to the phosphorus in order to prevent phosphorus evaporation. The ampoule is then put into the furnace and heated until the designated temperature profile is obtained. The pressure in the ampoule is held at a high phosphorus pressure (about 27.5 atm) by controlling the temperature of the phosphorus region. Temperature control is important since stoichiometry of the synthesized crystals strictly depends on the phosphorus pressure. The synthesis reaction proceeds as the ampoule is continuously drawn from the furnace. The growth rate depends on the temperature gradient at the solid-liquid interface. For example, a combination of a very high growth rate and a low temperature gradient will result in precipitation of indium due to supercooling. For the production process, the furnace is generally designed to realize a high temperature gradient at the growth interface, which permits a high growth rate. After the synthesis is completely finished, the furnace is cooled, and the ampoule is pulled out of the furnace. Then, the ampoule is broken, and the InP polycrystal is taken out.

2. GROWTH OF InP SINGLE CRYSTAL

a. Techniques of Single Crystal Growth

The LEC method is widely used for the growth of compound semiconductor single crystals having highly dissociative elements. Dissociation from the melt is prevented by covering the surface of the melt with a B_2O_3 encapsulant and by keeping a high pressure of inert gas in the chamber. Namely, single crystals are grown (using the Czochralski method) from the InP melt covered with a liquid B_2O_3 layer that, under the high pressure of inert gas, acts as a deterrent to the dissociation of InP. Figure 4(a) shows a typical structure and Fig. 4(b) a photograph of a high-pressure LEC furnace. The diameter of the crystal is calculated using the weight of the growing crystal, which is

5. InP SUBSTRATES: PRODUCTION AND QUALITY CONTROL 185

TABLE V
PRODUCTION PROCESS OF HB METHOD FOR POLYCRYSTAL SYNTHESIS.

No.	Process	Control point	Sampling	Measurement equipment or method
1	Etching quartz parts	Etching condition	Every etching time	Timer, thermometer
2	Baking quartz parts	Degree of vacuum Temperature	Every baking time	Ionization vacuum gauge Timer, thermometer
3	Inspection of indium and phosphorus	Surface oxidation Surface cleanliness Purity	Every ingot	Sensory evaluation
4	Weighing indium and phosphorus	Quantity and its allowance	Every weighing time	Digital balancer
5	Loading into ampoule	Kind of material Sequence of loading	Every loading time	Sensory evaluation
6	Evacuating and sealing ampoule	Degree of vacuum	Every evacuating time	Ionization vacuum gauge
7	Heating up the furnace	Heating speed	Every heating time	Thermometer
8	Synthesis	Temperature Growth rate	Every synthesizing time	Thermometer
9	Cooling down the furnace	Cooling speed	Every cooling time	Thermometer
10	Taking out from the ampoule	Furnace condition	Every taking out	Sensory evaluation

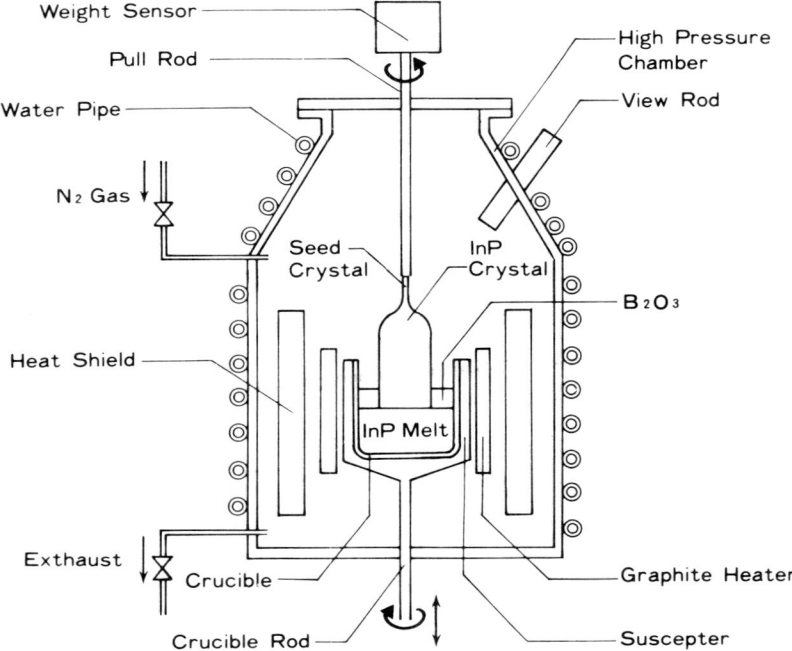

FIG. 4. (a) Schematic diagram of a high-pressure LEC furnace.

measured by a weight sensor equipped on the pull rod. The shape of the crystal can be controlled by adjusting both the temperature at the growth interface and the pulling rate of the crystal. Particular attention must be paid to the water content and the temperature of the B_2O_3. The water in the B_2O_3 affects the extent to which impurities are gettered from the InP melt, and the temperature of the B_2O_3 determines the viscosity, which seriously affects the conditions for crystal growth. In order to stably produce high-quality InP single crystals by the LEC method, it is necessary to realize a system having both mechanical and thermal stability, a suitable temperature profile around the growth interface, and a highly pure environment. Desired techniques for the system include automatic diameter control, solid-liquid interface shape control, and doping control. These conditions are established by repeated attempts at improving the growth process and by subsequent evaluations of the grown crystals.

b. Production Process of the LEC Method for Single Crystal Growth

The LEC method has been widely employed as a sophisticated technique for the production of III–V semiconductor crystals. Table VI shows a typical

5. InP SUBSTRATES: PRODUCTION AND QUALITY CONTROL

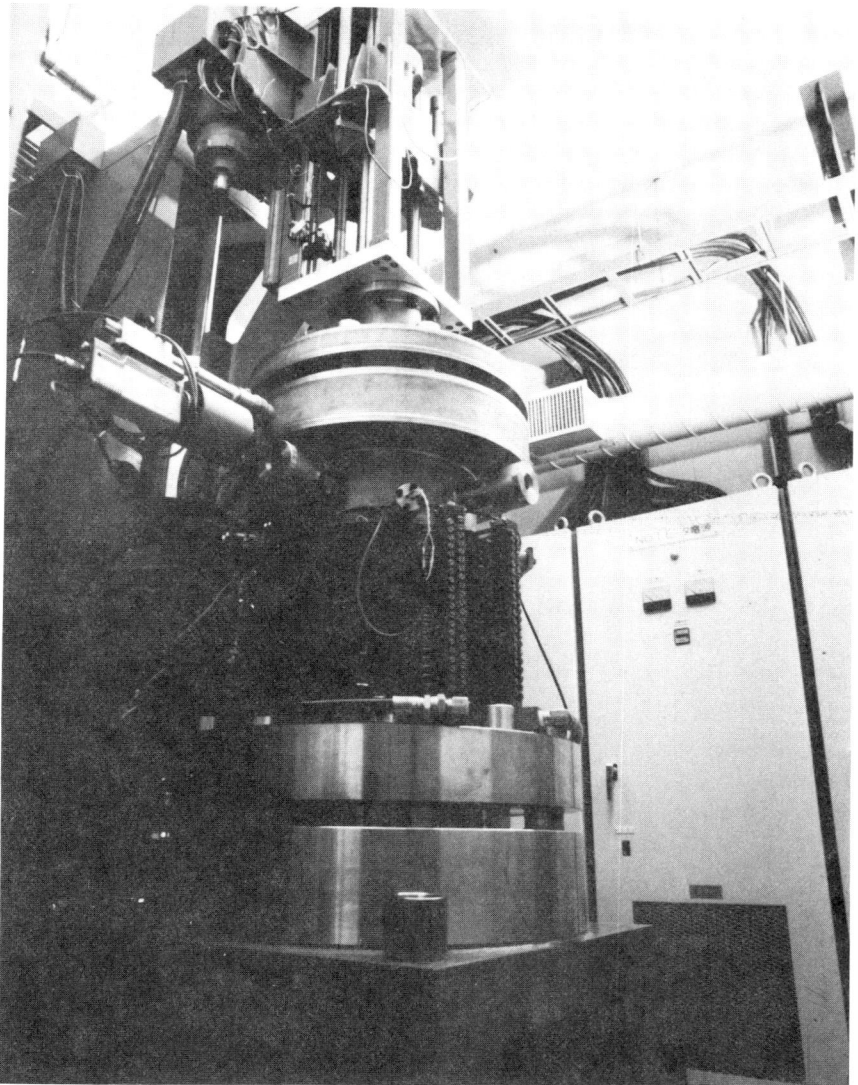

FIG. 4. (b) Photograph of the external view.

TABLE VI
PRODUCTION PROCESS OF InP SINGLE CRYSTAL GROWTH BY LEC TECHNIQUE.

No.	Process	Control point	Sampling	Measurement equipment or method
1	Etching quartz parts	Etching condition	Every etching time	Timer, thermometer
2	Inspection of seed crystal	Orientation Size	Every seed crystal	Orientation checker Slide calipers
3	Material preparation Doping design	Specification	Every ingot	
4	Weighing polycrystals	Quantity and its allowance	Every etching time	Digital balancer
5	Etching	Etching condition	Every etching time	Timer, thermometer
6	Inspection of polycrystal	Surface oxidation Surface cleanliness Purity Quantity and its allowance	Every ingot Every weighing time	Sensory evaluation Results of mass analysis Digital balancer
7	Inspection of dopant	Quantity and its allowance	Every weighing time	Digital balancer

5. InP SUBSTRATES: PRODUCTION AND QUALITY CONTROL 189

8	Loading into crucible	Kind of material Sequence of loading	Every loading time	Sensory evaluation
9	Setting seed crystal	Setting condition	Every setting time	Sensory evaluation
10	Crystal growth	Heating speed Temperature Crystal weight Cooling speed	Continuous record and feed-back to diameter control	Thermometer Temperature record Load cell
11	Taking out from the puller	Furnace condition	Every taking out	Sensory evaluation
12	Etching	Etching condition	Every etching time	Timer, thermometer
13	Sketching	Ingot no.	Every sketching time	Identification no. on ingot
14	Slicing for inspection	Orientation	Every slicing time	Light pattern
15	Inspection of EPD	Counting accuracy Ingot no. Shape of etch pit Etching condition	Every inspection wafer Every etching time	Magnification of microscope Identification no. on wafer Timer, thermometer
16	Hall effect measurement	Good contact to electrode Shape and size of sample Ingot no. and wafer no.	Every Hall sample	Confirmation by ohmic characteristics Sensory evaluation Slide calipers Identification no. on wafer

production process of InP single crystal growth by the LEC technique. In this process, it is important not to reduce the purity level of polycrystals. The contaminants in the furnace must be completely removed by suitable methods in advance. Since polycrystals used as source materials are usually covered with oxidized layers, they must be treated with etching and cleaned just before being loaded into the crucible. After contaminant removal, the treated polycrystals must be kept in an inert atmosphere. The temperature profile around the growth interface is liable to be influenced by the properties of the materials that constitute the hot zone as well as by the structure. Therefore, the operator must carefully check for any slight deviations of the hot zone from the standard alignment and must also check the state of the materials. Fundamentally, the CZ furnace is built in such a manner as to establish a temperature profile having cylindrical symmetry. Another point that must be checked is whether or not the center of rotation of the seed crystal coincides with that of the crucible. After setting up the hot zone, the InP seed crystal, which has been etched in aqua regia is attached to the seed holder. The seed crystal usually has an $\langle 001 \rangle$ direction. Etched polycrystals and dopants are loaded into the bottom of a pBN or quartz crucible and are then covered with an encapsulant of B_2O_3. After these preparations, the chamber is closed and evacuated down to $10^{-2} \sim 10^{-3}$ torr. The crucible is heated, under high pressure (30 $\sim$ 50 atm) inert gas (such as N_2 or Ar), to melt the B_2O_3 encapsulant and the source material. The temperature of the InP melt is stabilized at near-seeding temperature, whereupon the seed crystal is brought into contact with the surface of the melt. Shape control of the growing crystal is automatically performed by the computer control system. InP single crystal, 250 mm long and 50 $\sim$ 60 mm in diameter, can be grown using 3.0 kg of the source material. The cooling rate of the furnace after growth must be carefully determined, since rapid cooling increases the dislocation density in the crystal. On the other hand, rapid cooling is essential for production. The grown InP single crystal is taken out of the chamber and is sent on to wafer processing.

c. *Automatic Growth Control*

The diameter control of single crystals is important during crystal growth, because variations in the diameter affect not only the properties of the crystals but also the yield of wafers cut from them. Therefore, the automatic diameter control technique is especially essential for cost reduction in the production process. Figure 5 shows our automatic growth control system, which can automatically control growth conditions from immediately after the seeding process up to the process of cutting the crystal from the melt. In the LEC method, the diameter of the crystal is usually determined by the

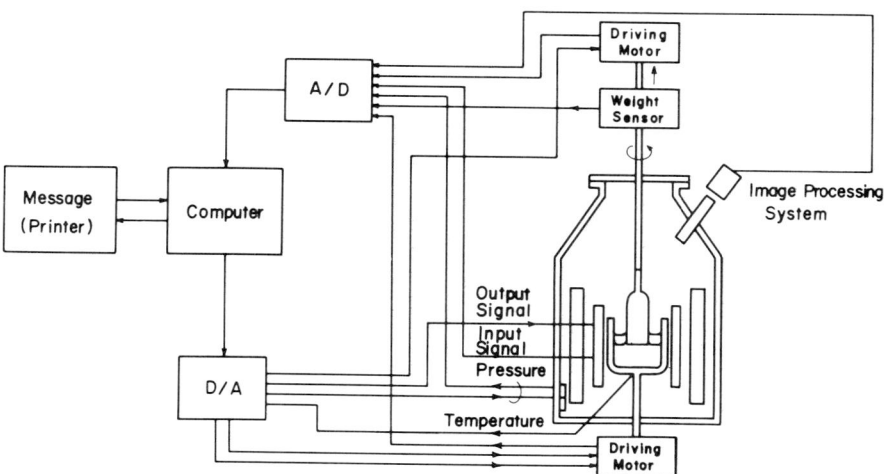

FIG. 5. A typical automatic growth control system of the LEC furnace. Several sensors are equipped to measure growth parameters such as temperature, pressure, crystal weight, pulling rate, and rotation rate of both the crystal and the crucible. The data are put into the computer and treated in real time.

amount of change in the crystal weight after making corrections in the buoyancy due to the B_2O_3 encapsulant. The several parameters, shown in Fig. 5, are put into the computer, treated in real time, and the signals are then fed back to both the heating and control systems. By using this system, we can control the diameter of a crystal 50–75 mm in diameter to within 2–3 mm. Such a crystal is shown in Fig. 6.

d. Suppression of Twinning

The generation of twinning is the most troublesome problem in growing InP single crystals (Iseler, 1981). Twinning occurs often in the early stages of growth, namely, in the period of crystal diameter change. Although the mechanism of twinning has not yet been clarified, it is empirically noted that the following conditions relate to the occurrence of twinning (Bonner, 1981):

1. deviation of the melt composition from the stoichiometric one,
2. shape of solid-liquid interface,
3. dissociation of phosphorus from the surface of the crystal,
4. scum on the surface of the melt, and
5. instability of the temperature at the solid-liquid interface.

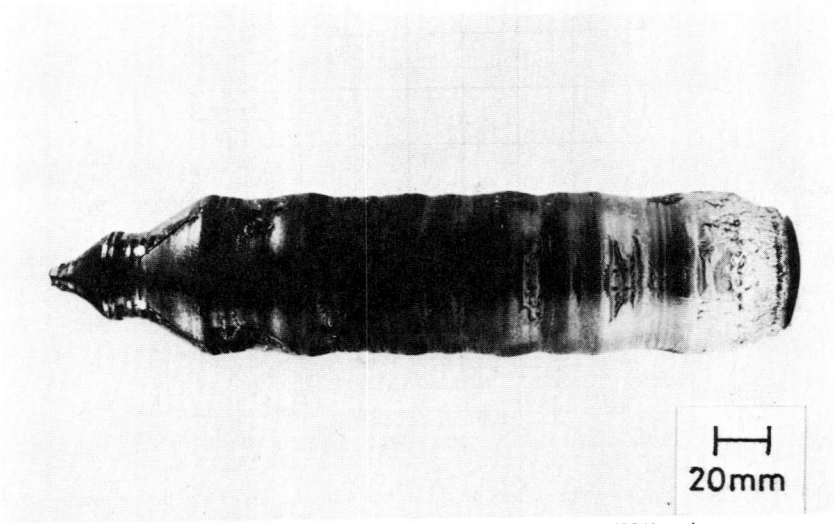

FIG. 6. An InP single crystal that was grown along a $\langle 001 \rangle$ axis.

Therefore, the following means are recommended for effectively preventing the twinning (Shinoyama et al., 1981):

1. using InP polycrystals of a stoichiometric composition as source material,
2. keeping the shape of the solid-liquid interface convex to the melt,
3. keeping the temperature of the B_2O_3 surface less than 600–700°C,
4. using B_2O_3, which includes an extremely small amount of water, and
5. suppressing the instability of the temperature at the solid-liquid interface.

The application of a magnetic field may also be effective in preventing the occurrence of twinning.

e. Doping Techniques

In practical usage, several kinds of impurities are added to the melt when producing InP single crystals, which controls the type of electrical conduction or reduces the dislocation density. The types and the amounts of impurities for particular devices depends on the specifications for the InP substrates. In optical device usage, Sn or S are added for n-type substrates and Zn is added for p-type substrates. Doping with S or Zn results in an impurity-hardening effect, and the concentration is increased to the extent that the effect becomes sufficiently large. High levels of doping, however, lower the optical transmittance due to an increase in free carrier absorption, which may degrade the characteristics of the devices. High levels of doping also cause out diffusion of the dopant in the process of device fabrication. The doping level of the impurity is determined by the device maker's requirements as well as by the limitations described previously. It is now impossible to obtain semi-insulation in undoped InP crystals, since excess donors, due to residual impurities, make them conductive. Semi-insulating crystals are obtained by doping with Fe, which can compensate for the excess shallow donors by forming deep acceptors levels. The amount of Fe, usually 1 to 2×10^{16} cm^{-3} at the seed end, depends on the amount of residual impurities. Reduction of the doping level is strongly required in order to suppress the thermal conversion during the device processing. This can be achieved by growing single crystals having higher purity. Since the segregation coefficients of these dopants are not equal to unity (one), the concentrations change along the growing direction of the crystal, which result in changes in the crystal properties there. Consequently, the yield of the available substrates decreases. In order to increase this yield, the dopant level must be reduced as much as possible. When a dopant has a high dissociation pressure, it is added in the form of In or P compound having relatively low dissociation pressure.

f. Magnetic Field Effects on the Melt

Prominent effects have not yet appeared in the growth of III–V semiconductors while applying a magnetic field, but some slight effects have been noticed in the case of GaAs. Magnetic field usage, however, has resulted in prominent effects during Si crystal growth. The magnetic field generally affects the convection of the melt in stabilizing the temperature (Terashima and Fukuda, 1983) and in changing the temperature profile (Osaka *et al.*, 1984) and the distribution of impurities in the melt (Kawase *et al.*, 1986). The degree of the effects depends on the overall properties of the melt, which are represented by the Hartman number. In InP, the magnetic field is expected to be effective at levels above 1,000 gauss. It has been reported that magnetic fields are useful both in the suppression of temperature fluctuation and in the improvement of controllability of the crystal diameter. For application in the production process, it is necessary to raise the extent of the effect by applying the magnetic field under an optimally thermal environment.

V. New Developments in InP Crystal Growth

InP single crystals have been industrially grown by the LEC method for optical and electrical devices (Morioka *et al.*, 1987). The characteristics of each device have not been explicitly related to the properties of single crystal at the present stage, so there is not a strong demand for improvement on particular properties of InP. However, it will become important for crystal growers to supply higher quality single crystals in order to realize the higher performance of future devices (Foyt, 1981). Several new developments have been tried in crystal growth. The main purpose of these developments is to achieve lower dislocation density and higher purity crystals with a larger size.

3. Preparation of Polycrystal with High Purity

A highly purified InP single crystal is considered valuable in realizing the high performance of both optical devices and high-speed microwave devices. In InP crystals there are no deep levels created by native defects such as EL2, which is observed in GaAs. (EL2 plays an important role in semi-insulating GaAs by compensating shallow impurity levels.) In the case of InP, a deep acceptor level is introduced by doping with Fe, whose concentration depends on the concentration ($N_{SD}-N_{SA}$) of the difference between the shallow donor and the shallow acceptor arising from residual impurities. Semi-insulating InP must be high purity since the high doping of Fe can result in a large change in resistivity caused by out diffusion of Fe during the device process (Kamada et al., 1984). Contamination can arise in both the synthesis and growth process of the InP crystals. Contamination is also a result of impurities in the raw materials. Crystal growers can now obtain higher pure In and P source materials, but they require even higher purity levels than they can presently achieve. We will not describe these in further detail here.

Polycrystal ingots of InP have usually been synthesized by the horizontal Bridgman method (Adamski and Ahern, 1985). In this method, silicon contamination from quartz chiefly occurs as is depicted in the following reaction.

$$10\ SiO_2\ (s) + P_4\ (g) \rightleftarrows 10\ SiO(g) + P_4O_{10}\ (g). \tag{1}$$

To prevent this contamination, the following improvements (Adamski, 1983; Wardill et al., 1983; Allred et al., 1981) were attempted. (1) the reduction of process temperature especially at the raw indium zone, (2) the reduction of pressure, and (3) the improvement of the purity of materials constituting the synthesis system.

The reduction of contamination, especially by silicon, is expected to occur by reducing the process temperature and the phosphorus pressure in accordance with previous reaction (Eq. 1). The indium zone and the phosphorus zone temperature are related to process temperature and pressure, respectively. In the conventional process, the operating temperature is about 1,100°C ($>T_M$) for the indium zone and about 500°C for the phosphorus zone. By this approach, Adamski (1983) obtained highly pure polycrystals with carrier concentration of $3 \times 10^{14}\ cm^{-3}$ and mobility of 1.385 ×

10^5 cm^{2}V^{-1}s^{-1}(77 K) at a relatively low temperature (1030°C) for the indium zone. The reductions of the process temperature and the phosphorus pressure, however, tend to induce deviation from the stoichiometry in synthesized polycrystals. If the reductions are too great, this process becomes similar to the solution growth method (or the SSD method). To prevent this deviation, the growth rate must be greatly reduced, and the temperature and pressure must be stabilized. In the present stage of production, synthesized polycrystals have a carrier concentration of $2 \sim 10 \times 10^{15}$ cm^{-3} and a mobility of $3.5 \sim 4.2 \times 10^3$ cm^2 V^{-1}s^{-1}(300 K). We have been trying to attain higher levels of purity by improving both the system and the process. The SSD method can be performed at temperatures as low as 900°C, and this produces highly pure polycrystals having a carrier concentration of 4×10^{14} cm^{-3} and a mobility of 1.26×10^5cm^2 V^{-1}s^{-1}(77 K) (Kubota and Katsui, 1987). The phosphorus injection method is the other technique. Phosphorus gas supplied from a red phosphorus reservoir is injected into liquid indium, which is contained in a crucible in the pulling furnace (Hyder and Holloway, 1983; Farges, 1982). This method enables in situ synthesis and growth in the puller, but as for the purification, good results have not yet been reported.

4. REDUCTION OF DISLOCATION DENSITY

To reduce dislocation density, there are generally two techniques that have already been attempted for GaAs single crystals grown by the pulling method. One method is to reduce the temperature gradient along both the axial and radial directions and the other is to utilize the impurity-hardening effect.

a. Reduction of the Temperature Gradient

Reduction of the temperature gradient is effective for reducing the thermal stress that causes evolution of dislocations if it exceeds some threshold value. Lowering the temperature gradient, however, raises the temperature of the InP single crystal at the surface, which results in a crucial damage because of the high dissociation pressure of InP. This drawback can be overcome by using our modified LEC or the FEC (fully encapsulated Czochralski) method.

We are now trying to grow InP single crystals with low dislocation density by our modified LEC method (Tada *et al.*, 1987). Figure 7 shows the concept of our modified LEC system. Even under conditions with the temperature gradient lower than 30°C/cm, the dissociation at the crystal surface can be prevented by introducing a small amount of phosphorus vapor into the space

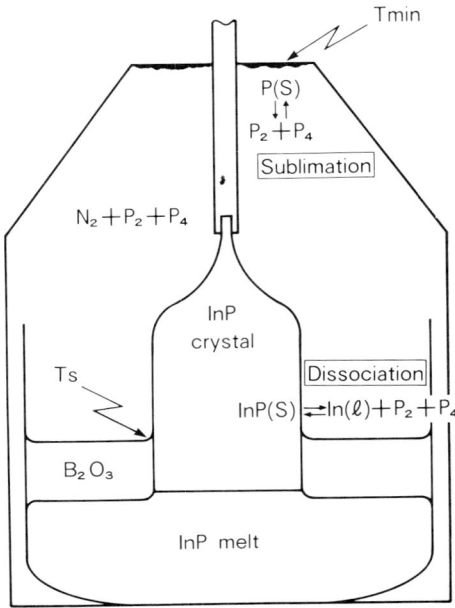

FIG. 7. Concept of the modified LEC system used for the growth of InP single crystal. This method enables growth in a low temperature gradient near the solid-liquid interface without dissociation at the crystal surface. Dissociation of InP crystal occurs at the highest temperature region of Ts. This is suppressed by phosphorus pressure, which is generated from the sublimation process at the lowest temperature region. Crystal growth is carried out under conditions in which the sublimation pressure of the red phosphorus is higher than the dissociation pressure at the crystal surface.

above the surface of the B_2O_3 encapsulant. In this condition, the phosphorus vapor pressure must be higher than the dissociation pressure (P_1) at the crystal surface. The dissociation pressure of the InP crystal is given by the following equation (Bachmann and Buehler, 1974)

$$\log(P_1/\text{atm}) = -17.8 \times 10^3/Ts + 13.72, \tag{2}$$

where $P_1 = P_{p_2} + P_{p_4}$, and T_s is the highest temperature of the crystal surface, which corresponds to the temperature at the surface of B_2O_3. The pressure within the vessel is determined by the lowest temperature (Tmin.) at which the phosphorus vapor is in equilibrium with solid phosphorus at the sublimation pressure (P_0). This is described by the following equation (Bachmann and Buehler, 1974)

$$\log(P_0/\text{atm}) = -6.99 \times 10^3/T\text{min} + 9.86. \tag{3}$$

5. InP SUBSTRATES: PRODUCTION AND QUALITY CONTROL

By satisfying the condition of $P_0 > P_1$, we have successfully grown InP single crystals under the low temperature gradient of 30°C/cm. The growth of InP single crystals was performed using InP polycrystals (about 1 kg) contained in a 100 mm diameter crucible, which is in a high-pressure Czochralski puller with multizone heaters. The pressure of the dry nitrogen (35 ~ 40 atm) in the puller was balanced against the total pressure of both the nitrogen and the phosphorus in the vessel. The crystals were pulled along an $\langle 001 \rangle$ direction at a constant rate of 5–6 mm/h. The axial temperature profile in this system is shown in Fig. 8 together with that in the conventional LEC. The temperature gradient along the axial direction in the B_2O_3 layer is about 30°C/cm, which is much lower than the temperature gradient of 100–150°C/cm in the conventional LEC. The temperature of the B_2O_3 at the surface is 950°C in the modified LEC system and is 730°C in the conventional LEC system. Figure 9 shows a grown semi-insulating InP single crystal doped with Fe of 2×10^{16} cm^{-3} at the seed end. The crystal is 50 mm in diameter and 120 mm in length and has a surface of metallic luster. The wafer of 40 mm in diameter, which is cut from the middle part of the crystal, has an average EPD of 2×10^3 cm^{-2} with a W-shaped profile. The value is one order of magnitude less than those of crystals grown by the conventional LEC method (Fig. 10). This method also gives Fe-doped crystals which have very few inclusions in

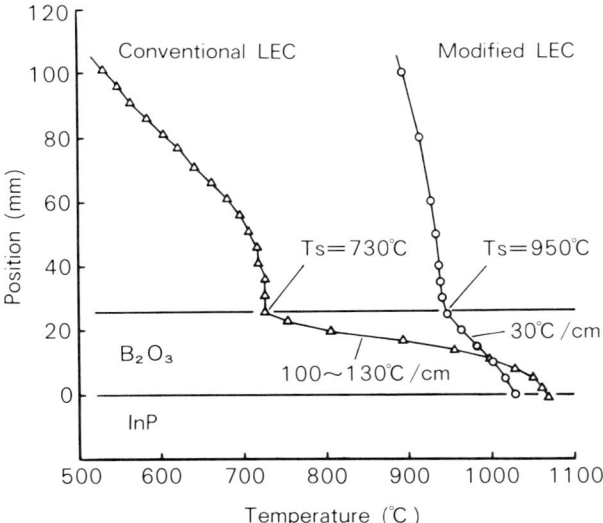

FIG. 8. Temperature profiles along the axial direction above the InP melt in both the conventional and modified LECs. The main differences are observed for temperatures at the B_2O_3 surface and for temperature gradients in the B_2O_3.

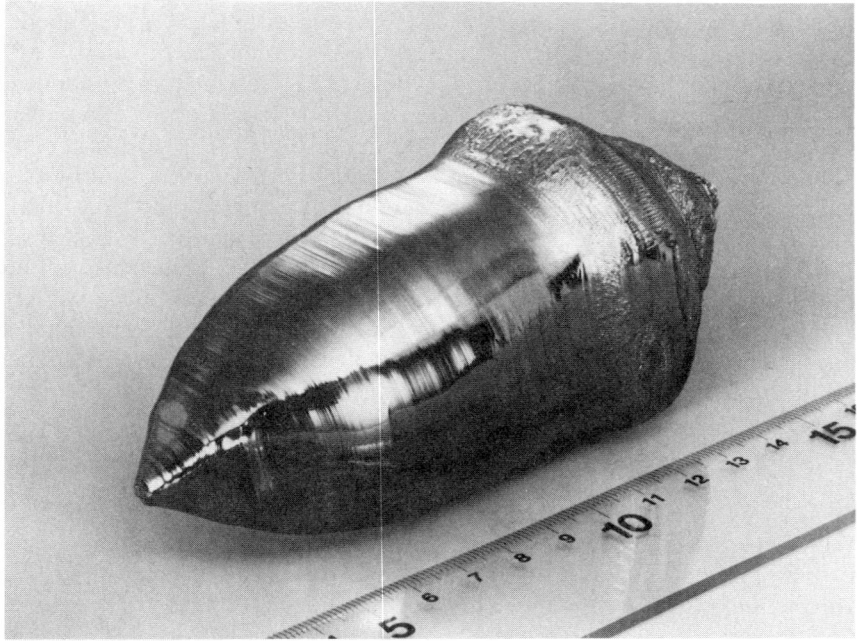

FIG. 9. An Fe-doped InP single crystal grown by the modified LEC method. The amount of doped Fe at the seed end is 2×10^{16} cm^{-3}. The crystal was grown along a $\langle 001 \rangle$ axis and is 50 mm in diameter and 120 mm in length.

comparison with those grown by the conventional LEC method. The inclusions are observed as scatterers in experiments of infrared light-scattering tomography (see VI6d), of which the results are shown in Fig. 11. By using our modified LEC method, we also obtained slip-free (dislocation-free) S-doped InP crystals and low dislocation density Sn-doped crystals (less than 2×10^3 cm^{-2}) (Fig. 12).

The FEC method is the other method of growth under a low temperature gradient in which the dissociation of the crystal at the surface is suppressed by fully encapsulating the single crystal with B_2O_3. However, controlling the crystal diameter is very difficult because of the variation in B_2O_3 height that occurs while pulling the crystal and also because of the fluctuation of the thermal conduction through the B_2O_3 layer. In FEC GaAs, a diameter-controlled crystal has been obtained by applying a magnetic field, but it has not yet been reported in InP. Another problem of this method is twinning, which often occurs in crystal growth under a low temperature gradient. The twinning is considered to occur mainly due to temperature fluctuations and

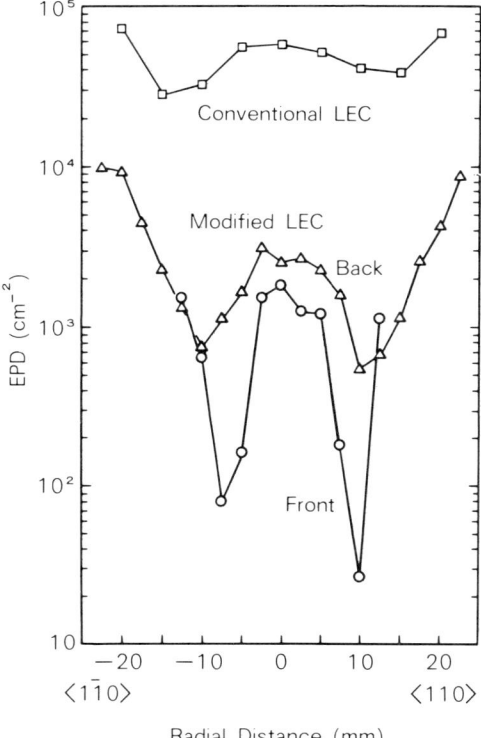

FIG. 10. EPD profiles of the (001) InP wafers obtained from modified and conventional LEC semi-insulating crystals. The EPDs were estimated using the number of etch pits in the 1.9 mm square region.

mechanical vibrations. Overcoming this problem is essential for employing this method in the production of InP single crystals.

b. Impurity Hardening

Adding a specific impurity to an InP crystal can also suppress the occurrence and the propagation of the dislocations in the crystal. The cause of this effect is that dislocations are bound by impurity atoms and become relatively immobile, which is known as the pinning effect. Seki et al., (1978) explained that the pinning effect arises when the bond energy between a substituted impurity atom and a host atom exceeds the bond energy between the two host atoms at the lattice site. Table VII shows the bond energy for various kinds of impurity atoms and the results of experiments on their

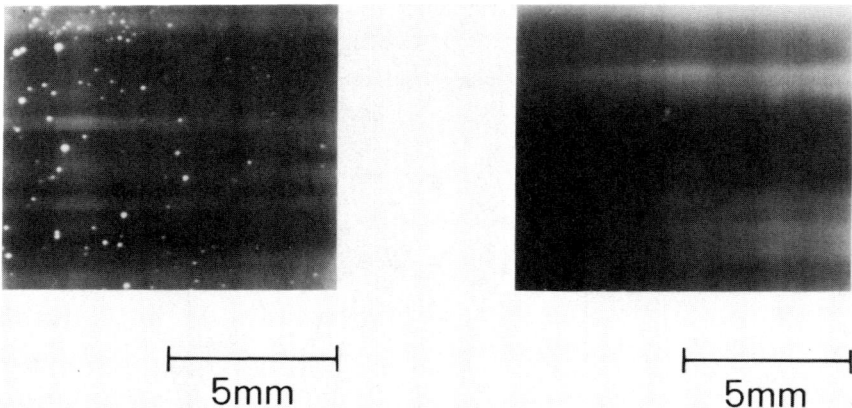

FIG. 11. Images of infrared light scattering from Fe-doped InP crystals, which are grown by (a) the conventional LEC and (b) the modified LEC methods. The bright spots in the photograph are due to scattered infrared light from microprecipitations in the InP crystals.

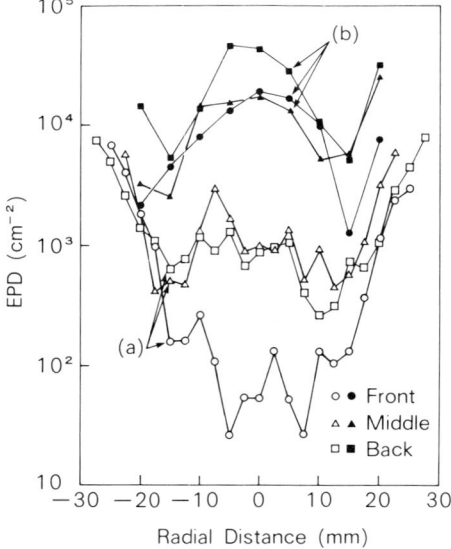

FIG. 12. An EPD profile of a Sn-doped InP single crystal grown by (a) the modified LEC and (b) the conventional LEC methods. The amount of Sn at the seed end is 1×10^{18} cm^{-3}. In this method, the Sn-doped InP single crystal displays impurity-hardening effect. That is, the EPD becomes lower than that of an undoped crystal.

TABLE VII

BOND ENERGIES OF IMPURITIES IN AN InP SINGLE CRYSTAL AND EXPERIMENTAL RESULTS ON THEIR IMPURITY-HARDENING EFFECTS.

Impurity (site)	Bond energy $\left(\dfrac{kcal}{mol}\right)$	Results of experimental EPD
Zn(In)	94.2	○
N(P)	85.1	—
Al(In)	72.8	—
S(P)	68.7	○
Ga(In)	58.2	○
Te(P)	47.0	△
Cd(In)	46.8	—
InP	46.8	×
Ge(In)	45.2	○
Sn(In)	43.5	△
Sb(P)	37.6	△
As(P)	36.0	△

○: DF
△: $10^2 \sim 10^3$ cm^{-2}
×: $10^4 \sim 10^5$ cm^{-2}

impurity-hardening effects (Tohno and Katsui, 1986b; Cockayne et al., 1986; Katsui and Tohno, 1986a; Shimizu et al., 1986). As shown in the results for Ge, Sn, Sb, and As, Seki's model fails to explain the effect due to impurity hardening. The extent of the reduction of the dislocation density depends on the amounts of the added impurity, and the effect becomes remarkably useful in crystal growth when sufficiently low temperature gradients are used. The same situation has been encountered in the growth of the dislocation-free single crystal of In-doped GaAs, which was first achieved under an improved thermal environment. Figure 13 displays the effects of impurity hardening in S-doped InP crystals grown under axial temperature gradients of 40°C/cm and 100°C/cm. This crystal, grown in a low temperature gradient, has a dislocation-free region of about 30 mm in diameter without slip dislocation. The effective impurities, for the reduction of the dislocation density, such as S and Zn are conductive impurities and contribute to the increase of the carrier concentration. Therefore, it is difficult to independently control the dislocation density and the free carrier concentration, both of which effect the properties of the InP optical devices. To overcome the previous problem, isoelectronic impurities were selected as dopants from group III or group V

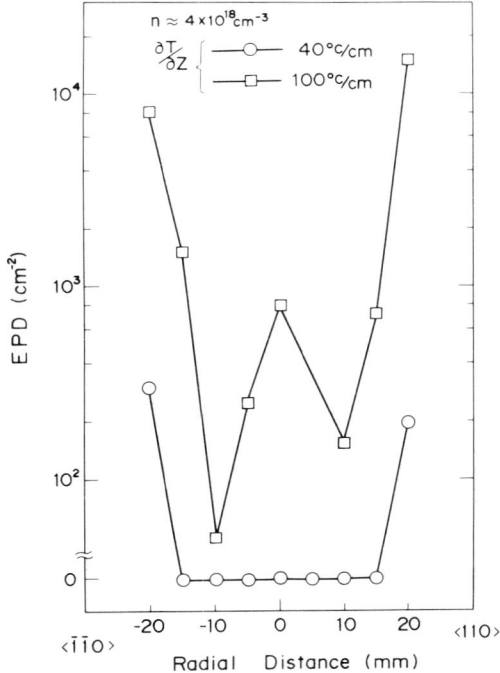

FIG. 13. S-doped InP crystals' impurity-hardening effects under different temperature gradients (40 °C/cm, 100 °C/cm). The amount of S is 4×10^{18} cm^{-3} at the seed end. By reducing the temperature gradient, W-shaped EPD distributions tend to become U-shaped ones.

in the periodic table, although the effect of their impurity hardening is less than in the case of S or Zn. The needed amounts of dopant in the crystal are more than 10^{19} cm^{-3}, which is one order of magnitude higher than that of S or Zn. The concentration of the impurity in the crystal constantly changes along the direction of growth. The distribution depends on the magnitude of the segregation coefficient, especially if it is more or less than unity (one). The distribution influences the uniformities of the crystal properties, which include electrical and optical properties, dislocation density, and lattice constant. Such influences are reduced by codoping in which more than two kinds of impurities have different segregation coefficients. Also, the total concentration must be more than some critical value to result in the impurity-hardening effect throughout the crystal. The combinations of the codoping are reported for Ga + Sb, Ga + As, Ga + Sb + As, etc. Katsui et al., (1986b) have grown InP single crystals codoped with Ga and Sb, which have a dislocation free region that is 25 mm in diameter throughout the 45 mm diameter crystal. The total doping level is held at more than 1×10^{19} cm^{-3}. We have also succeeded in reducing the dislocation density to a level of $1 \sim 2 \times 10^3$ cm^{-2} by doping Ga and As with a concentration of $5 \sim 10 \times 10^{19}$ cm^{-3}. We grew a single crystal (50 mm in diameter) under the low temperature gradient of 35–45°C/cm. Figure 14 shows the x-ray topograph on the wafer cut from the crystal doped with Ga and As in addition to Sn. We also confirmed that this crystal had a lattice constant (throughout the crystal) that nearly matched that of the GaInAs epitaxial layer. As for the semi-insulating crystal, about 10^{16} atoms cm^{-3} of Fe were added in order to create a deep acceptor level, and Ga and As, a total concentration of more than 8×10^{19} cm^{-3}, were added to reduce the dislocation density. The EPD of this crystal was about 2×10^3 cm^{-2}, the resistivity was more than $10^7 \Omega$cm, and the electron mobility was $2.5 \sim 3.0 \times 10^3$ cm^2 V^{-1}s^{-1}. For practical usage, however, further investigations of the effect of high doping on the properties of the devices are necessary.

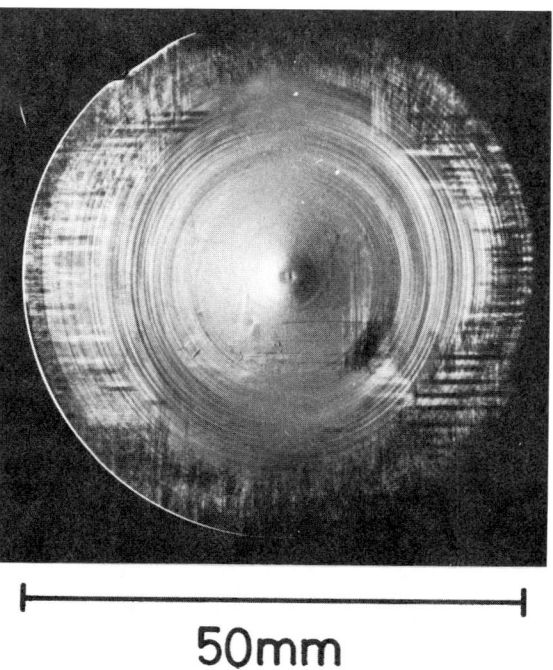

FIG. 14. X-ray topograph of an InP single crystal doped with Ga and As (in addition to Sn). The concentrations of Ga and As are 8×10^{18} cm^{-3} and 8×10^{19} cm^{-3}, respectively.

VI. Evaluations

Crystal growers need to understand the single crystal's properties and its uniformity, and they must also see whether or not the crystal's growth process is satisfactory. There are many ways to evaluate semiconductor crystals. The kind of evaluation to be performed depends on the purpose for evaluation. Such purposes include the following: observing the growth processes, developing new methods or new processes, and replying to customers' demands. Table VIII shows some of the evaluation methods that we have usually carried out. We can classify them for convenience into two kinds, that is, on-line evaluations and off-line evaluations. On-line evaluations are routinely performed at each phase of growth in order to control the quality of crystals. We can then confirm whether or not the production process has proceeded stably. The resulting feedback is used for the next growth. Therefore, the evaluations must be comprehensive, and they must

TABLE VIII

EVALUATION TECHNIQUES.

	Item	Method or instrument
Crystallographic evaluation	EPD X-ray diffraction	• Chemical etching • X-ray topography (XRT) • X-ray double crystal rocking curve method
Chemical evaluation	Impurity	• Instrumental analysis (GDMS, SSMS, SIMS, GFA, ICP, etc.)
	Matrix	• Coulometric titration
Electrical evaluation	Hall effect	• Hall effect analysis system
	Deep level	• DLTS
	Resistivity	• ρ mapping technique
Optical evaluation	PL Raman scattering Light absorption/transmittance	• Spectrophotometer detectors
	Infrared light scattering	• IR tomography/topography
Evaluation for practical usage	Annealing effects Characteristics or real devices	• Annealing process • Device process and operation
Theoretical analysis (calculation)	Shear stress	• Computer simulation

5. InP SUBSTRATES: PRODUCTION AND QUALITY CONTROL

totally reflect the crystal's properties and the conditions of the processes. It is also important that these evaluations be quickly and routinely performed. Off-line evaluations are often performed on particular properties of special cases, namely, in troubleshooting and improving the growth process, in developing better crystals, and in responding to customers' demands. In the following sections, we introduce typical evaluations and results generally obtained from InP single crystals.

5. ON-LINE EVALUATIONS

Here we carry out a characterization of the crystal perfection and an investigation of the electrical properties of an InP single crystal. We adopt the etch pit density (EPD) measurement for the former. Etch pit density corresponds to dislocation density, and its evaluation provides a useful guide to crystal quality. Measurements for the electrical properties of semiconductors are fundamentally important since they provide information concerning the crystal's purity, stoichiometry, and perfection. For these purposes, the Hall effect measurement is considered the most suitable method.

We routinely carry out these measurements on two kinds of wafers cut from both the seed side and the tail side of a crystal. Using such wafers enables us to check the variation along the pulling direction.

a. Etch-Pit Density (EPD) Measurement

Etch-pit density (EPD) is one of the best indicators of the quality of III–V semiconductor crystals. In this measurement, a wafer is chemically etched to reveal etch pits, which are then counted under an optical microscope. Of course, we must know in advance the correlation of EPD to the crystallographic properties found using other measurements. These properties include x-ray topography, x-ray double crystal rocking curve, and photoluminescence measurements. In addition to the etch pits, other defects revealed on the etched surface, such as inclusion, precipitate, and microtwinning, must be carefully observed. The EPD measurement process is briefly outlined here.

(1) (001) wafers, cut perpendicularly to the direction of growth, are polished.
(2) The polished wafers are etched for a few minutes with an $H_3PO_4 - HBr$ room temperature aqueous solution (Huber and Linh, 1975).
(3) The number of etch pits on each wafer is carefully counted under an optical microscope. Usually, this counting process is carried out at many points on the wafer in order to learn the EPD distribution and to evaluate its uniformity.
(4) The number of etch pits is converted into an EPD (number of etch pits per 1 cm^2) value.
(5) From the results, the single crystals are classified into various grades.

Figure 15 shows an optical micrograph of etch pits revealed on an InP wafer. A typical distribution of EPD on a Sn-doped InP wafer along the radial direction is shown in Fig. 16 (Morioka et al., 1987). The radial distributions are ordinarily W or U shaped, depending on the temperature profile around the growing crystal. This method also produces other useful information about the secondary structures of the etch pit distribution, which include slips, network structures, and lineages.

This evaluation has been performed routinely, but the counting process is very time consuming, inaccurate, and fatiguing for the operator. Therefore, this process has been replaced by the automatic counting system (Fig. 17) (Toyoda et al., 1985). Using this system, we can quickly obtain various forms of precise results on EPD. Figure 18 shows an example, which was obtained using this system, of the EPD distribution on a Sn-doped InP wafer. It takes less than 30 minutes to obtain the final results of measurements for the whole region of a two-inch wafer with a 3 mm grid pitch. On the other hand, using visual measurement, it would take four and a half hours.

b. Hall Effect Measurement

Using this measurement, we can determine conduction type and three fundamental quantities of semiconductors, namely, electric resistivity, carrier concentration, and (Hall) mobility. From these results, we can confirm how much residual impurity is included in the undoped crystal and determine

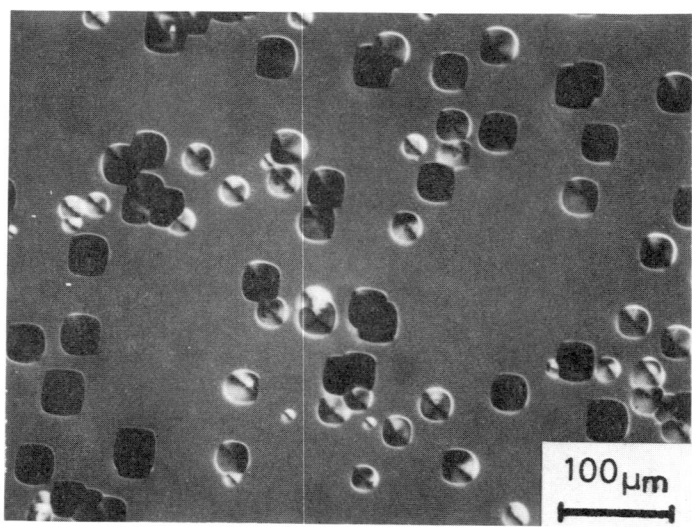

FIG. 15. Optical micrograph of etch pits on an (001) InP surface.

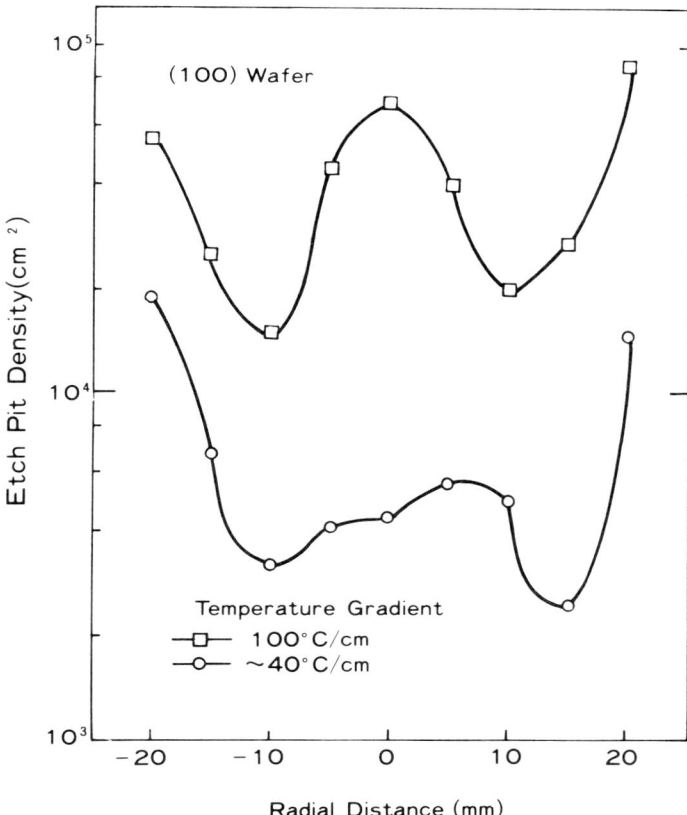

FIG. 16. EPD profiles of (001) Sn-doped wafers. The amount of Sn is 1×10^{18} cm^{-3} at the seed end. The EPD level and profile are greatly affected by the temperature gradient during growth.

whether or not the crystal is properly doped. Generally, the Hall effect measurement is carried out by measuring the voltage induced when mutually perpendicular electrical and magnetic fields are applied to the sample. In the actual Hall effect measurement for compound semiconductors, the van der Pauw method (Pauw, 1958; 1958/59) is widely employed because it can accommodate flat samples of any shape. This practical procedure is indicated next.

(1) Pretreatment: Wafers are polished and etched.
(2) Preparation of samples: The wafer is cut into square chips (4×4 mm², in our case). For *n*-type InP, Au-Ge-Ni alloy (in the form of electrodes) is

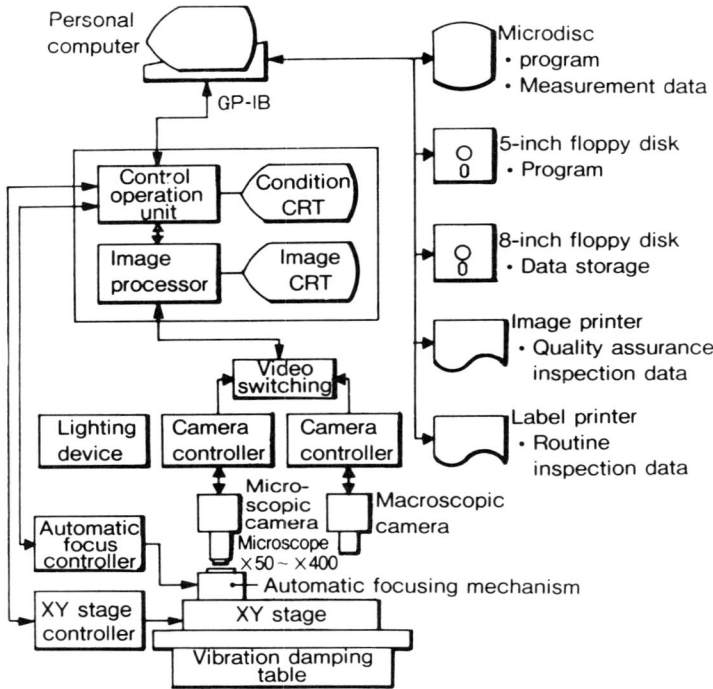

FIG. 17. Block diagram of the EPD automatic counting system.

evaporated at the chip's four corners. Au-Zn alloy is used for *p*-type InP. After this, the chips are annealed in an atmosphere of nitrogen for 20 minutes. The annealing temperature is 450°C.

(3) Measurement: Each of the chip's electrodes is connected electrically to a Hall effect measurement instrument.

The measurements are then performed using the van der Pauw method. Recently, measurement and data processing have been carried out automatically, as shown in Fig. 19. Figure 20 shows the relationship between electric resistivity and carrier concentration in InP crystals. This data can be applied to estimate the amount of added impurity or the amounts of the residual impurities in crystals. Figure 21 shows the carrier concentration dependence of EPD (Morioka et al., 1987), which is the impurity-hardening effect of S and Zn. As mentioned previously, the evaluations of EPD and Hall effects can provide information concerning crystallographic and electrical properties, which are representative properties of an InP single crystal.

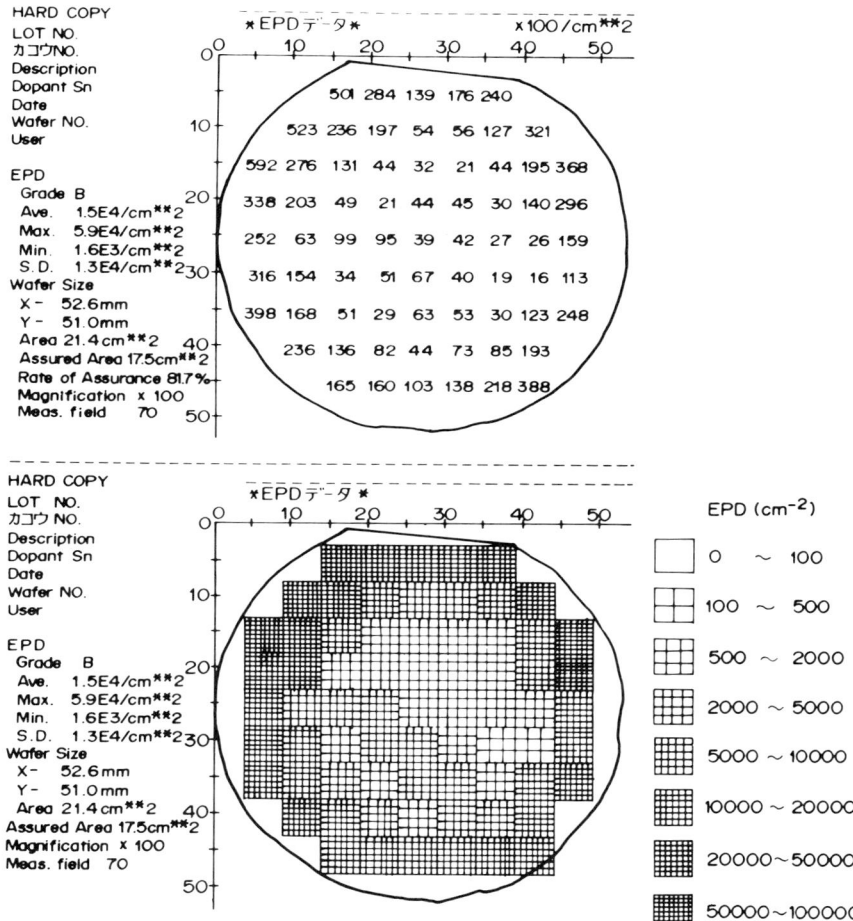

FIG. 18. Examples of output data using the EPD automatic accounting system. The crystal used was Sn-doped InP.

6. OFF-LINE EVALUATIONS

Off-line evaluations are carried out when a special need is revealed. There are many kinds of evaluation methods; hence, it is important to employ only that which is the most suitable for the purpose. Evaluations must provide clues that suggest which growth technique should be modified. In many cases, evaluations concerning the conditions of crystal growth are those that have already been established. Therefore, the results of these established evaluations can reveal the status of present crystal growth. They can also be

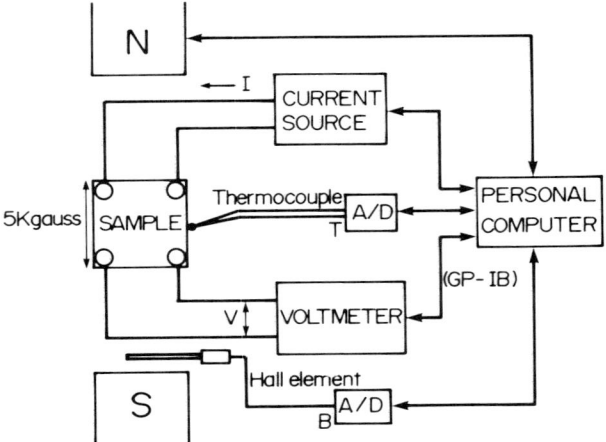

Magnetic Installment

FIG. 19. Block diagram of the Hall effect measurement system. Both measurement and data processing are automatically carried out in this system.

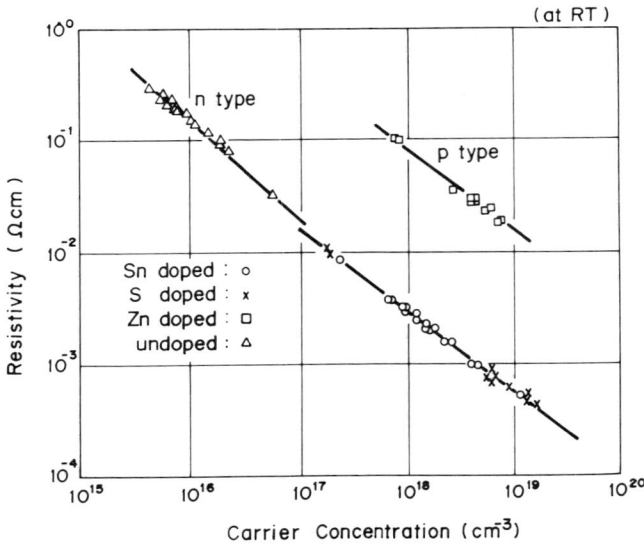

FIG. 20. Relationships between electric resistivity and carrier concentration in InP.

5. InP SUBSTRATES: PRODUCTION AND QUALITY CONTROL

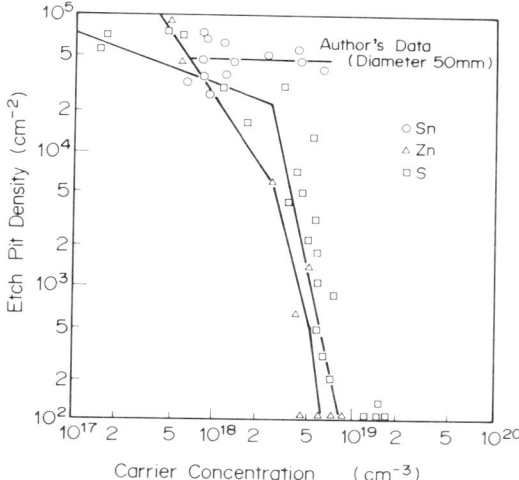

FIG. 21. Dependence of the etch pit density on the carrier concentrations (impurity-hardening effect). In this case, variations in carrier concentration are due to doped impurities such as Sn, Zn, and S.

correlated to the results of the evaluations. However, we sometimes attempt to develop new evaluation methods in order to obtain better information. In this section, we will introduce representative evaluation methods.

a. Crystallographic Characterization

Evaluation of crystal perfection is essential since the properties of devices (fabricated on the crystal) strongly depend on the structural perfection of the single crystal. The EPD measurement is one of these evaluation methods, but it gives only an approximation of the perfection. Detailed information is obtained by using x-ray or electron microscopy techniques. The x-ray topography technique reveals imperfections such as dislocations, subgrain-boundaries, striations, and lattice strains (Tohno and Katsui, 1986a; Uemura et al., 1981). This technique can be carried out easily and has several modified forms, such as the Berg Barrett, the Lang, and the double crystal diffraction techniques. Figure 22 shows the x-ray topographs of the Sn-doped and S-doped InP single crystals. The x-ray double crystal rocking curve method is useful in that it gives the quantitative measure of crystal perfection. However, the surface of the crystal must be carefully treated, since this method only provides information for the areas, usually on the surface, that are affected by the x-ray beam. This, in turn, depends on the beam's penetration depth. Recently, lattice constants of single crystals have been precisely determined

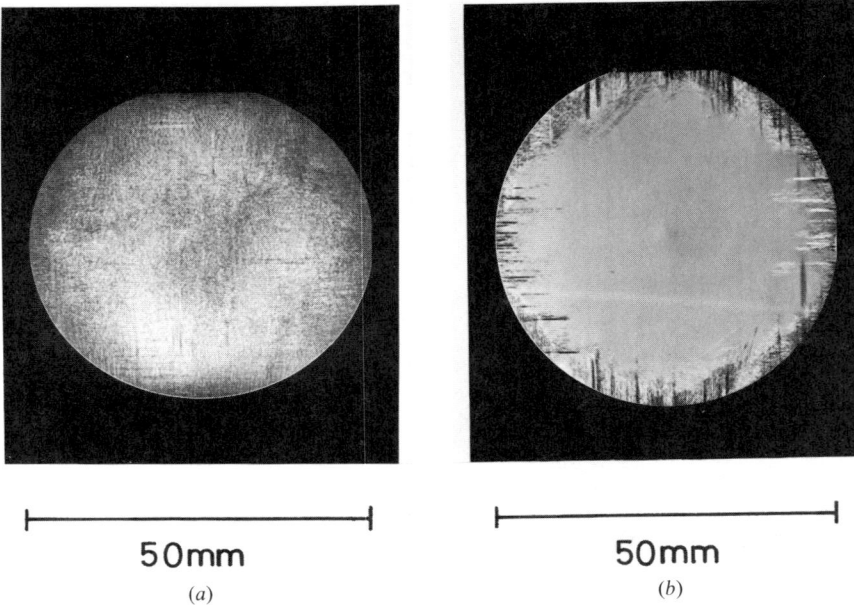

FIG. 22. X-ray topographs taken, under a $\langle 220 \rangle$ diffraction vector with $Mo_{k_{\alpha 1}}$, radiation, by the Lang transmission method. The wafers used were (a) Sn-doped and (b) S-doped InP crystals having an $\langle 001 \rangle$ direction.

by the x-ray diffraction method, which provides information concerning crystal stoichiometry.

b. *Characterization of Impurities*

Impurities seriously affect the characteristics of semiconductor materials and their devices. Therefore, it is essential to identify the kinds of impurities present and to determine their amounts. Many kinds of instrumental analyses are applied to these purposes (Millet, 1980; Mullin, 1985). In many cases, calibration measurements must be carried out using standard samples whose concentrations have already been determined by other methods. These results are useful in understanding both the electrical and optical properties of crystals. In order to do this, it is necessary to determine the concentration of very low levels of impurity, typically 10^{-2}–10^{-3} parts per million of atoms (ppma). Such characterization has been recently achieved by advancements in the sensitivity of analytical instruments. Techniques most frequently used are mass spectroscopy, atomic absorption spectrometry, and atomic emission spectrometry, and a description of each follows.

5. InP SUBSTRATES: PRODUCTION AND QUALITY CONTROL

(1) *Mass spectroscopy.* The spark source mass spectrometry (SSMS) technique is often used to determine roughly impurity amount. In this technique, many kinds of ion species, which are made from a discharged sample, are separated according to their m/e (mass/charge) by using a combination of electrostatic and magnetic analyzers. The typical sensitivity for most of the elements is in the range of 0.01 to 0.1 ppma. For a more precise analysis, the secondary ion mass spectrometry (SIMS) technique is usable. Its features include high spatial resolution with high sensitivity. It can also analyze the depth profile of an element's amount. Figure 23 shows a depth profile of Fe concentration in an annealed Fe-doped InP crystal. Identification limits for the impurities in InP crystals have been reported in detail by Tanaka *et al.*, (1988). Recently, the glow discharge mass spectrometry (GDMS) technique has gained attention due to its high sensitivity (Harrison *et al.*, 1986). Table IX shows our results using the GDMS technique on an Fe-doped InP crystal.

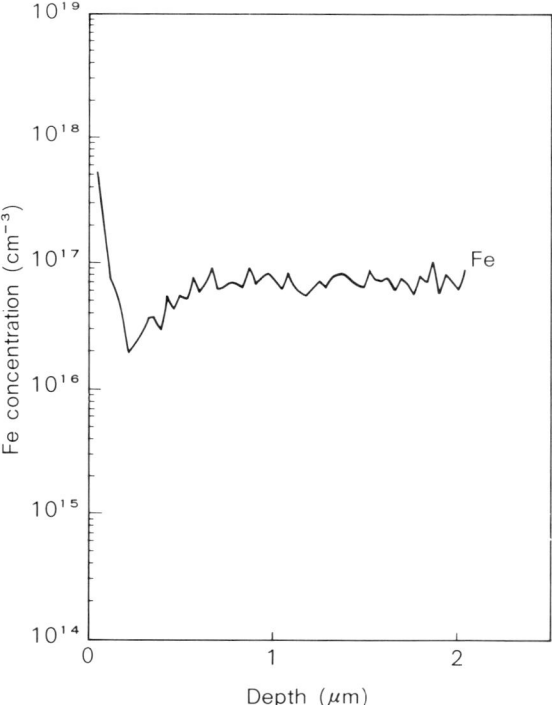

FIG. 23. Depth profile of Fe concentration, determined by the SIMS analysis. The abscissa represents depth from the InP wafer surface. The wafer was annealed at 750 °C for 30 min. This profile shows the out diffusion of Fe occurred at the surface.

TABLE IX

GDMS ANALYSIS DATA OF AN FE-DOPED InP.

Peak	Atomic-ppm	Atoms cm^{-3}
B 11	<0.000738	2.85 E13
Ma 23	0.003808	7.00 E14
Mg 24	<0.000761	2.93 E13
Al 27	0.007897	4.73 E14
Si 28	0.017564	1.20 E15
S 32	0.032489	1.99 E15
Ti 48	<0.000814	3.12 E13
V 51	<0.000606	2.31 E13
Cr 52	<0.000715	2.75 E13
Mn 55	<0.000598	2.30 E13
Fe 56	0.969549	4.16 E16
Ni 58	<0.000883	3.40 E13
Co 59	<0.000598	2.30 E13
Cu 63	<0.001094	4.32 E13
Zn 64	<0.001229	4.71 E13
Sb121	<0.001053	4.04 E13
I 127	<0.000601	2.30 E13

<: Peak rejected or count rate limited.

(2) *Atomic absorption spectrometry/atomic emission spectrometry*. These techniques are the graphite furnace atomic absorption spectrometry (GFA) technique and the inductively coupled plasma atomic emission spectrometry (ICP) technique. They are often used in determining the amounts of doping elements or other particular element in crystals. The distribution of Fe in an Fe-doped InP crystal, which was obtained using our automated GFA technique (Shibata *et al.*, 1985), is shown in Fig. 24. In this case, the identification limit of Fe was very low (0.01 wt ppm).

c. Electrical Evaluations

Electrical or optical properties chiefly depend on the energy band structure, especially on its impurity levels. These are clarified by total evaluations including the DLTS, PL (emission/excitation spectroscopy), and ESR methods in addition to the impurity analysis.

Practically, the important values concerning electrical properties include those of the carrier concentration, mobility, and resistivity, which result from the energy band structure. These quantities are obtained by the Hall effect measurements.

Recently, device makers have begun to require that the electrical properties

5. InP SUBSTRATES: PRODUCTION AND QUALITY CONTROL

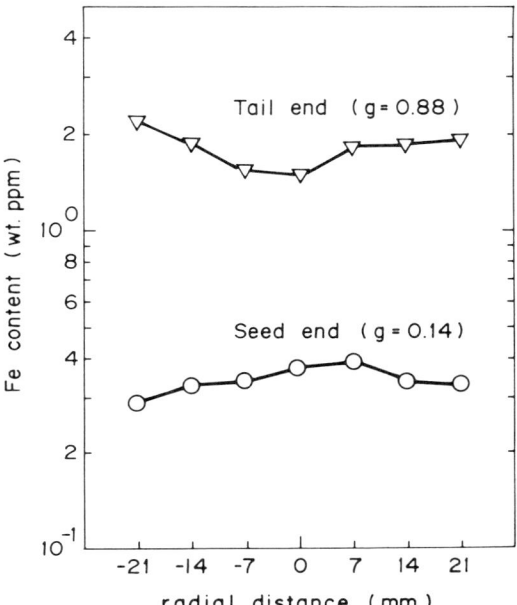

FIG. 24. Radial distribution of Fe in an Fe-doped InP crystal. These analyses were performed using the GFA method.

of the wafer's whole region be microscopically uniform. Advancements in evaluation techniques have enabled us to perform microscopic measurements of the properties described previously. As an example, the microscopic resistivity distribution measurement is introduced next (Matsumura et al., 1985; Obokata et al., 1984). The uniformity of resistivity in a wafer can be determined by using a technique having a high spatial resolution (about 70 μm). This technique is shown schematically in Fig. 25. This method is carried out by the following procedures:

(1) Patterns of microelectrodes (Fig. 25) are fabricated on one face of a wafer by the vacuum evaporation technique. The opposite face is entirely covered with an electrode.

(2) Probes (with one being used for the guard electrode) are kept in contact with a set of electrodes. Then, the current is measured by applying voltage to both sides of the wafer. The resistivity value at this point is calculated from the voltage and the current.

(3) For the next pattern, the same procedure is carried out. Results that were obtained in this way are shown in Fig. 26. They show the radial distribution of resistivity in an Fe-doped InP wafer.

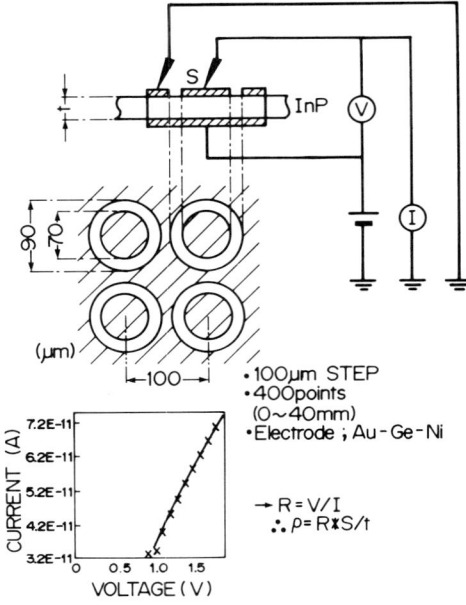

FIG. 25. An electric circuit used for resistivity distribution (μ-ρ) measurements. These measurements are conducted for 100 μm step intervals.

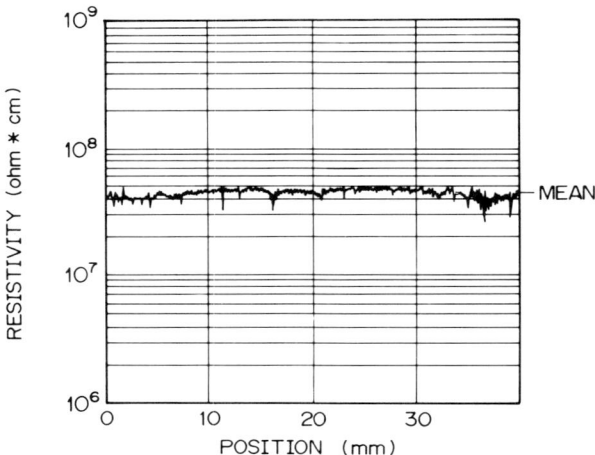

FIG. 26. Radial distribution of resistivity in an Fe-doped (001) wafer. The measurements were performed along an $\langle 011 \rangle$ direction at 100 μm step intervals.

d. Optical Evaluations

Optical measurements are a nondestructive method and generally have high sensitivity and high spatial resolution. Optical properties of semiconductors are strongly related to the band structure, which includes band gap and energy levels formed by impurities. Therefore, optical measurements are useful tools for understanding electrical properties. They also provide information about defects in a single crystal. It is essential to fully understand the optical properties of InP single crystals, since they are widely applied in various optical devices. Descriptions of photoluminescence, optical transmittance, and infrared light scattering measurements follow.

(1) *Photoluminescence (PL) technique.* Photoluminescence is emitted when excess electrons and excess holes are recombined after their separation due to the crystal's surface being exposed to light whose energy is greater than the band gap energy. The PL measurements are often employed to identify impurities in crystal (Temkin *et al.*, 1982). Figure 27 shows one of the PL spectra at 4.2 K, which was measured on an Fe-doped InP crystal. This crystal's excitation resulted from an Ar ion laser beam ($\lambda = 514.5$ nm). In this figure, a 1.378 eV peak, which is presumed to be related to Zn, can be observed in addition to the main peak, which results from the free or bound exciton. Photoluminescence measurements also provide information concerning defects, strains, and their positional uniformities in a crystal. This technique, however, has two drawbacks: (1) it is difficult to analyze its results quantitatively; and (2) it is useless in the detection of nonradiative impurity centers.

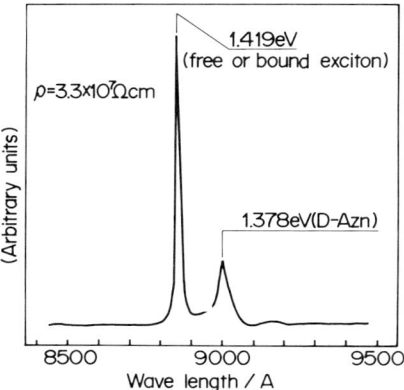

FIG. 27. PL spectrum of an Fe-doped InP at 4.2 K excited by an Ar ion laser beam ($\lambda = 514.5$ nm).

(2) Transmittance measurement. When InP single crystals are utilized for optical devices, high transmittance through the substrate is necessary. Optical transmittance in the infrared region depends on the free carrier absorption, that is, on the carrier concentration. Fig. 28 shows the dependence of the transmittance of a 1.3 μm wave on the carrier concentration (Morioka et al., 1987). It was measured by a double beam infrared spectrophotometer.

(3) Infrared light scattering measurement. This technique has recently been developed and is useful in investigating small-size inclusions, aggregations of defects, and precipitates (Ogawa and Nango, 1987; Fillard et al., 1987). The measuring system is shown schematically in Fig. 29. Using this system, we can obtain the scattering pattern, shown in Fig. 30, from undoped and Fe-doped InP single crystals.

e. *Evaluations for Practical Usage*

It is very important for both crystal suppliers and device makers to correlate the qualities of the devices and their processes to the properties of the crystal. Furthermore, as device productions have become more developed, customers now tend to require not only guarantees of the crystal

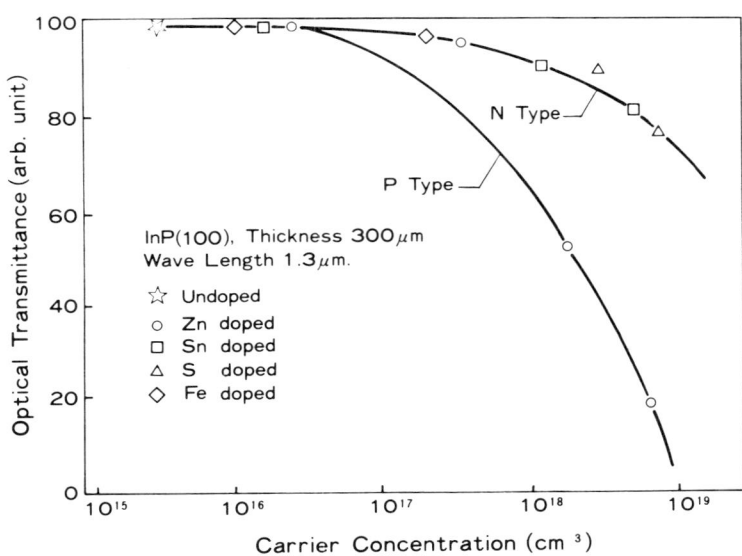

FIG. 28. Dependence of a 1.3 μm wave transmittance. Absorptions at high carrier concentration levels, free carrier absorption occurs.

5. InP SUBSTRATES: PRODUCTION AND QUALITY CONTROL

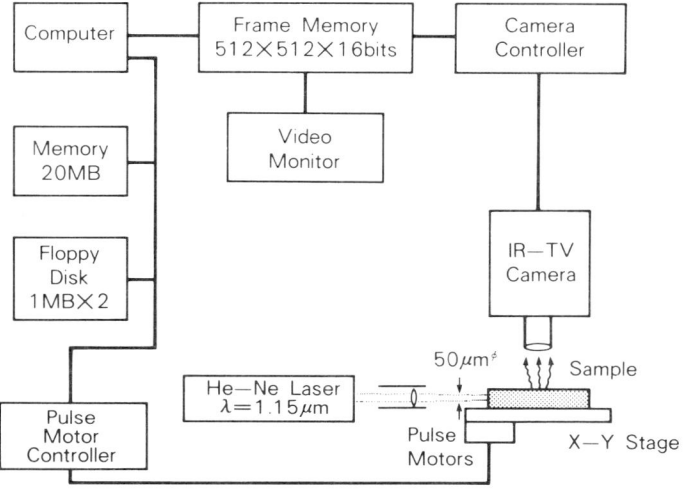

FIG. 29. Infrared light-scattering measurement system. A crystal's scatterer distribution can be two dimensionally displayed on a video monitor.

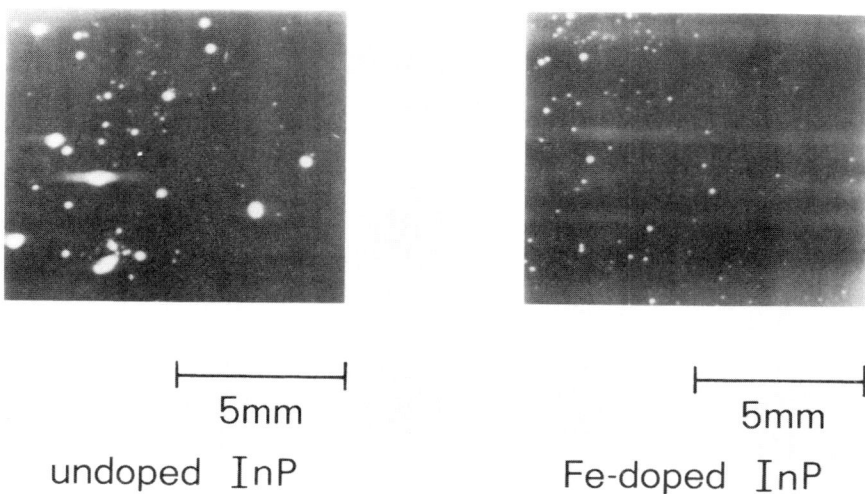

FIG. 30. Infrared light-scattering tomographs of undoped and Fe-doped InP.

properties, but also guarantees of the fundamental properties of the devices. In the case of GaAs LEC crystals for IC substrates, the uniformity of threshold voltages of FETs is required throughout the crystal. On the other hand, such demands have not yet appeared for InP crystals. There are mainly evaluation methods for ensuring a strong correlation between crystal property and device quality.

(A) Annealing evaluation. This evaluates the thermal stability of substrates, which is required in the device processes. Usually, the evaluations are carried out for crystal properties, such as resistivity, photoluminescence, and dopant's distribution. They are conducted both before and after annealing (Paris et al., 1983; Kamada et al., 1984; Farley et al., 1987; and Pande et al., 1984.

(B) Quasi-device evaluation. This evaluates the properties of the devices, which are fabricated in the form of a simplified structure. Recently, this type of evaluation has been performed on GaAs FETs (Takebe et al., 1985).

(C) Real device evaluation. This is often conducted in collaboration with the device makers.

For crystal suppliers, these evaluations are very useful, since they provide information concerning which crystal properties should be controlled to ensure high-quality devices.

7. SIMULATIONS

A crystal grower's ideal of crystal growth is described next. Crystal growers can completely simulate both the temperature profile of their furnaces and the growth conditions within them. They accomplish this by using high-speed, high-capacity supercomputers. They can, on the basis of the predictions of their simulations, employ both an appropriate hot zone structure and the most suitable growth conditions. As a result, they can obtain high-quality single crystals from the first growth attempt, without repeatedly trying troublesome crystal growth. Unfortunately, computer simulation cannot strictly reproduce such growth situations at the present time, though many crystal growers are still trying to realize this dream. This is not only due to the fact that there are too many unknown factors in controlling crystal growth. In addition, the growth temperature data on the physical properties of materials, which constitute the growth system, is insufficient. However, it is useful to understand roughly the conditions of crystal growth. Without using the complex three-dimensional analysis, useful information can still be obtained using a simple, one dimensional heat flow analysis. Using this method, we achieved long GaAs single crystals having high uniformity of properties along the pulling direction. Other simulations are concerned with the shear stress within a crystal, which is induced by thermal strain. From

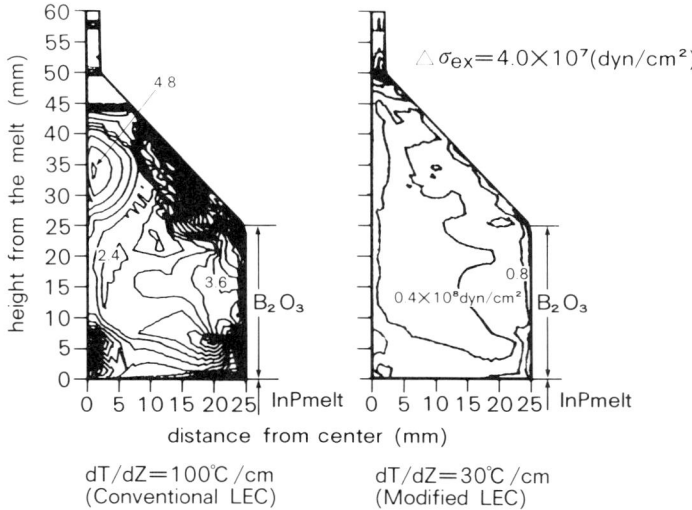

FIG. 31. Estimated contour lines of iso-excess shear stress in two InP single crystals.

these simulations, we can now roughly learn about both dislocation densities and their distribution. Figure 31 shows an example of contour lines, which were caused by estimated iso-excess shear stress in the crystals (Tada *et al.*, 1987).

VII. Quality Control of Crystal Growth

The fundamental purposes of quality control are to ensure the constant quality of developed products during production and to continue to produce them with high yield and low cost. Moreover, continuous quality control is essential in order to ensure these qualities. It is desired that the product's fundamental quality and production process have been established during development (prior to production). In most cases, these are further improved during the production process. Single crystal growth, such as that of InP, has many unknown factors that are complicatedly interrelated. Therefore, it is very difficult to control perfectly both the processes and the quality using only limited control parameters, while at the same time strictly correlating the results of the crystal evaluations to the growth conditions. Scientific approaches, however, have recently progressed in this field by developing the following control and evaluation techniques. In addition, both growth experiments under varying conditions and evaluations on individual crystals during production have been conducted. As a result, crystal growers have come to understand the control parameters and evaluations necessary for growing high-quality single crystals. Some of the quality control methods carried out in the InP production process are briefly described next. The control parameters and conditions are indicated by numerals, and the crystal properties related to them follow in parentheses.

The polycrystal synthesis process; in this process, it is essential that the highly pure InP polycrystals be stoichiometrically synthesized quickly. The Hall effect measurement is, therefore, effective for this purpose.

(1) temperature at the lowest temperature region, where phosphorus pressure is controlled (stoichiometry)
(2) temperature at the highest temperature region, where InP polycrystals are synthesized (purity, Hall effect)
(3) temperature gradient near the solid-liquid interface (precipitation of indium, stoichiometry)
(4) rate of movement of the quartz ampoule, which correspond to the growth rate (precipitation of indium, stoichiometry)
(5) purity of raw materials (purity)

The single crystal growth process; in this process, it is essential to achieve a high yield of diametrically controlled InP single crystals having both very few defects and high property uniformity. Therefore, attentions must be paid to both the temperature profile and the temperature fluctuation in the furnace.

(1) temperature profile and fluctuation (near the growth interface), which relate to various conditions, including the state of the hot zone, the heating power, the pulling rate, the crystal and crucible rotations, and the pressures in the furnace (defects, inclusions and uniformity of electrical properties)
(2) crystal diameter control (defect)
(3) purity in the furnace (purity)

Recent developments in computer systems have enabled crystal growers to effectively control these parameters and to fully understand the varying conditions of the entire process in real time. In the future, artificial intelligence (AI) will be introduced to this field. It is expected that even better quality control of InP single crystal will then be achieved.

VIII. Wafer Processing

InP single crystals have been used as the substrates for epitaxial layers to fabricate long wave optical communication devices and microwave devices. The substrates are required to have a high quality polished surface. Recently, the device fabrication technologies have become more complicated and precise. Also, in addition to the popular methods, LPE or VPE, methods such as OMVPE and MBE have begun to be used to grow thin crystalline layers. These tendencies have strengthened the requirement of high quality polished wafers with flatness, no contamination, and no damage. InP material is one kind of III–V compound semiconductor crystal like GaAs. Wafer processing of these materials is fundamentally the same as that of Si semiconductor crystal. However, III–V compound crystals have different physical and chemical properties than those of Si crystals, such as crystallographic structures, bonding strengths, and hardness. It is very important to optimize the wafer processing conditions that fit the material's properties and to construct the quality control system to meet various customer's requirements. In this section, before InP wafering process techniques are discussed in detail, InP material properties are summarized, and the fundamental functions of the InP wafering process are analyzed in connection with those properties. Quality evaluations and inspections executed to maintain a high quality polished wafer are also discussed.

8. Fundamental Properties of InP Single Crystal

InP single crystals have a zinc blende structure. This structure is the same as those of the other III–V compound crystals but is different from the diamond lattice crystal structure of Si. In shaping a single crystal, it is necessary to consider orientation dependency according to the crystallographic structure, which is different from materials that are amorphous or polycrystal with no specific orientation. The cleavage plane of Si crystal is the $\{111\}$ plane that combining strength between neighboring planes, is the weakest because it has the longest plane interval. On the other hand, the zinc blende crystal structure possesses not only covalent bonding but also ionic bonding. As a result, its cleavage plane is not the $\{111\}$ plane with the longest plane interval, but $\{110\}$ plane, which has electrical neutrality. The disagreement between the $\{111\}$ slip plane and the $\{110\}$ cleavage plane is a feature of III–V compounds, which results in the mechanical properties being different from those of Si. In III–V compounds, the mechanical properties show an orientation dependency. The hardness of a particular plane depends upon the angle difference against cleavage plane. Therefore, in the wafer processing of III–V compounds such as InP crystal, the orientation dependency must be considered carefully. For crystal with a diamond lattice structure, the (111) and ($\bar{1}\bar{1}\bar{1}$) faces appear identical, but for crystals with the zinc blende structure shown in Fig. 32, this is not true, since each double layer of the stacking sequence consists of one sheet of group III component atoms and one sheet of group V component atoms. Therefore, the face polarity must be considered in chemical treatments such as etching and polishing. In addition,

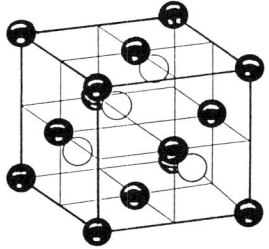

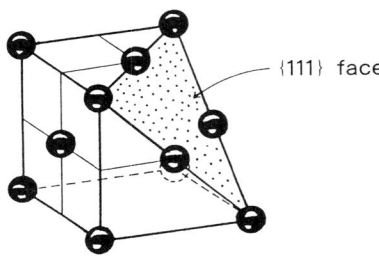

FIG. 32. Crystallographic structure of zinc blende type crystals.

for wafers with a {100} face, the etching characteristics of one face are different than those of another face. The Knoop hardness values for III–V compound crystals are less than those for Si crystals as shown in Fig. 33 (Gornynova et al., 1968). Among the crystals with the same crystallographic structure, Goldshmid's relationship (log $H = K_1 - K_2 \times \log r$) may be applied between crystal hardness H and intraatomic length r. Therefore, the longer the intraatomic length is, the smaller the hardness. The hardness of InP crystal is about half the hardness of Si crystal and softer than GaP and GaAs. In the InP polishing process, the greatest care must be taken in selecting and using the correct polish pad and polishing solution in order to prevent scratches from occurring.

9. FUNDAMENTAL SPECIFICATION FOR InP SUBSTRATES

Table X classifies the required specifications of III–V compound semiconductor substrates in accordance with the device applications. InP substrates are mainly utilized for the fabrication of laser diodes and photo detectors used in long distance communication. Surface orientations of InP polished wafers used as substrates for the epitaxial layers of laser diodes must be controlled with strict precision because of strong correlation between orien-

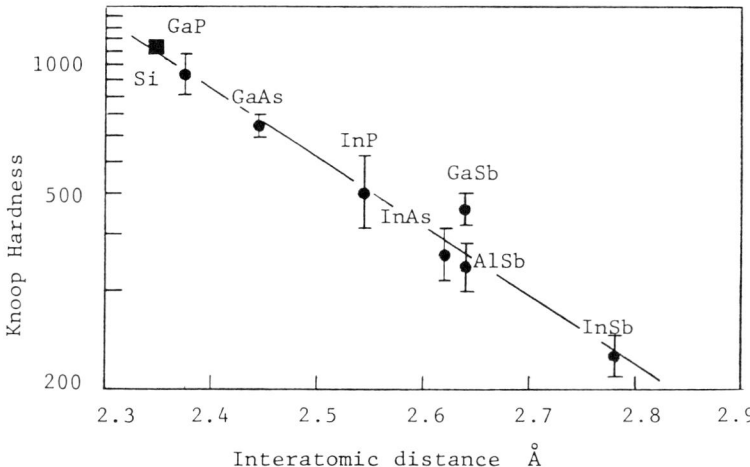

FIG. 33. Knoop hardness of III–V compound crystals (and also of Si crystal). The Knoop hardness of InP crystal is less than half of that of the Si crystal.

tation and epitaxial layer morphology. Recently, OMVPE technology has been developed as the epitaxial process for optical device fabrication. As the new epitaxial process has been developed, the requirements for surface cleanliness and subsurface damage have been growing. Surface flatness requirements for InP polished wafers are not so strong because optical devices and the substrate wafers are small. In the future, with device size enlargement, surface flatness requirement for InP wafers will be growing at a rate similar to that of GaAs.

10. WAFER PROCESSING FOR InP SUBSTRATES

The required specifications for InP substrates are modified in various ways with the development of the new epitaxial growth method and the extension of the device's applied field. To meet various requirements properly, required specifications must be classified according to the fundamental function of the wafer processing. Table XI shows the fundamental properties of the substrates and the fundamental wafer processing. The basic wafer processing of InP substrates is not different from that of GaAs or Si. These required specifications are basically the same because the processes from wafering through device fabrication are basically the same. Figure 34 shows the wafer processing for InP substrates to ensure the substrate characteristics. Under each process, fundamental function and intermediary product quality are checked, and, as a result, the polished wafers, which are final products, are

TABLE X

REQUIRED SPECIFICATIONS OF III–V COMPOUND SEMICONDUCTOR SUBSTRATES IN ACCORDANCE WITH DEVICE APPLICATIONS.

Devices	Infrared LED	Laser diode	Optical detecter	FET
Fabrication method	Epitaxial	Epitaxial	Epitaxial	Epitaxial, (I^2)
Fundamental Characteristics				
Crystallographic Orientation	○	○	○	○
Wafer thickness variation	○	○	○	○
Wafer size	△	○	○	○
Subsurface damage	△	○	○	○
Cleanliness of polished wafer	△	○	○	○
Planarity	△	○	○	○
Wafer number (Ingot position)	△	△	△	○
Mesa orientation				
Clarification of sides	△	○	○	○

TABLE XI

FUNDAMENTAL PROPERTIES OF SUBSTRATES AND FUNDAMENTAL WAFER PROCESSING.

Fundamental functional process		Parameter							Objective characteristics	Equipments
		Surface orientation	Mesa orientation	Wafer size	Thickness variation	Planarity	Subsurface damage	Flatness	Cleanliness		
▽	(Ingot)										
○	Decision of orientation	○	○							Orientation accuracy	X-ray diffraction Laser beam
○	Cutting		○	○	○	○	△			Thickness variation Work damage Planarity	Slicer
○	Shaping		○	○			△			Size accuracy	Grinding machine Edge-contouring machine Scriber
○	Removal of damage					○	○			Work damage	Etching instrument Polisher
○	Regulation of thickness				○	○	○			Thickness variation	Lapping machine
○	Polishing				○	○	○	△		Thickness variation Work damage planarity Flatness, cleanliness	Polisher
○	Cleaning							○	○	Thickness variation Work damage, planarity Cleanliness	Polisher
△	(Polished wafer)								○		Organic solvent Detergent

Fundamental Characteristics	Common characteristics
	1 Wafering number (χ/L)
	2 Packaging

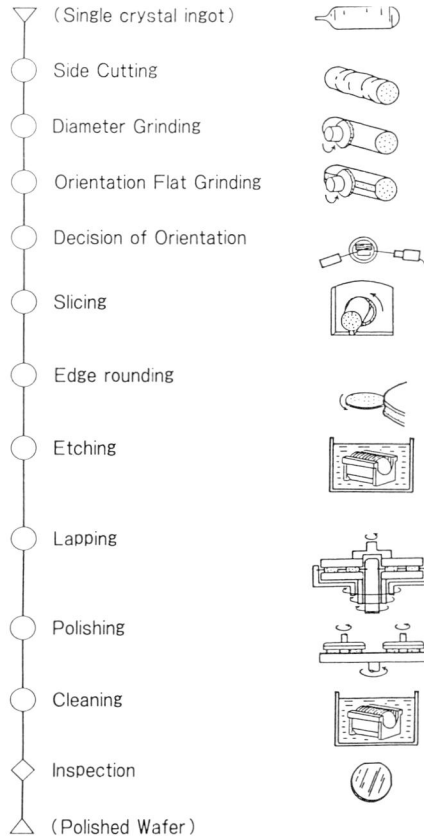

FIG. 34. Wafer processing of InP substrates.

correctly manufactured. Next, the wafering technique of InP single crystal substrates is presented in detail.

a. Grinding

Crystallographic properties and electrical properties of the crystal ingots grown by the LEC method are investigated for the seed and tail ends. The outside of the crystal ingots, which come up to the standard, are ground by a three-dimensional cylindrical grinding machine to define the diameter of the material. A rotating cutting tool makes multiple passes down a rotating ingot until the chosen diameter is attained. Following diameter grinding, two flats are ground along the length of the ingot. The orientation flats serve several

5. InP SUBSTRATES: PRODUCTION AND QUALITY CONTROL

purposes for III–V compound substrates. They serve as identifiers on both sides of the wafers and also as mechanical locators in automated processing equipment. In III–V compounds, the etching rate depends on the surface orientation. The etching rate of the (111) A face with a group III atom surface is slower than that of the other faces. The dependency of the etching rate on the surface results in different etched structures ("dovetail" and "V" groove) between the $(0\bar{1}\bar{1})$ cleavaged face and the $(0\bar{1}1)$ cleavaged one, as shown in Fig. 35. These etched surface characteristics are important factors that influence the device's electrical properties and the precision of the fine patterns required by device fabrication processes. The figures of the etch pits that reveal the crystal dislocation are also dependent on surface orientation. Prior to generating the identifying flats, the ground single crystal ingots are etched and inspected for the etched surface characteristics. On the $(0\bar{1}\bar{1})$ face, primary flat is identified, and the secondary flat is identified on the $(0\bar{1}1)$ face.

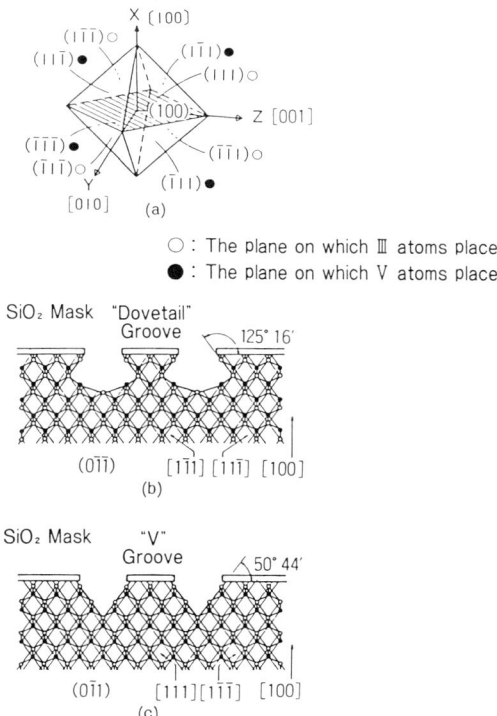

FIG. 35. (a) Elements (III or V atom) appeared on {111} crystal plane. Etched structure of (b) an $(0\bar{1}1)$ cleaved surface and (c) an $(0\bar{1}\bar{1})$ cleaved surface.

Once the previous operations are completed, the ingot is usually ready to be converted to a wafer geometry, as shown in Fig. 36.

b. *Slicing*

(1). Orientation measurement. Slicing is an important process because it determines four wafer parameters: surface orientation, thickness, taper, and bow. Among these parameters, crystal surface orientation has an effect on the epitaxial layer's uniformity, the surface etching rate, and the other device characteristics. InGaAs or InGaAsP epitaxial layers grown on InP substrates

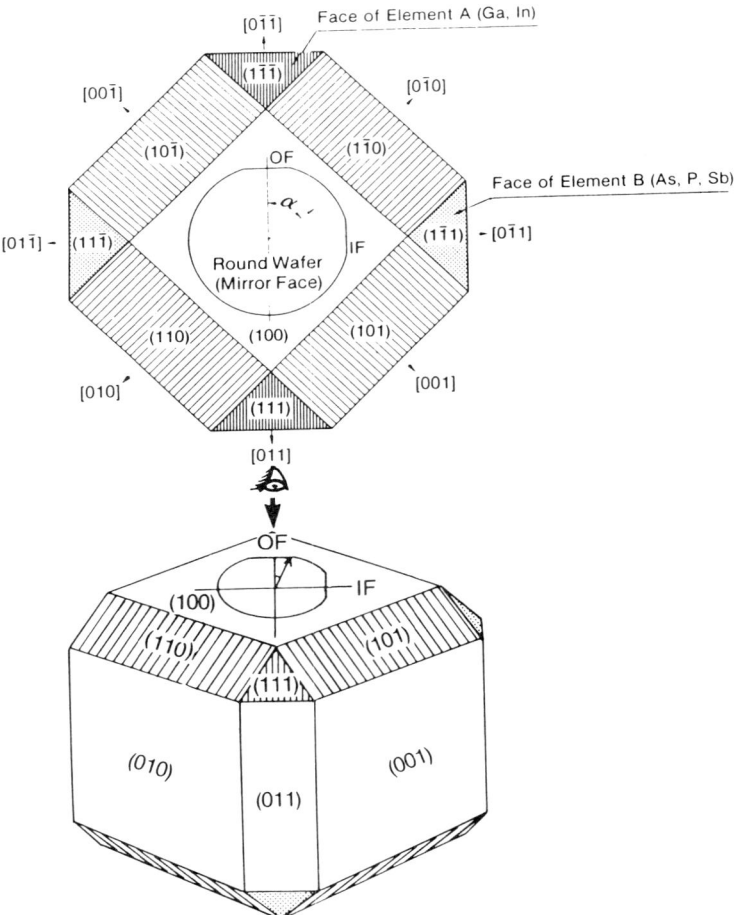

FIG. 36. Definitions of both OF and IF.

are required for uniformity of layer thickness and the surface morphology. Therefore, the required accuracy of the substrate's surface orientation is strict and is very strict for laser application. Orientation for sawing can be determined by cutting several wafers, measuring the orientation, and then resetting the ingot axis until the correct orientation is achieved. Measuring methods are mainly divided into the x-ray method and the laser evaluation method. These methods are summarized in Fig. 37. For the strict precision of the surface orientation, the x-ray method is utilized.

(2). Cutting. The conversion of ground ingots into slices is typically accomplished by inner diameter (ID) diamond sawing. This is done because wire sawing or band sawing is unsuitable for slicing to the strict orientation precision required especially in InP substrates. The ingot is prepared for slicing by mounting it in wax or epoxy, placing it on a support, and then positioning the support on the saw. The saw blade is a thin sheet of stainless steel with diamonds bonded on the inner rim. This blade is tensioned in a collar and then mounted on a drum that rotates at high speed on the saw. The key points of quality control in the cutting process are as follows:

(1) maintaining the machine performance, for example, parallelium between blade face and cutting direction, and the deflection of the blade
(2) controlling the sharpness of a blade, for example, diamond size, diamond thickness, stainless steel's stiffness, cutting loss, and blade life
(3) maintaining a suitable cutting condition, for example, the regularity of blade tensioning, blade-dressing techniques, and the blade translation speed

These items are checked and controlled in order to attain quality characteristics of the sliced wafers.

(3). Edge rounding. Sliced wafers have the square-cornered slice periphery, which tends to be laden with microcracks from slicing and is easy to break. Edge-rounded wafers have the following advantages:

(1) fewer chips are developed during device fabrication,
(2) the wafer is more fracture resistant,
(3) the epicrown growth is reduced, and
(4) the buildup of photoresist at the wafer edge is controlled.

This edge rounding step is done in cassette-fed equipment, which uses a multistep sequence to grind the outside diameter, top edge, and bottom edge on a cam-controlled profile grinder.

(4). Etching. The previously described shaping operations leave the surface and edges of the wafer damaged and contaminated. In addition, a bow

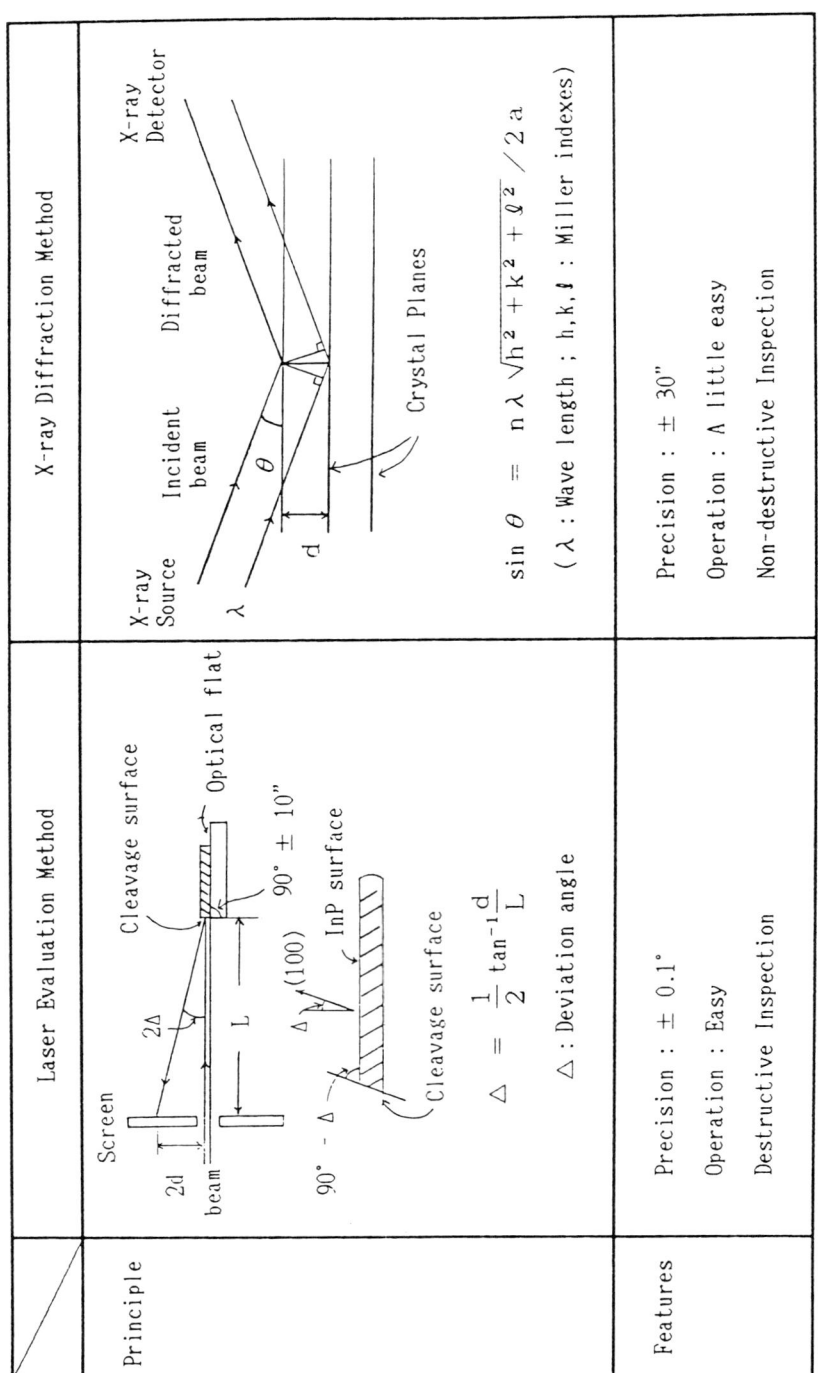

FIG. 37. Methods of orientation measurements for crystal cutting.

induced by nonsymmetrical surface stresses on opposing faces may still exist. The application of etching media to as-cut wafers or as-lapped wafers is effective in removing the surface layers that have residual damage and contamination, and in reducing the excessive bow of the wafer. The damaged and contaminated region of InP as-cut wafers is about 10 μm deep and can be removed by chemical etching, which uses mixed bromine and methanol, or nitric acid, hydrochloric acid, and, deionized water. Etching conditions must be carefully controlled because acid etching is prone to cause nonuniform and nonplanar wafer surface due to the variation of the etching conditions and the condition of the surface before etching.

c. *Lapping*

Lapping has the following operating functions: (1) the regulation of wafer thickness, (2) the reduction of wafer thickness variation, (3) the improvement of the dimensional uniformity, and (4) the improvement of surface flatness.

Lapping can be conducted as a single-side or double-side process. Alumina abrasive powders, typically of 5 to 10 μm particle size, remove 30–45 μm of stock per side, with their motion between wafers and lapping plates. A double-side lapping process is typically used for Si or GaAs. For an InP substrate, however, the single-side lapping process is typically used because the double-side process has been proved to cause the edge chips and cracks on the edges of InP wafers, which are softer and more brittle, and typically used with thinner thicknesses. In the future, stricter adherence to surface flatness will be required with the development of epitaxial growth methods such as OMVPE or MBE, and the enlargement of device size. The double-side lapping methods suitable for InP substrates must be developed.

d. *Polishing*

Polishing is the final step of the wafer processing. Sub-surface damage, which is presently the most important characteristic for InP substrates, is controlled in this process. The required characteristics in this step are as follows: (1) wafer thickness and its variation, (2) wafer planarity, (3) surface smoothness of polished face, (4) residual damage layer on the polished surface, and (5) cleanliness of the polished surface. These characteristics are determined by the polishing method and the materials used. Polishing can be accomplished by one-side polishing or double-side polishing, which is similar to lapping. For InP wafers, the one-side polishing is typically used because it can control the surface quality more easily. Polishing slurrys are divided into the following three types: mechanical abrasive, mechano-chemical slurry, and chemical solution. Si crystals are typically polished by caustic colloidal-silica slurry with ~ 100 Å diameter particles, which removes 10–30 μm from each side. This process can produce a highly reflective planar surface with a very thin damaged layer of several tens Å thickness, free from microscratches and haze. For InP wafers, the similar chemical-mechanical system results in a thicker damaged layer of 2–3 μm thickness and easily causes microscratches because of the relative softness of InP. Chemical polishing processes, which do not use abrasives are needed for III–V compound materials.

InP wafer polishing is typically accomplished by a $Br_2 - CH_3OH$ solution and a rotating pad made of artificial fabric, such as polyurethane laminate, polyester felt. Typical processes remove 25–35 μm stock of InP. In the polishing process, the following parameters must be controlled to maintain the important polished-wafer characteristics such as flatness, taper, and quality of surface finish:

(1) composition, flow rate, radial distribution, and temperature of the polishing solution
(2) surface speed, temperature distribution, relative diameters, rigidity, and geometrical contours of the platen on which the wafers are mounted
(3) wafer mounting procedures
(4) compressibility and polishing solution retention of pads
(5) life of polishing pads
(6) unit content pressure and ambient conditions.

e. Wafer Cleaning

The purposes of the wafer cleaning process are as follows: (1) the removal of residual material such as wax and contamination on the wafer backside, and (2) the removal of residues such as particles, contamination, and stains on the wafer frontside.

It is important to decide the cleaning steps according to the kind of residual contaminants. Typical cleaning steps are accomplished by using deionized clean water, organic solvents, detergents, and acid or alkaline etching solutions according to contaminations. For InP polished wafers, an organic solvents cleaning step is typically used. In the wafer cleaning process, not only is the control of the cleaner very important, but the control of other parameters such as cleaning tank, etching basket, and ambient environment are also very important. If only one parameter lacks among these parameters, polished wafers with cleanliness cannot be constantly obtained. Another cleaning step must be added in order to remove contamination once the polished wafers are contaminated. It can also be the causing parameter of other contaminations.

11. WAFER CHARACTERISTICS AND THEIR EVALUATIONS

Evaluation technologies for the substrate's wafering characteristics are important for the sake of improvement in the wafering process, quality control in the wafering process, and quality assurance for products. The evaluation technologies of wafer characteristics, including surface defects, sub-surface damage, planarity, and cleanliness are explained.

a. Evaluation of Surface Defects

Two kinds of the unaided eye inspection methods are usually accomplished according to defects on polished wafers. One of those inspections is with the use of the unaided eye under a fluorescent light source with an intensity of more than 2500 luxes, which is executed to detect chips, cracks, crows' feet, dimples, grooves, orange peels, mounds, pits, saw marks, and striations. Another inspection also uses the unaided eye with the polished surface under a narrow strong beam light source with a 1 − 2 cm focused light intensity of more than 65,000 luxes, which is used to detect contamination and scratches.

b. Evaluations of Damaged Layers

The evaluation of the damaged layers formed during the wafering processes are accomplished through the etching or the x-ray double-crystal diffraction methods. The etchant for inspection of the damaged layers on an InP polished wafer is typically a mixed solution with H_3PO_4 and HBr, which reveals some etch patterns corresponding to scratches on a polished surface by etching for 10–20 seconds, as shown in Fig. 38. This occurs because the only portion of the wafer on which mechanical work damages remain is etched more selectively than the other portion. The occurrence of these kinds of work damage can be suppressed by the optimization of the polishing process and the prevention of particles intruding into the polishing environment. The other evaluation utilizes the double crystal x-ray diffraction measurement method. InP wafers polished by $Br_2 - CH_3OH$ solution were evaluated as they were and were evaluated after etching for various periods. For the double crystal x-ray diffraction, a deeply etched S-doped dislocation-free InP (100) wafer was used as the first crystal. The (311) asymmetric parallel setting was applied for the measurement in order to evaluate the damage extremely near the surface by setting the illumination beam near parallelism for the (100) surface. Figure 39 shows the decrease of the full width at half maximum of the rocking curve as a function of the etching depth for the S-doped polished wafer. When more than 0.4 μm of the surface was etched off, no variation of the full widths of the half maximum was observed.

FIG. 38. Scratches on the surface of an InP wafer that was polished using the mechanochemical polishing method. Both abrasive slurry and a chemical solution were used.

It can be deduced that the damaged layer on the polished wafer is less than 0.4 μm.

c. Evaluations of Flatness

Flatness specification is important because it has a major influence on the practical limit for shrinking line width and increasing circuit complexity in Si or GaAs device fabrication. InP substrates, however, are mainly used for the fabrication of discrete devices such as laser diodes, light-emitting diodes, photo detectors, and high-speed FETs. The research and development for the integration of these devices will be a more important subject in the future. Typical TTV (total thickness variation) value of InP polished wafer is ~ 10 μm for two inches in diameter. The requirement for flatness will be more stringent. For flatness inspection, several different types of equipment are in use, each with different principles of measurement according to the customer's specifications.

d. Evaluations of Surface Cleanliness

In addition to the unaided visual inspection previously presented, the surface cleanliness on the polished wafers are evaluated by a surface

5. InP SUBSTRATES: PRODUCTION AND QUALITY CONTROL

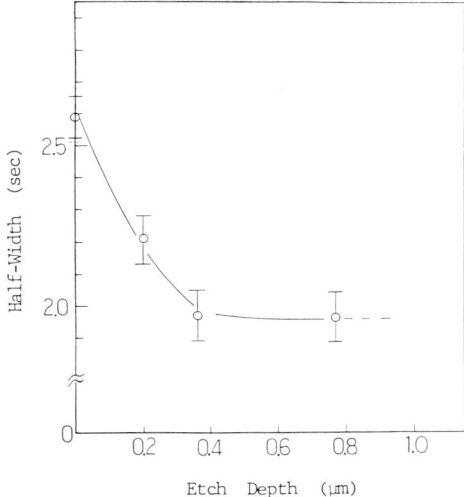

FIG. 39. Damaged layer thickness evaluation of InP wafers that were polished using the $Br_2 - CH_3OH$ solution. The evaluation was performed using the x-ray double crystal diffraction measurement method.

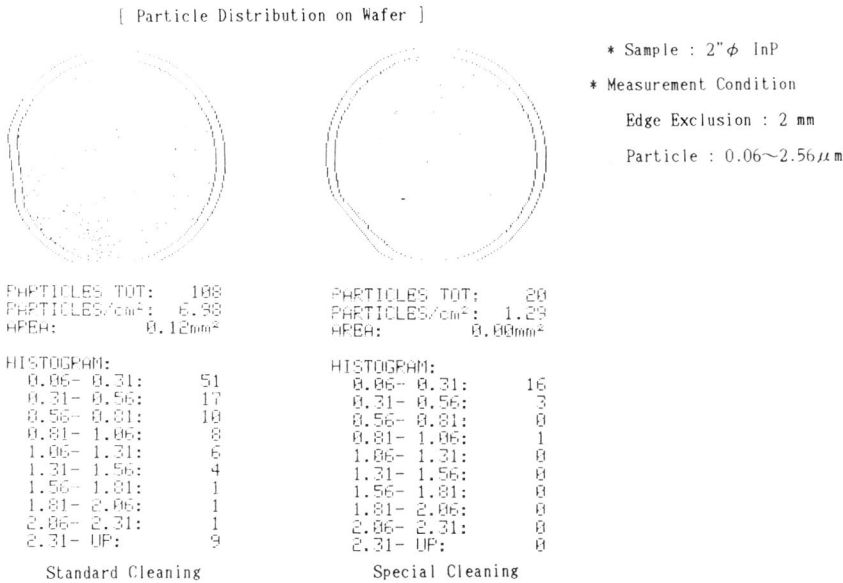

FIG. 40. Evaluation of surface cleanliness of an InP polished wafer. These evaluations were performed using the Surf Scan instrument.

measurement system that utilizes a laser-beam scan with a solid-state sensor to determine the number of particles and the degree of contamination. Figure 40 shows the comparison of surface cleanliness between typical standard products and the improved wafers cleaned by special treatment. The number of particles with a size greater than 0.2 μm on typical two inch InP polished surface is about 100 particles. On the other hand, the particles of the improved wafers diminished to about 20 particles.

IX. Summary

The technology of indium phosphide single crystal growth has recently shifted to the development of crystals having lower dislocation densities and higher purity. The commercially available conductive single crystal wafers now have relatively lower dislocation density ($2 \times 10^4 \sim 1 \times 10^5$ cm^{-2}). The dislocations in indium phosphide cause not only low efficiency of the light-emitting diodes but also degradation of the photodiode's properties. The latter result is due to an increase in dark current. Although the effect of the dislocations on electronic devices is not yet clear, there is a continuing demand for defect reduction. There are two methods for reducing dislocation densities. In the first method, dislocations are reduced by obtaining a low-temperature gradient near the solid-liquid interface during crystal growth. In the second method, the generation and propagation of dislocations are suppressed by doping with certain impurities. Electronically inactive elements, such as Ga or Sb, are used in reducing the dislocation density without affecting electric conductivities of the substrates. This method is also effective for obtaining low dislocation density semi-insulating substrates. Regarding thermal stability against changes in electric conductivity during device processing, there is the problem of residual impurities, which arises due to the impurities' thermal diffusion. Thus, the purification of both the raw materials and the polycrystal materials is very important and still requires development efforts. Techniques of crystal processing including slicing, polishing and etching, and surface treating are not well developed at present. Improvements in wafer surface roughness and cleanliness are necessary. Such techniques will provide an indispensable basis for the development of progressive applications for these materials in the areas of high-frequency microwave devices, optoelectronic devices, and future optoelectronic integrated circuits and high-speed digital devices.

REFERENCES

Adamski, J. A. (1983). Synthesis of indium phosphide. *J. Crystal Growth* **64**, 1-9.
Adamski, J. A., and Ahern, B. S. (1985). Rapid synthesis of indium phosphide. *Rev. Sci. Instrum.* **56**, 716-718.

Allred, W. P., Burns, J. W., and Hunter, W. L. (1981). Synthesis of indium phosphide in a pressure balanced system. *J. Crystal Growth* **54**, 41–44.

Bachmann, K. J., and Buchler, E. (1974). Phase equibria and vapor pressures of pure phosphorus and of the indium/phosphorus system and their implications regarding crystal growth of InP. *J. Electrochem. Soc.: Solid-State Science and Technology* **121**, 835–846.

Bimberg, D., Blachnik, R., Cardona. M., Deam, P. J., Grave, T., Harbeke, G., Hübner, K., Kaufman, U., Kress, W., Madelung, O., von Munch, W., Röseler, U., Schneider, J., Schulz, M., and Skolnick, M. S. (1982). "*Landolt-Börnstein III/17a*" (Madelung, O., ed.). Springer-Verlag, Berlin, Heidelberg, New York.

Bonner, W. A. (1981). InP synthesis and LEC growth of twin-free crystal. *J. Crystal Growth* **54**, 21–31.

Boomgaard, J., and Schol, K. (1957). The P-T-x phase diagrams of the systems In-As, Ga-As and In-P. *Philips Res. Rept.* **12**, 127–140.

Cameron, D. C. (1984). The correlation of channel mobility with interface state measurements on InP MOSFET structures. *Solid-State Electronics* **27**, 305–309.

Cockayne, B., Bailey, T., and MacEwan, W. R. (1986). Dislocation-free LEC growth of InP doped with Ge and S. *J. Crystal Growth* **76**, 507–510.

Donnay, J. D. H. (1963). "*Crystal Data, Determination Tables*," Second Ed., American Crystallographic Association, April, 1963, ACA Monograph No. 5.

Farges, J. P. (1982). A method for the "in-situ" synthesis and growth of indium phosphide in a Czochralski puller. *J. Crystal Growth* **59**, 665–668.

Farley, C. W., Kim, T. S., Streetman, B. G., Lareau, R. T., and Williams, P. (1987). Encapsulation and annealing studies of semi-insulating InP. *Thin Solid Films* **146**, 221–231

Fillard, J. P., Gall, P., Baroudi, A., George, A., and Bonnafe, J. (1987). Defect structures in InP crystals by laser scanning tomography. *Jpn. J. Appl. Phys.* **23**, L1255–L1257.

Foyt, A. G. (1981). The electro-optic applications of InP. *J. Crystal Growth* **54**, 1–8.

Gault, W. A., Monberg, E. M., and Clemans, J. E. (1986) A novel application of the vertical gradient freeze method to the growth of high quality III-V crystals. *J. Crystal Growth* **74**, 491–506.

Goryunva, N. A., Borshchevskii, A. S., and Tretikakov, D. N. (1968). Hardness. In "*Semiconductors and Semimetals: vol. 4,*" *(Physics of III–V Compounds)* (Willardson, R. K., and Beer, A. C., eds). 3–34, Academic Press, New York.

Gottschalk, H., Patzer, G., and Alexander, H. (1978). Stacking fault energy and ionicity of cubic III-V compounds. *Phys. stat. sol. (a)* **45**, 207–217.

Harrison, W. W., Hess, K. R., Marcus, R. K., and King, F. L. (1986). Glow discharge mass spectrometry. *Anal. Chem.* **58**, 341A–356A.

Huber, A., and Linh, N. T. (1975). Révélation métallographique des défauts cristallins dans InP. *J. Crystal Growth* **29**, 80–84.

Hyder, S. B., and Holloway, S. B. (1983). In-situ synthesis and growth of indium phosphide. *J. Electron. Mater.* **12**, 575–585.

Iseler, G. W. (1981). Liquid-encapsulated Czochralski growth of InP crystals. *J. Crystal Growth* **54**, 16–20.

Jordan, A. S. (1985). Some thermal and mechanical properties of InP essential to crystal growth modeling. *J. Crystal Growth* **71**, 559–565.

Jordan, A. S., Caruso, R., Von Neida, A. R., and Nielsen, J. W. (1981). A comparative study of thermal stress induced dislocation generation in pulled GaAs, InP, and Si crystals. *J. Appl. Phys.* **52**, 3331–3336.

Kamada, H., Shinoyama, S., and Katsui, A. (1984). Redistribution of Fe in thermally annealed semi-insulating InP (Fe): Determination of Fe diffusion coefficient in InP. *J. Appl. Phys.* **55**, 2881–2884.

Kaneko, K., Ayabe, M., Dosen, M., Morizane, S., and Watanabe, N. (1973). A new method of growing GaP crystals for light-emitting diodes, *Proc. IEEE* **61**, 884–890.

Katsui, A., and Tohno, S. (1986a). LEC growth and characterization of Ga doped InP crystals. *J. Crystal Growth* **74**, 221–224.

Katsui, A., and Tohno, S. (1986b). Growth and characterization of isoelectronically double-doped semi-insulating InP(Fe) single crystals. *J. Crystal Growth* **79**, 287–290.

Kawase, T., Kawasaki, A., and Tada, K. (1986). Improvement in the uniformity of electric properties of semi-insulating $3''\phi$ In-mixed GaAs single crystal by high magnetic field applied liquid encapsulated Czochralski technique. *In "Proc. of 13th Int. Symposium on GaAs and Related Compounds,"* (Lindley, W. T., ed.). 27–32, Las Vegas.

Kubota, E. (1984). Growth of InP single crystals by growth-rate controlled synthesis, solute diffusion technique. *J. Crystal Growth* **68**, 639–643.

Kubota, E., and Katsui, A. (1987). Growth of high purity InP crystals under high growth temperature by synthesis, solute diffusion technique. *J. Crystal Growth* **82**, 573–577.

Matsumura, T., Obokata, T., and Fukuda, T. (1985). Two-dimensional microscopic uniformity of resistivity in semi-insulating GaAs. *J. Appl. Phys.* **57**, 1182–1185.

Millet, E. J. (1980). Progress in the analysis of crystalline solids. *J. Crystal Growth* **48**, 666–682.

Morioka, M., Tada, K., and Akai, S. (1987). Development of high-quality InP single crystals. *Ann. Rev. Mater. Sci.* **17**, 75–100.

Mullin, J. B. (1985). Characterization of III–V compounds. *In "Crystal Growth of Electronic Materials,"* (Kaldis, E., ed.). 371–378, North-Holland.

Obokata, T., Matsumura, T., Terashima, K., Orito, F., Kikuta, T., and Fukuda, T. (1984). Improved uniformity of resistivity distribution in LEC semi-insulating GaAs produced by annealing. *Jpn. J. Appl. Phys.* **23**, L602–L605.

Ogawa, T., and Nango, N. (1986). Infrared light scattering tomography with an electrical streak camera for characterization of semiconductor crystals. *Rev. Sci. Instrum.* **57**, 1135–1139.

Osaka, J., Kohda, H., Kobayashi, T., and Hashikawa, K. (1984). Homogeneity of vertical magnetic field applied LEC GaAs crystal. *Jpn. J. Appl. Phys.* **23**, L195–L197.

Pande, K. P., Nair, V. R. K., and Aina, O. (1984). Capless annealing of InP for metal-insulator-semiconductor field-effect transistor applications. *Appl. Phys. Lett.* **45**, 532–534.

Paris, J. C., Gauneau, M., L'Haridon, H., and Pelous, G. (1983). Effects of annealing on unintentionally doped LEC InP. *J. Crystal Growth* **64**, 137–141.

Richman, D. (1963). Dissociation Pressures of GaAs, GaP and InP and the nature of III–V melts. *J. Phys. Chem. Solids.*

Seki, Y., Watanabe, H., and Matsui, J. (1978). Impurity effect on grown-in dislocation density of InP and GaAs crystals. *J. Appl. Phys.* **49**, 822–828.

Shibata, M., Yoshida, H., Shima, H., Sasaki, M., and Fijita, K. (1985). Personal computer-controlled automatic analytical system for graphite furnace atomic absorption spectrometry. *Sumitomo Electric Technical Review* **24**, 176–182.

Shimizu, A., Nishine, S., Morioka, M., Fujita, K., and Akai, S. (1986). Low dislocation crystal growth of semi-insulating InP through multi-heater LEC technique and co-doping of Ga and As. *In "Semi-Insulating III–V Materials"* 41–46, Ohmsha.

Shinoyama, S., Uemura, C., Yamamoto, A., and Tohno, S. (1981). Growth and crystal quality of InP crystals by the liquid encapsulated Czochralski technique. *J. Electron. Mater.* **10**, 941–956.

Tada, K., Tasumi, M., Nakagawa, M., Kawase, T., and Akai, S. (1987). Growth of low dislocation-density InP by the modified CZ method in the atmosphere of phosphorus vapor pressure. *In "Proc. of 14th Int. Symposium on GaAs and Related Compounds,"* (Crete) (to be published).

Takebe, T., Shimazu, M., Kawasaki, A., Kotani, T., Nakai, R., Kikuchi, K., Murai, S., Tada, K., Akai, S., and Suzuki, T. (1985). Characterization of SI GaAs substrate uniformity for IC application. *Sumitomo Electric Technical Review* **24**, 158–165.

Tanaka, T., Homma, Y., and Kurosawa, S. (1988). Secondary ion mass spectrometric, ion yield and detection limits of impurities in indium phosphide. *Anal. Chem.* **60**, 58–61.

Temkin, H., Dutt, B. V., Bonner, W. A., and Keramidas, V. G. (1982). Deep radiative levels in InP. *J. Appl. Phys.* **53**, 7526–7533.

Terashima, K., and Fukuda, T. (1983). A new magnetic field applied pulling apparatus for LEC GaAs single crystal growth. *J. Crystal Growth* **63**, 423–425.

Tohno, S., and Katsui, A. (1986a). X-ray topographic study of twinning in InP crystals grown by the liquid encapsulated Czochralski technique. *J. Crystal Growth* **74**, 362–374.

Tohno, S., and Katsui, A. (1986b). Single crystal growth and characterization of isoelectronically-doped InP. *J. Crystal Growth* **78**, 249–256.

Toyoda, N., Aota, Y., and Takahashi, N. (1985). Development of an automatic etch pit density inspection system. In *"Defect Recognition and Image Processing in III–V Compounds,"* (Fillard, J. P., ed.). 141–147. Elsevier Science Publishers B. V., Amsterdam.

Uemura, C., Shinoyama, S., Yamamot, A., and Tohno, S. (1981). LEC growth and characterization of undoped InP crystals. *J. Crystal Growth* **52**, 591–596.

Van der Pauw, L. J. (1958). A method of measuring specific resistivity and Hall effect of discs of arbitrary shape. *Philips Res. Repts.* **13**, 1–9.

Van der Pauw, L. J. (1958/59). A method of measuring the resistivity and Hall coefficient on lamellae of arbitrary shape. *Philips Technical Review* **20**, 220–224.

Wardill, J. E., Dowling, D. J., Brunton, R. A., Crouch, D. A. E., Stockbridge, J. R., and Thompson, A. J. (1983). The preparation and assessment of indium phosphide. *J. Crystal Growth* **64**, 15–22.

CHAPTER 6

LP-MOCVD Growth, Characterization, and Application of InP Material

Manijeh Razeghi*

THOMSON-C.S.F.
LABORATOIRE CENTRAL DE RECHERCHES
ORSAY, FRANCE

	LIST OF ACRONYMS.	244
I.	INTRODUCTION	245
II.	ENERGY BAND STRUCTURE OF InP	248
III.	GROWTH TECHNOLOGY	252
	1. *LPE*	254
	2. *VPE*	255
	3. *MBE*	256
	4. *MOCVD*	257
	5. *New Non-equilibrium Growth Technique*	259
	6. *ALE*	261
	7. *MEE*	263
IV.	LOW PRESSURE METALORGANIC CHEMICAL VAPOR DEPOSITION (LP-MOCVD)	264
	8. *Experimental Details*	264
	9. *Starting Materials*	265
	10. *Reactor Design*	269
	11. *Growth Procedure*	270
	12. *Flow Patterns*	271
V.	InP-InP SYSTEMS	272
	13. *Growth and Characterization of InP Using TEIn*	272
	14. *Orientation Effects*	274
	15. *Source-Purity Effects*	275
	16. *Characterization*	276
	17. *Interfaces*	293
VI.	GROWTH AND CHARACTERIZATION OF InP USING TMIN	295
VII.	INCORPORATION OF DOPANTS	301
VIII.	MICROWAVE APPLICATIONS	307
	18. *Gunn Diodes*	307
IX.	LP-MOCVD GROWTH OF InP ON ALTERNATIVE SUBSTRATES	313
	19. *Growth Procedure*	314
	20. *Growth of GaInAs–InP Multiquantum Wells on Garnet $(GGG = Gd_3Ga_5O_{12})$*	323

*Parts of this chapter are reprinted with permission from "The MOCVD Challenge," published by, IOP Publishing Ltd., Techno House, Redcliffe Way, Bristol, U.K.

	21. GaInAsP/InP Heterostructures Grown by LP-MOCVD on Si Substrates	327
	22. Growth of GaAs on Si Substrate, Using GaAs/$Ga_{0.49}In_{0.51}P$ Superlattices as a Buffer Layer	335
X.	OPTOELECTRONIC APPLICATIONS	339
	23. GaInP/GaInAs/InP MESFETs	341
	24. Planar Monolithic Integrated Photoreceiver for 1.3–1.55 μm Wavelength Applications using GaInAs–GaAs Heteroepitaxies	341
	25. Monolithic Integration of a GaInAs/GaAs Photoconductor with a GaAs FET for 1.3–1.55 μm Wavelength Applications.	346
	26. Monolithic Integration of a Schottky Photodiode and a FET Using a $Ga_{0.49}In_{0.51}P/Ga_{0.47}In_{0.53}As$ Strained Material	349
XI.	CONCLUSION	351
	ACKNOWLEDGMENTS	351
	REFERENCES	351

List of Acronyms

AES	Auger electron spectroscopy
ALE	atomic layer epitaxy
BRS	buried ridge structures
CBE	chemical beam epitaxy
DEZ	diethyl zinc
DH	double heterostructure
DLTS	deep-level transient spectroscopy
EPD	etch-pit density
ESCA	emission spectroscopy chemical analysis
FET	field-effect transistors
FME	flow rate-modulation epitaxy
GGG	gallium-gadolinium-garnet
GSMBE	gas source molecular beam epitaxy
hh	heavy hole
IC	integrated circuits
IHS	integral heat sink
InP	indium phosphide
LDS	low-dimensional structures
LEED	low energy electron diffraction
lh	light hole
LPE	liquid phase epitaxy
LP-MOCVD	low pressure metalorganic chemical vapor deposition
MBE	molecular beam epitaxy
MEE	migration-enhanced epitaxy
MOCBD	metalorganic chemical beam deposition
MOCVD	metalorganic chemical vapor deposition
MOMBE	metalorganic molecular beam epitaxy
MQW	multiquantum well
PL	photoluminescence
QSE	quantum size effect

RHEED	reflection high-energy electron diffraction
SCE	saturated calomel electrode
SIMS	secondary-ion mass spectroscopy
SL	superlattices
TDEG	two-dimensional electron gas
TDHG	two-dimensional hole gas
TEGa	triethyl gallium
TEIn	triethyl indium
TMIn	Trimethyl indium
UHV	ultr-high-vacuum
VPE	vapor phase epitaxy
XPS	x-ray photoelectron spectroscopy
YAG	yttrium aluminum garnet

I. Introduction

The early work on InP was stimulated by the need for long wavelength lasers as light sources, with the lower loss and lower dispersion at $\lambda = 1.3$–$1.55\,\mu m$ in silica fibers. Future telecoms systems will require narrow spectrum $1.55\,\mu m$ lasers and very sensitive detectors, and this is forcing a significant research push on the technology of InP and related compounds. For microwave devices, with the exception of mobility, which is higher in GaAs, the other characteristics favor InP over GaAs in terms of superior Gunn device performance. The main use for InP and related compounds appears to be in the new field of semiconductor quantum devices using heterojunction, multiquantum wells and superlattices. The discovery of these devices has been a direct result of basic research in InP and related compounds. Knowledge of the bond structures, the electrical properties, optical properties, and the carrier recombination mechanisms led a number of physicists to conclude that the InP and related compound multilayers is the best candidate for the future optoelectronic and microwave devices.

The low pressure metalorganic chemical vapor deposition (LP-MOCVD) technique is well adapted to the growth of the entire composition range of InP and related compounds with uniform thickness and composition over a large area of substrate. This results first from the ability of the process to produce abrupt composition changes and second from the result that the composition and growth rate are generally temperature independent. It is a versatile technique; numerous starting compounds can be used, and growth is controlled by fully independent parameters.

Growth by MOCVD takes place far from a thermodynamic equilibrium, and growth rates are determined generally by the arrival rate of material at the growing surface rather than by temperature-dependent reactions between the gas and solid phases. In contrast to liquid phase epitaxy (LPE) growth, it has been found that during MOCVD growth of double heterostructure (DH), InP can be grown directly on GaInAsP with no disturbance of the active layer; i.e., there is no effect equivalent to melt-back.

One of the key reasons for the usefulness of this method is the possibility of obtaining high purity and, therefore, high-mobility $Ga_xIn_{1-x}As_yP_{1-y}$ lattice matched to InP. As long-wavelength 1.0–1.65 μm GaInAsP-InP electro-optical devices become more widely used motivated by low-fiber absorption and dispersion, high transmission through water and smoke, and greatly enhanced eye safety at wavelengths greater than 1.4 μm LP-MOCVD offers advantages of smooth uniform surfaces, sharp interfaces (lower than 10 Å for GaInAsP/InP), uniformly low background doping density, and economy of scale for large-area devices.

Using this technique, Razeghi showed the feasibility of using various metalorganic sources of group III elements with hydride sources of group V species for the following:

InP–InP System

(1) The growth of high-quality InP epilayer with carrier concentration as low as 3×10^{13} cm^{-3} and electron Hall mobility as high as 6000 cm^2 V^{-1} s^{-1} at 300 K and 200,000 cm^2 V^{-1} s^{-1} at 50°K. Photoluminescence measurements at 2°K showed that this is the purest InP that has yet been reported in the literature with zero compensation ratio (Razeghi *et al.*, 1987a).

These materials have been used for the fabrication of Gunn diodes, and have been transferred from development to the production department at Thomson (Poisson *et al.*, 1985; Razeghi *et al.*, 1982).

GaInAs–InP System

(2) The growth of high-quality $Ga_{0.47}In_{0.53}As$–InP heterojunction, multi-quantum wells (MQW) and superlattices (SL). The carrier concentration is as low as 3×10^{14} cm^{-3} with electron Hall mobility as high as 13,000 cm^2 V^{-1} s^{-1} at 300 K and 700,000 cm^2 V^{-1} s^{-1} at 2°K, and photoluminescence linewidth at 2°K less than 2 meV has been measured (Razeghi *et al.*, 1987b).

(3) The observation of two-dimensional electron gas (2DEG) and quantum Hall effect with one, two, and three subbands filled with 2DEG (Guldner *et al.*, 1982; Razeghi *et al.*, 1986a).

(4) The observation of two dimensional Hole gas with Hall mobility of 10,500 cm^2 V^{-1} s^{-1} at 2°K (Razeghi *et al.*, 1986b).

(5) The observation of room temperature negative resistance in a double barrier resonant-tunnelling structure (Razeghi *et al.*, 1987c).

(6) The observation of room temperature excitons in GaInAs–InP superlattices (Razeghi et al., (1986c).

(7) The growth of monoatomic layer (ALE) epitaxy of $(InAs)_n(GaAs)_n$–InP MQW (Razeghi et al., 1987d).

(8) The PIN photodetector with dark current density as low as 10^{-6} A cm^{-2}. The technology of PIN photodetector has been transferred to production (Razeghi, 1984a; Poulain et al., 1985).

(9) The growth and fabrication of junction FET with excellent characteristic (Raulain et al., 1987).

(10) The growth of very low loss optical waveguide (Delacourt et al., 1987).

$Ga_xIn_{1-x}As_y$-P_{1-y}–InP System

(11) The growth of the entire compositional range of $Ga_xIn_{1-x}As_y$-P_{1-y} lattice matched to InP with carrier concentration as low as 6×10^{14} cm^{-3} (for GaInAsP $= \lambda = 1.3$ μm), electron Hall mobility of 6,400 cm^2 V^{-1} s^{-1} at 300°K and 36,000 cm^2 V^{-1} s^{-1} at 77°K (Razeghi and Duchemin, 1984).

(12) The observation of 2DEG in $Ga_{0.25}In_{0.75}As_{0.5}P_{0.5}$–InP, HJ, MQW and SL (Razeghi et al., 1985).

(13) The observation of QHE (Razeghi et al., 1987e), in a GaInAsP/InP heterojunction.

(14) The GaInAsP–InP, double heterostructure laser emitting at 1.3 μm with threshold current density as low as 430 A/cm^2 (Razeghi et al., 1983a).

(15) GaInAsP–InP laser emitting at 1.55 μm with threshold current density of 500 A/cm^2 (Razeghi et al., 1983b).

(16) BRS laser emitting at 1.3 μm and 1.5 μm with threshold current as low as 6 mA (Razeghi et al., 1985).

(17) BRS–DFB laser emitting at 1.55 μm with threshold current of 10 mA (Razeghi et al., 1984a; Razeghi et al., 1987b).

(18) High-power phase-locked laser arrays emitting at 1.3 μm with output power of 300 mW without optical coating (Razeghi et al., 1986g).

(19) 1.5 μm GaInAsP–InP waveguide (Bourbin et al., 1987).

(20) SCH GaInAsP–GaInAs–InP MQW laser (Razeghi, 1985; Nagle et al., 1986).

Strained layer epitaxy

(21) Growth of GaInAs–InP on GGG substrate (Razeghi et al., 1986).

(22) Growth of GaInAsP–InP on Si substrate (Razeghi et al., 1987h).

(23) Growth of GaAs–GaInP on Si substrate (Razeghi 1984b; Razeghi et al., 1987i).

(24) Growth of InP on GaAs and InAs substrates (Razeghi et al., 1987j).

(25) GaInAsP–InP laser emitting at 1.3 μm on GaAs substrate (Razeghi et al., 1984b).

(26) Monolithic integration of a GaInAs/GaAs photoconductor with a GaAs F.E.T. for 1.3–1.55 μm wavelength applications (Razeghi et al., 1987k).

(27) Monolithic integration of a Schottky photodiode and a F.E.T. using a $Ga_{0.49}In_{0.51}P/Ga_{0.47}In_{0.53}As$ strained material (Razeghi et al., 1987l).

(28) Monolithic integration of planar monolithic integrated photoreceiver for 1.3–1.55 μm wavelength applications using GaInAs heteroepitaxies (Razeghi et al., 1986d).

$GaAs$-$Ga_{0.49}In_{0.51}P$ System

(29) The growth of high quality: $GaAs$-$Ga_{0.49}In_{0.51}P$, HJ, MQW and SL, with electron Hall mobility of 8,000 $cm^2 V^{-1} s^{-1}$ at 300°K and 800,000 $cm^2 V^{-1} s^{-1}$ at 2°K (Razeghi et al., 1989).

(30) The observation of 2DEG in $GaAs$-$Ga_{0.49}In_{0.51}P$, HJ, MQW, and SL (Razeghi et al., 1986e).

All of these results are firsts and most of them are also the best independent of the growth technique.

This chapter describes the recent progress made on the growth, characterization, and application of InP and related compounds grown by low pressure metalorganic chemical vapor deposition growth technique. Part II describes the energy-band structure of InP. Part III introduces epitaxial growth techniques. In Part IV, low pressure metalorganic chemical vapor deposition (MOCVD) growth is treated in detail. Part V describes the growth and characterization of InP using tetraethyl In (TEIn) and Phosphine (PH_3). In Part VI, the growth of InP using trimethyl In (TMIn) is described. The incorporation of dopants is the subject of Part VII. The application of InP epitaxial layers in the fabrication of Gunn diodes is detailed in Part VIII. The use of silicon and gallium-gadolinium-garnet (GGG) substrates for the MOCVD growth of InP and GaAs is described in Part IX, while Part X is devoted to optoelectronic applications utilizing layers grown by LP-MOCVD.

II. Energy Band Structure of InP

Indium phosphide (InP) has the zinc blende crystal structure. Its lattice may be considered as two interpenetrating face-centered-cubic sublattices, one made up of indium atoms, and the other of phosphorus atoms. When more than one element from groups III or V is distributed randomly on indium or phosphorus lattice sites III–III–V (such as: $Ga_xIn_{1-x}P$ or

$Ga_xIn_{1-x}As$) or III–V–V (such as: $InAs_yP_{1-x}$) ternary alloys, or III–III–V–V (such as: $Ga_xIn_{1-x}As_yP_{1-y}$), III–III–III–V (such as: $Al_xGa_yIn_{1-x-y}P$, $Al_zGa_xIn_{1-x-z}As$), quaternary alloys can be achieved.

The Bloch theorem states that, in a regular lattice, the allowed wave functions for an electron may be written as a free electron wave function modified by a function $U_{n,k}(\mathbf{r})$, which is periodic with the lattice:

$$\phi_{n,k}(\mathbf{r}) = e^{i\mathbf{k}\cdot\mathbf{r}} U_{n,k}(\mathbf{r}), \tag{1}$$

where

$$U_{n,k}(\mathbf{r}+\mathbf{R}) = U_{n,k}(\mathbf{R})$$

(n is the band index and $\mathbf{r}$ a lattice vector).

Information relating to the energy band structure is usually presented by plotting the energy of the electron, E for values of the wave vector, $\mathbf{k}$, limited to within the first Brillouin zone.

In a one-dimensional model with lattice periodicity a_o, a change of Brillouin zone corresponds to changing the value of k by $\pm 2\pi/a_o$, which will not change the lattice periodicity of the Bloch function.

$E(k)$ is periodic with a period $2\pi/a_o$. Thus, one may map E versus k curves into any interval of width $2\pi/a_o$ without loss of any information.

Referring to the E versus k curve for the weak binding model in a one-dimensional model, it is generally agreed that the first set of allowed energy levels ranging from $K = \pm \pi/a$ is called the first Brillouin zone. The energy levels lying between the intervals $K = -2\pi/a, -\pi/a$ and $K = +2\pi/a, +\pi/a$ make up the second Brillouin zone (see Fig. 1 for a square two-dimensional array).

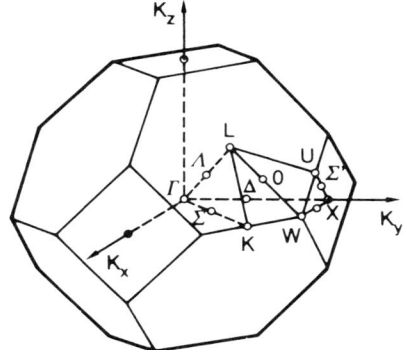
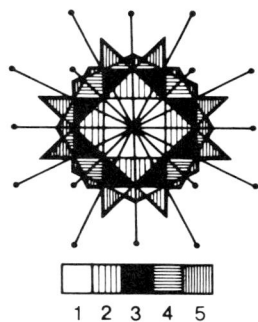

FIG. 1. First Brillouin zone for the InP-based materials (Zinc-blende lattice), including important symmetry points and lines.

The construction of the Brillouin zones in reciprocal space is attributed to Léon Brillouin, 1946. The first Brillouin zone in k space (or reciprocal lattice space) is found in the following manner: In two dimensions, the first Brillouin zone is the area enclosed by lines that are the bisectors of the smallest K vectors of the reciprocal lattice.

Figure 1 shows the Brillouin zone of the InP crystal (zinc-blende lattices) and indicates the most important symmetry points and symmetry lines, such as the center of the zone $[\Gamma = 2\pi/a\ (0, 0, 0)]$, the $\langle 111 \rangle$ axes (Λ) and their intersections with the zone edge $[L = 2\pi/a\ (\frac{1}{2}, \frac{1}{2}, \frac{1}{2})]$, the $\langle 100 \rangle$ axes (Δ) and their intersections $[X = 2\pi/a\ (0, 0, 1)]$ and the $\langle 110 \rangle$ axes (Σ) and their intersections $[K = 2\pi/a\ (3/4, 3/4, 0)]$.

The E - k diagram of InP is shown in Fig. 2. The energy E varies with k - k_o, where k_o is the value of k at the Γ extremum. The expression relating E with k-k_o has a parabolic form near the Γ point minimum. The relation is

$$E = \frac{\hbar^2 k^2}{2m_e^*}. \qquad (2)$$

The constant m_e^* is the effective mass of electron (Kane, 1966). The E-k relation for the minima lying on the Λ or Δ directions is of the form:

$$E = \frac{\hbar^2}{2}\left(\frac{k_\parallel^2}{m_\parallel^*} + \frac{k_\perp^2}{m_\perp^*}\right) \qquad (3)$$

where $m_\parallel^*$ is the effective mass for the $\langle 111 \rangle$ direction and $m_\perp$, that for a perpendicular direction. Effective mass: At the E-k relation, the coefficient of

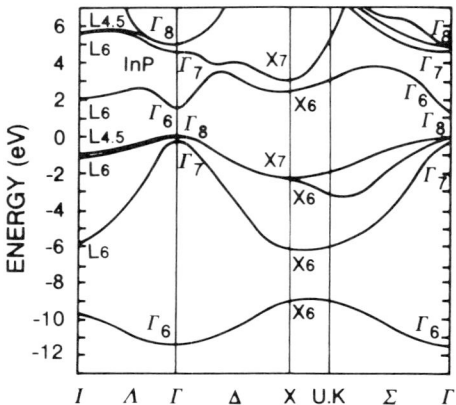

FIG. 2. The energy band structure of InP.

k^2 determines the curvature of E versus k. Turned about, we can say that $1/m^*$, the reciprocal mass, determines the curvature.

InP and related compounds have a covalent bond structure. When a covalent bond is produced between two unlike atoms, such as In and P, it is called a heteropolar bond. In InP semiconductor crystals, the bond orbitals are constructed from SP^3 hybrids (see Fig. 3) (a linear combination of P orbital and S orbital called an SP hybrid, 3 indicates there is three times as much probability to find an electron in a P state than to find it in an S state). The In has the electronic configuration of $4d^{10}$, $5S^2$, $5P^1$ and P has the electronic configuration of $3S^2$, $3P^3$. After hybridization there are two atoms, and hence eight electrons, per primitive unit cell in the InP crystal. There are one 5S level and three 5P levels in the In atom and one 3S level and three 3P levels in the P atom, since each gives rise to a bond in the crystal, this gives a total of four bonds arising from the atomic energy levels containing the valence electrons of InP. These four bonds are called the valence bonds of InP. We now consider the allocation of the eight valence electrons of InP among the four valence bonds, since two in primitive unit cells of the InP lattice. There are a total of eight electrons to put into four valence bonds. Consider, for example, the (100) and (111) directions near the center of the Brillouin zone, the point in K space (reciprocal lattice) at which a valence electron has the highest energy is the Γ point at the center of Brillouin zone. We, therefore, say that the valence band maximum occurs at the center of the Brillouin zone. At $K = 0$, all the valence band states are occupied by electrons, and all the higher bands of the InP band structure are empty. At temperature above $T = 0K$, a few electrons will be thermally excited from the highest valence band into the vacant next higher band.

The next empty band above the highest valence band is called the conduction band, because electrons thermally excited into it will find empty

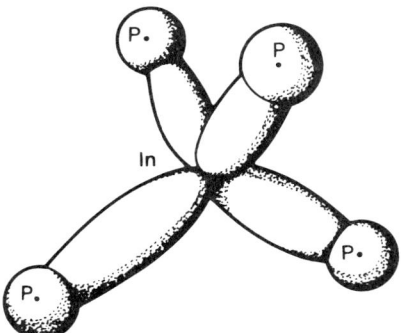

FIG. 3. SP^3 hybride bond orbitals.

TABLE I

EFFECTIVE MASSES OF ELECTRONS AND HOLES IN InP CRYSTAL.

Electron mass effective m^*_e/m_o	Heavy Hole m^*_{hh}/m_o	Light Hole m^*_{lh}/m_o	Split Off Hole m_{soh}/m	Spin Orbit $\Delta(eV)$
0.073	0.4	0.078	0.15	0.11

states available for the electrical conduction process. The conduction band minimum is at the Γ point at the zone center in III–V compounds (the conduction band minimum is often called the conduction band edge, and the valence band maximum is called the valence band edge). The minimum energy gap E_g is the energy difference between the conduction band minimum (E_c) and the valence band maximum (E_v), so the magnitude of E_g of the energy gap in InP is given by the relation:

$$E_g = E_c - E_v,$$

InP has a direct gap with band edges at the center of the Brillouin zone (see Fig. 2). The conduction band edge is spherical with effective mass m_e^*:

$$E_c = E_g + \frac{\hbar^2 k^2}{2m_e^*}.$$

The valence bands are characteristically threefold near the edge, with the heavy hole (hh) and light hole (lh) bands degenerate at the center, and a band spin-orbit hole split off by the spin-orbit splitting Δ:

$$E_v(hh) \equiv -\frac{\hbar^2 k^2}{2m_{hh}^*}$$

$$E_v(lh) \equiv -\frac{\hbar^2 k^2}{2m_{lh}^*}$$

$$E_v(soh) \equiv -\Delta - \frac{\hbar^2 k^2}{2m_{soh}^*}$$

The values of the mass parameters of InP are given in Table I.

III. Growth Technology

Figure 4 shows the x-y compositional plane for ternaries and quaternaries alloys lattice matched to InP substrate. Each corner has a binary III–V material. The mixture of binaries at each side gives the ternaries III–III–V or

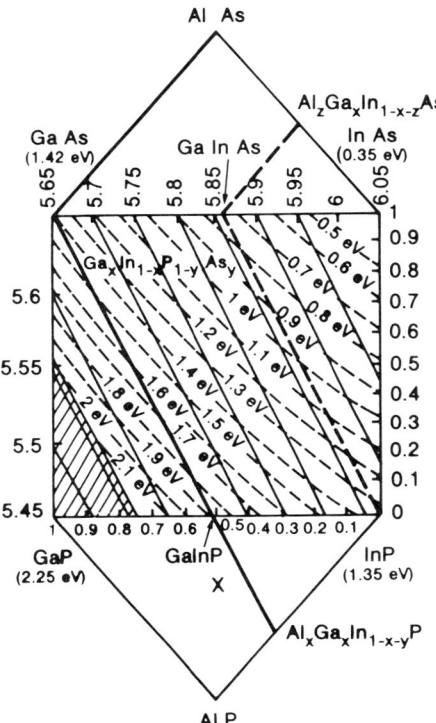

FIG. 4. The $x - y$ compositional plane for quaternaries III–V alloys at 300 K. The $x - y$ coordinate of any point in the plane gives the compositions. The solid lines are lattice parameters. The broken lines are direct energy gap values.

III–V–V alloys, and the combination of binaries or ternaries inside gives the quaternaries alloys. The dotted lines represent equal energy gap, and the solid line represent equal lattice parameter. Two systems are very interesting for optoelectronics and microwaves devices. The first system is the ternaries and quaternaries lattice matched to GaAs substrate, where the energy gap varies between 1.43 eV up to 2.4 eV, and all of the ternaries and quaternaries have an energy gap higher than the GaAs binary crystal, so the electrons and holes are confined in the binary alloys.

The second system is composed from the ternaries and quaternaries lattice matched to InP substrate, where the energy gap varies between 0.75 eV up to 1.35 eV. In this system, all of the ternaries and quaternaries have an energy gap lower than InP, so electrons and holes confine in ternary or quaternary alloys. InP and related compounds with energy gap less than 2 eV have been considered as narrow gap semiconductor materials.

During the past ten years, extensive theoretical and experimental studies of low-dimensional structures (LDS) of InP and related compounds in two distinct contexts have been performed: (1) the low-dimensional meaning less than three, and the properties of the dimensional electron or hole gas in different III–V semiconductor heterojunctions is now a mature, even aging subject. (2) Low has come to small, as in the physcis of semiconductors, where the feature size of the sample is small compared with intrinsic quantum length scale associated with carriers in semiconductors. This is given by de Broglie wavelength defined as $\lambda = h/p$, where h is Planck's constant and p the carrier momentum, typically given by $p^2/2m^* = KT$.

Low-dimensional structures have become essential elements in modern device technology.

There is now a need for a wide range of semiconducting materials, especially InP and related compounds, to be grown as thin, single crystal films. The thickness has to be controlled more and more accurately, and the uniformity of thickness can be vital. Epitaxial layers as thin as 10 Å or less are now needed. Excellent homogeneity and purity is required. It is also desirable to have no misfit dislocations present and to have a very sharp boundary between the substrate and epitaxial layers. A major goal of solid-state physics and solid-state technology is the perfection of materials and substrates in which charge carriers have long time, low scattering, high mobilities and controlled densities. Such materials have allowed the elucidation of the electronic structure of solids and the development of semiconductor electronics and photonic devices.

For growing heterojunctions, quantum wells (QW), and superlattices of InP-based materials, there are several different growth techniques.

1. LIQUID PHASE EPITAXY (LPE)

Growth of InP–GaInAsP has been started mainly by LPE, with the LPE reactor consisting of a horizontal furnace system and a sliding graphite boat. InGaAsP lattice matched on InP were grown for the first time by Antypas *et al.*, (1973). Since it was reported that InGaAsP–InP heterostructures were of potential use for laser (Bogatov *et al.*, 1975; Hirch, 1976), light-emitting diode (Pearsall *et al.*, 1976), photocathode (James *et al.*, 1973) Escher and Sankaran, 1976, Escher *et al.*, 1976), and photodiode (Wieder *et al.*, 1977) applications. In the LPE-growth technique, the composition of the layer formed on the substrate depends mainly on the equilibrium phase diagram and, to a lesser extent, on the orientation of the substrate. There are basically three parameters that influence the growth in LPE: the melt composition, the growth temperature, and the growth time. The first reported growth apparatus was developed by Nelson in 1963. The advantages of LPE for the growth

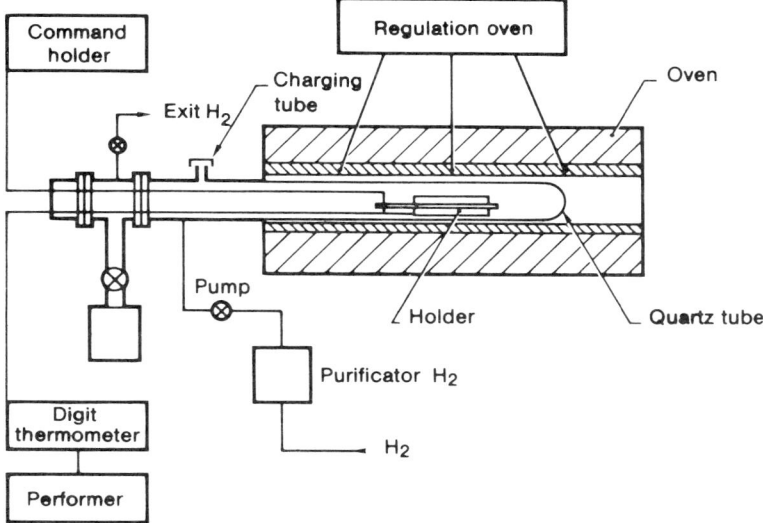

FIG. 5. Principle of LPE system.

of InP-based materials are: simplicity of equipment, and higher deposition rates and elimination of parasitic reactions due to use of reactive gases and their reactive products, which are often toxic and explosive.

The limitations of LPE technology include the nonuniformity of the layer thickness, roughness of surface morphology, and smaller wafer sizes. Owing to the high growth rate and melt-back effect, it is very difficult to grow multiquantum well (MQW) or superlattice structures with $Lz \leqslant 100$ Å, using LPE technology (Fig. 5).

2. VAPOR PHASE EPITAXY (VPE)

The VPE method was first demonstrated by Tietjenend and Anik, 1966, for the growth of GaAsP alloys. The growth of InP-based materials can be obtained in a fused-silica reactor composed of two zones set at different temperatures by using a multielement furnace that surrounds the reactor. Hydrogen is often used as a carrier gas, arsine (AsH_3) and phosphine (PH_3) for arsenic (As) and phosphorus (P) sources. Pure indium and gallium metal can be used as group III element sources. In the first zone of the reactor, called the source zone and held at a temperature T_s laying in the range 750–800°C, the gaseous species to be transported are synthesized, following the reaction:

$$In + HCl \rightarrow InCl + \tfrac{1}{2}H_2$$
$$PH_3 \rightarrow \tfrac{1}{4}P_4 + \tfrac{3}{2}H_2.$$

In the second zone, called the deposition zone, with a temperature T_D in the range 650–750°C, growth occurs

$$\tfrac{1}{4}P_4 + InCl + H_2 \rightarrow InP + HCl.$$

The VPE is a thermodynamic equilibrium growth technique (like LPE). The advantages of VPE over LPE are a high degree of flexibility in introducing dopant into the material as well as control of composition gradients by accurate flow metering. The disadvantages include the potential for hillock and haze formation, and interfacial decomposition during the preheat stage. It is very difficult to grow (MQW) or superlattice structures with $Lz \leqslant 50$ Å using VPE technology. One of the major advantages of VPE over LPE is the possibility doing localized epitaxy. If one grown InP-based material on a InP substrate consists of stripe of SiO_2 or Si_3N_4, there exists monocrystal growth on the substrate but no growth on the oxide (SiO_2 or Si_3N_4) surface, which is very important for the monolithic integration circuit on InP substrate. Figure 6 shows the schematic illustrations of a VPE reactor.

3. Molecular Beam Epitaxy (MBE)

The MBE process involves the reaction of one or more thermal beams of atoms and molecules of the III–V elements with a crystalline substrate surface held at a suitable temperature under ultra-high-vacuum (UHV) conditions. In 1958, using multiple beams, Gunther described the growth of III–V materials, Gunther's films were grown on glass substrates and hence

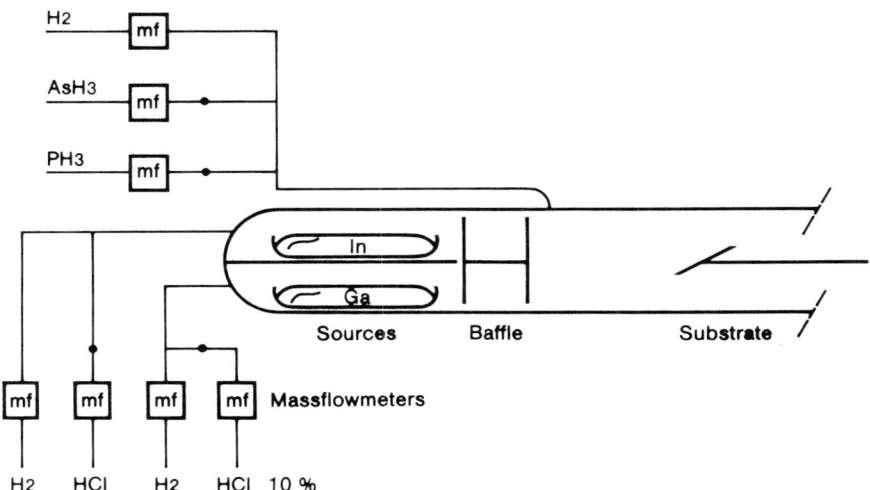

Fig. 6. Schematic illustrations of a VPE reactors.

were polycrystalline. In 1968, Davey and Pankey, Arthur grew monocrystalline GaAs films with MBE.

InP and GaInAs lattice matched to InP were prepared by MBE and their properties studied by (McFee et al., 1977; Miller and McFee, 1978; Kawamura et al., 1981, Chang et al., 1981; Davies et al., 1983; Olego et al., 1982; Lambert et al., 1983, and Tsang et al., 1982).

Since the MBE is essentially a UHV evaporation technique, the growth process can be controlled in situ by the use of equipment such as a pressure gauge, a mass spectrometer, an electron-diffraction facility located within the MBE reactor. The MBE growth chamber can contain other components for surface analytical techniques including the following: reflection high-energy electron diffraction (RHEED); (the information about the dynamics of film growth by MBE have been obtained by Neave and Joyce (1983) by studying the RHEED oscillations); Auger electron spectroscopy (AES) (another nondestructive analytical method for characterization of the initial substrate surface, verification of surface accumulation of dopant elements during epitaxial growth, and determination of the relative change in the relative ratio of the constituent elements of reconstructed surface structures), X-ray photoelectron spectroscopy XPS; low energy electron diffraction (LEED) emission spectroscopy chemical analysis ESCA; secondary-ion mass spectroscopy (SIMS) (a powerful surface and bulk material-composition analysis tool for pre- and post-deposition analysis during MBE growth); and elipsometry, which can be used as an in situ surface diagnostic technique during MBE growth, thanks to the UHV growth conditions. MBE has attracted a lot of interest as an excellent crystal growth technology, especially for GaAs-based multilayer structures, because of its extremely precise control over layer thickness and doping profile, and the high uniformity of the epitaxial layer over a large area of a substrate (>3 inches diameter).

The disadvantages of MBE are that it is expensive, and that difficulties have been reported with P-type doping by Naganuma and Takahashi, 1975, and with the growth of phosphorus-bearing alloys such as InP and GaInAsP. Figure 7 shows the schematic illustration of a MBE reactor.

4. METALORGANIC CHEMICAL VAPOR DEPOSITION (MOCVD)

Metalorganic chemical vapor deposition (MOCVD) has established itself as a unique and important epitaxial crystal growth technique yielding high-quality LDS for fundamental semiconductor physics research and useful semiconductor devices, both electronic and photonic. The growth of semiconductor III-V compounds results from introducing metered amounts of the group III alkyls and the group V hybrides into a quartz tube, which contains a substrate placed on an RF-heated carbon susceptor. The hot

FIG. 7. Photograph of a MBE reactor.

susceptor has a catalytic effect on the decomposition of the gaseous products, and the growth, therefore, primarily takes place on this hot surface. MOCVD is attractive because of its relative simplicity compared to other growth methods. It can produce heterostructures, MQW and superlattices (SL) with very abrupt switch-on and switch-off transitions in composition as well as in doping profiles in continuous growth by rapid changes of the gas composition in the reaction chamber. The technique is attractive in its ability to grow uniform layers, low-background of doping density, sharp interfaces, and the potential for commercial application. MOCVD can prepare multilayer structures with thickness as thin as a few atomic layers. This allows the study and device applications of two-dimensional electron gas (TDEG) (Razeghi et al., 1985; Delahaye et al., 1985; Guldner et al., 1986; Razeghi et al., 1986), two-dimensional hole gas (TDHG) (Razeghi et al., 1986; Rogers et al., 1986), and transport and quantum size effect (QSE) (Razeghi et al., 1983) in a variety of III–V compound semiconductors, heterojunctions, and multilayers (Razeghi et al., 1986; Nicholas et al., 1985). It also makes it possible to 'engineer the band gap' by growing a predetermined alloy composition and doping profile (Razeghi et al., 1986). As a result, an entirely new class of electronic and photonic devices are realized (Razeghi et al., 1987). Another recent advance is the ability to grow strained layers, i.e. superlattices, in which the crystal lattices of the two materials are not very closely matched. In that case there is a built-in strain in each layer (see for example: Osbourn, 1985). The disadvantage of the MOCVD growth technique is in using high quantities of poisonous gases such as AsH_3 and PH_3. In comparison to MBE, it is very difficult to do in situ characterization during the MOCVD growth technique (see Fig. 8).

5. New Non-equilibrium Growth Technique

One part: for the growth of InP-based material with MBE, on the other part: for doing in situ characterization with MOCVD. In the past few years, there have been several modifications to the conventional nonequilibrium MBE and MOCVD growth techniques, which have enhanced its versatility and increased its potential for the growth of InP and related compounds with extremely thin layers and abrupt interfaces.

In the case of MOCVD, the use of a low-pressure MOCVD for the growth of InP by Duchemin (1978) showed that for silicon growth, the use of low pressure reduced the absorption of hydrogen on the growing surface, thus permitting the growth of good-quality silicon at a lower temperature than is possible at atmospheric pressure. Fraas (1981) reported that epitaxial GaAs, GaAsP, and GaInAs films with good surface morphologies and electrical

FIG. 8. Photograph of a MOCVD reactor.

properties could be grown by reacting combinations of triethylgallium (TEGa), triethylindium (TEIn), arsine, and phosphine in a high vacuum system.

In the case of MBE, Panish (1980) replaced the conventional condensed sources for the group V element, usually As and P, with sources that decompose AsH_3 and PH_3, and called it gas source molecular beam epitaxy (GSMBE). Vodjdani et al., (1982) and Tsang (1984) replaced the group III elements, such as In and Ga used as the condensed source in conventional MBE, with simple organometallic compounds. These extensions of GSMBE were called metalorganic molecular beam epitaxy (MOMBE), chemical beam epitaxy (CBE), and metalorganic chemical beam deposition (MOCBD).

The MOCBD reactor employed the gas handling system of MOCVD and growth chamber of MBE. The main advantages of this combination in comparison with the conventional MBE include the following:

(1) The capability of growing high-quality InP-based material (Panish, 1986; Tsang, 1986).
(2) The elimination of oval defects even at high growth rates.
(3) High-quality InP-based material at high growth rates.
(4) The flexibility of sources changes.
(5) The versatility of using different alkyl and hybride sources.
(6) Good homogeneity and composition uniformity over a large area of substrate, thanks to a single group-III (and group V if desired) beam.

Compared to MOCVD growth the MOCBD technique

(1) uses at least 100 times less PH_3 for the growth of InP-based material,
(2) eliminates the parasitic reactions in the gas phase, thanks to UHV condition,
(3) makes use of different in situ surface diagnostic techniques,
(4) improves homogeneity and composition uniformity and reproducibility of InP-based materials over large areas of substrate,
(5) provides in situ etching and the removal of oxides during the growth of InP on Si substrates,
(6) reduces memory effects during the P-type doping,
(7) reduces the boundary layer thickness on the hot substrate surface,
(8) compatible with other high-vacuum thin-film processings such as plasma etching, metal evaporation, ion beam etching and ion implantation and
(9) makes it possible to do localized epitaxy.

No deposition was observed on a SiO_2 film (Cho, 1975, Takahashi, 1984), which is in contrast with conventional MBE or MOCVD, where polycrystalline GaAs deposits are on the SiO_2 masked region. It was found that monocrystalline GaAs was grown on the bare substrate with the narrow-stripe patterns and that the growth did not take place on the large parts of the SiO_2-covered area. This phenomena has already been observed with VPE. The growth mechanism, which causes this remarkable feature, is unclear at present, but it is possible that this feature is due to the surface catalyzed growth process (see Fig. 9).

6. Atomic Layer Epitaxy (ALE)

For future integrated circuits (ICs) with a monolayer controlled multilayer structure, one needs a more advanced epitaxial technique, which has atomic scale, high uniformity of thickness and composition, perfect selective epitaxy, free from any defects such as splitting, dislocations, or hillocks. Atomic layer epitaxy is a new growth technique with control at the monolayer level. ALE was originally proposed by Suntola et al., 1980. The concept of ALE was proposed by Suntola, 1984, and a general review was made by Goodman and Pessa, 1986, and Watanale and Usui, 1986. In the conventional epitaxial methods, such as LPE, VPE, MBE, and MOCVD, the growth rate increases linearly with time, where the gradient is a function of analog quantities such as temperatures of source and substrate, pressures and flowrates of source gases, and growth time. On the other hand, the thickness in ALE increases stepwise. It is possible to divide ALE into two categories (Watanale, 1986), analog ALE and digital ALE. In analog ALE, the surface coverage θ depends on the analog quantities such as, substrate temperature T, adsorbant vapor phase pressure P, and flowrate F, $\theta \neq f(T, P, F)$. In digital ALE, $\theta \neq f(T, P, F)$, and the dependence of the surface coverage (θ) of adsorbed vapor species on its partial pressure P has been interpreted by several different formula such as Henry's equation $\theta \propto P$, Freundlich's equation $\theta \propto p^{1/n}$ (n is an integer), BET's equation ($\theta > 1$ at $P > P_c$, where P_c is a critical pressure above which double-layer adsorption occurs), or Langmuir's equation $\theta = KP/(1 + KP)$, where K is the adsorption constant [$K = \exp(-\Delta H P T \Delta S/RT)$ with ΔH and ΔS are enthalpy and entropy changes due to the adsorption, and R is the gas constant].

The Langmuir type results in a digital process in which a grown thickness is determined only by the growth cycle (see Watanale, 1986). When $KP \gg 1$ in the Langmuir equation, $\theta = 1.0$ is independent of T, P, and F. Digital ALE needs no control of the analog quantities, it only needs the cycle of the procedure.

a. MO–ALE

In the MO–ALE technique, alkyles and hybrides are used as the starting material. Nishizawa et al., 1986 and Doi et al., 1986 used MO–ALE for the growth of GaAs and they showed the GaAs layers were P-type, with hole

FIG. 9. Photograph of a MOCBD reactor.

concentrations of $10^{17} \sim 10^{19}$ cm^{-3} and mobilities of $60 \sim 200$ cm^2/Vs. Kohayashi et al., (1985) introduced a small amount of AsH$_3$ during the TEG-exposure step. They called this technique flowrate-modulation epitaxy (FME). They succeeded in the growth of n-type GaAs having carrier concentrations of $10^{14} \sim 10^{16}$ cm^{-3} and mobilities as high as 42,000 cm^2 V^{-1} s^{-1} at 77°K. The introduction of additional AsH$_3$ during TEG suppresses carbon contamination and explains the excellent properties (Makimoto et al., 1986).

b. Chloride-ALE

In this technique, chloride and hybride are used as starting materials. GaAs layers grown by GaCl–ALE are n-type with a carrier concentration of 6×10^{15} cm^{-3} and a mobility of 16,000 cm^2 V^{-1} s^{-1} at 77°K. Matsumoto and Usui (1986) showed that the layers are compensated by a carbon acceptor using photoluminescence at 5°K. The electron trap EL$_2$ was not detected in the GaCl–ALE sample. This is in contrast with conventional VPE or MOCVD samples, which exhibit EL$_2$ in the concentration range of $10^{14} \sim 10^{15}$ cm^{-3}.

The major advantages of ALE (especially a digital ALE, where grown thickness is insensitive to any analogue quantities such as source gas pressure, growth temperature, and growth time) are obtaining large-diameter integrated-circuit wafers containing monolayer level-controlled, three-dimensional confined devices.

7. Migration-Enhanced Epitaxy (MEE)

Rapid migration of the evaporated materials on the growing surface is essential to the growth of high-quality epitaxial layers. In conventional MBE growth of GaAs and AlAs layers, migrating materials on the growing surface include Ga–As and Al–As molecules rather than Ga and Al atoms, respectively, because these layers are grown under arsenic-stabilized conditions. The migration of these molecules on the surface is very slow, especially at low temperatures. Therefore, lowering the substrate temperature considerably deteriorates the crystal quality of grown layers.

Horikoshi et al., 1986 proposed a new mode of MBE growth, which makes it possible to grow high-quality GaAs and AlAs layers at very low substrate temperatures ($\sim 200°$C) by enhancing the migration of the materials evaporated on the growing surface. They called this method migration-enhanced epitaxy (MEE), which is based on the very rapid migration of Ga and Al atoms on the growing surface and on the alternate supply of Ga(Al) atoms and arsenic molecules to the growing surface. RHEED intensity observation revealed that Ga and Al atoms migrate on the GaAs surface much more rapidly than Ga-As and Al-As molecules, and that they migrate very actively even at temperatures as low as 200°C. They showed that high-quality GaAs and AlAs could be grown at very low temperatures by alternately supplying Ga(Al) atoms and arsenic molecules to the substrate surface. By applying the MEE method, 1-2 μm thick GaAs layers were grown at a substrate temperature of 200°C. These layers showed efficient PL due to the band-edge excitons at 4.2 K. AlAs-GaAs single quantum well structures with 3-6 nm widths were also grown at 200°C by MEE. These structures showed PL due to electronic transitions between quantized levels in the wells indicating the reasonable quality of AlAs (Horikoshi et al., 1985).

IV. Low Pressure Metalorganic Chemical Vapor Deposition (LP-MOCVD)

8. Experimental Details

The InP-based semiconductor alloys defined as having an energy gap $E_g < 2.0$ eV, which are discussed in this chapter, are listed in Table II along with the properties relevant to their crystal growth and some of their physical parameters.

They all melt at high temperatures and have a strong tendency to decompose well below their melting points. Consequently, controlled melt growth of these compounds is difficult, and most efforts have been directed toward vapor growth as a means for producing high-quality heterojunctions, quantum wells, and superlattices.

Elemental vapor-phase epitaxy usually requires growth temperatures within the range of 700-900°C. It has been increasingly recognized that growth at high temperature thermodynamically favors the formation of a multitude of impurity-defect complexes that act as electron traps and pin the

TABLE II

Some of the Physical Parameters of InP-based Materials.

Compound	Energy gap (eV)	Lattice parameter (Å)	Refractive index (n)	Electron effective mass m^*_e	Heavy hole effective mass m^*_h
InP	1.35	5.869	3.45	0.08	0.56
$Ga_{0.25}In_{0.75}As_{0.5}P_{0.5}$	0.95	5.869	3.52	0.053	0.5
$Ga_{0.40}In_{0.6}As_{0.85}P_{0.15}$	0.80	5.869	3.55	0.045	0.5
$Ga_{0.47}In_{0.53}As$	0.75	5.869	3.56	0.041	0.5

Fermi level far from both the conduction and valence band edges of these compounds. Hence, they often exhibit compensating properties. The lower growth temperature (500–630°C) for LP-MOCVD of InP-based materials, a feature shared with molecular beam epitaxy (MBE), has stimulated substantial interest in this preparative route as a potential way of either eliminating or controlling detrimental impurities and defects. This control is particularly important when the material is the active medium in optoelectronic devices.

A review by Ludowise (1985) presented practical information about the apparatus and technique as well as more fundamental details concerning the interrelation of the several parameters that control the deposition and material quality.

LP-MOCVD was first applied to the growth of InP using the metal alkyls triethyl indium ((C_2H_5)$_3$In, TEI) and phosphine PH_3 as the source of phosphorus (P) by Duchemin et al., (1979). A review by Razeghi, 1984, described advances in the field prior to 1984. The developments since 1984 in the growth of InP-based materials for optoelectronic and microwave devices form the basis of this chapter.

9. Starting Materials

The compounds used as sources of the group III and group V elements for this study are listed in Table III. Their vapor pressures as functions of temperature are given in Fig. 10. Their chemical properties are given in Tables IV, V, VI, VII and VIII.

Triethyl indium (TEI) and triethyl gallium (TEG) have been used as group III sources. Hydrides pure Arsine (AsH_3 and pure phosphine (PH_3) have been used as group V sources. Diethyl zinc (DEZ_n) is used for P-type doping, and sulphides H_2S or silane H_4Si have been used for n-type doping [16]. Pure hydrogen (H_2) and pure nitrogen (N_2) have been used as carrier gases. The presence of N_2 is necessary in order to avoid the parasitic reaction between

TABLE III

Starting Materials.

Group III sources	Group V sources	P-type dopant sources	n-type dopant sources
$Ga(C_2H_5)_3$	AsH_3	$(C_2H_5)_2Zn$	SH_2
$Ga(CH_3)_3$			SiH_4
$In(C_2H_5)_3$	PH_3		
$In(CH_3)_3$			

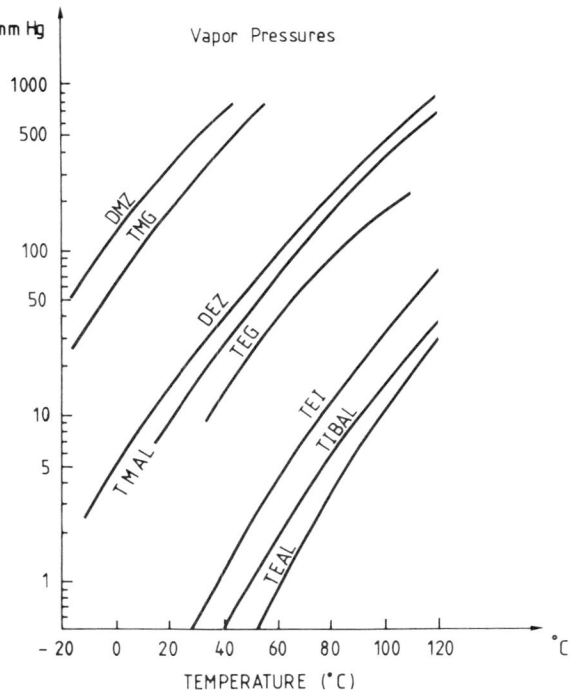

FIG. 10. The vapor pressure of alkyl sources of group III as a function of temperature.

TABLE IV

THE DETAILS OF CHEMICAL PROPERTIES OF TRIETHYL INDIUM.

Acronym	TEI
Formula	$(C_2H_5)_3In$
Formula weight	202.01
Metallic purity	99.9999 wt% (min) indium
Appearance	Clear, colorless liquid
Density	1.260 g/mL@20 °C
Melting point	-32 °C
Vapor pressure	1.18 mm Hg@40 °C
	4.05 mm Hg@60 °C
	12.00 mm Hg@80 °C
Behavior toward organic solvents	Completely miscible, without reaction, with aromatic and saturated aliphatic and alicyclic hydrocarbons. Forms complexes with ethers, thioethers, tertiary amines, -phosphines, -arsines and other Lewis bases.
Stability to air	Ignites on exposure (pyrophoric).
Stability to water	Partially hydrolyzed; loses one ethyl group with cold water.
Storage stability	Stable indefinitely at ambient temperatures when stored in an inert atmosphere.

TABLE V

THE DETAILS OF CHEMICAL PROPERTIES OF TRIETHYL GALLIUM.

Acronym	TEG
Formula	$(C_2H_5)_3Ga$
Formula weight	156.91
Metallic purity	99.9999 wt% (min) gallium
Appearance	Clear, colorless liquid
Density	1.0586 g/mL at 20 °C
Melting point	−82.3 °C
Vapor pressure	16 mm Hg @ 43 °C
	62 mm Hg @ 72 °C
	760 mm Hg @ 143 °C
Behavior toward organic solvents	Completely miscible, without reaction, with aromatic and saturated aliphatic and alicyclic hydrocarbons. Forms complexes with ethers, thioethers, tertiary amines, tertiary phosphines, tertiary arsines and other Lewis bases.
Stability to air	Ignites on exposure (pyrophoric).
Stability to water	Reacts vigorously, forming ethane and Et_2GaOH or $[(Et_2Ga)_2O]_x$.
Storage stability	Stable indefinitely at room temperatures in an inert atmosphere.

TABLE VI

THE DETAILS OF CHEMICAL PROPERTIES OF TRIETHYL INDIUM.

Acronym	TMI
Formula	$(CH_3)_3In$
Formula weight	159.85
Metallic purity	99.999 wt% (min) indium
Appearance	White, crystalline solid
Density	1.568 g/mL at 19 °C
Melting point	89 °C (192.2 °F)
Boiling point	135.8 °C (276.4 °F)/760 nm Hg 67 °C (152.6 °F)/12 mm Hg
Vapor pressure	15 mm Hg @ 41.7 °C (107 °F)
Stability to air	Pyrophoric, ignites spontaneously in air.
Solubility	Completely miscible with most common organic solvents.
Storage stability	Stable indefinitely when stored in an inert atmosphere.

TABLE VII
THE DETAILS OF CHEMICAL PROPERTIES OF DIETHYL ZINC.

Acronym	DEZ
Formula	$(C_2H_5)_2Zn$
Formula weight	123.49
Metallic purity	99.9999 wt% (min) zinc
Appearance	Clear, colorless liquid
Density	1.198 g/mL at 30 °C
Melting point	-30 °C
Vapor pressure	3.6 mm Hg @ 0.0 °C
	16 mm Hg @ 25.0 °C
	760 mm Hg @ 117.6 °C
Behavior toward organic solvents	Completely miscible, without reaction, with aromatic and saturated aliphatic and alicyclic hydrocarbons. Forms relatively unstable complexes with simple ethers, thioethers, phosphines and arsines, but more stable complexes with tertiary amines and cyclic ethers.
Stability to air	Ignites on exposure (pyrophoric).
Stability to water	Reacts violently, evolving gaseous hydrocarbons, carbon dioxide and water.
Storage stability	Stable indefinitely at ambient temperatures when stored in an inert atmosphere.

TABLE VIII
THE DETAILS OF CHEMICAL PROPERTIES OF TRIMETHYL GALLIUM.

Acronym	TMG
Formula	$(CH_3)_3Ga$
Formula weight	114.82
Metallic purity	99.9999 wt% (min) gallium
Appearance	Clear, colorless liquid
Density	1.151 g/mL at 15 °C
Melting point	-15.8 °C
Vapor pressure	64.5 mm Hg @ 0.0 °C
	226.5 mm Hg @ 25.0 °C
	760 mm Hg @ 55.8 °C
Behavior toward organic solvents	Completely miscible, without reaction, with aromatic and saturated aliphatic and alicyclic hydrocarbons. Forms complexes with ethers, thioethers, tertiary amines, tertiary phosphines, tertiary arsines, and other Lewis bases.
Stability to air	Ignites on exposure (pyrophoric).
Stability to water	Reacts vigorously, forming methane and Me_2GaOH or $[(Me_2Ga)_2O]_x$.
Storage stability	Stable indefinitely at ambient temperatures when stored in an inert atmosphere.

TEI and AsH_3 or PH_3. The presence of H_2 is necessary in order to avoid the deposition of carbon. TEI and TEG are contained in a stainless steel bubbler and are held in controlled-temperature baths at 31°C and 0°C, respectively. An accurately metered flow of nitrogen (N_2) for TEI and purified H_2 for TEG is passed through the appropriate bubbler. To ensure that the source material remains in vapor form, the saturated vapor that emerges from the bottle is immediately diluted by a flow of H_2.

The mole fraction, and thus the partial pressure of the source species, is lower in the mixture and is prevented from condensing in the stainless steel pipe. The flow rates of the hybrides, H_2 and N_2, were controlled by mass-flow-controllers within 0.2%. The metal alkyl or hydride flow can be either injected into the reactor or into the waste line by using the three-way valves. In each case, the source flow is first switched into the waste line to establish the flow-rate and then switched into the reactor.

10. REACTOR DESIGN

The most important part of the growth apparatus is the deposition chamber. There are two basic reactor types, a vertical design originated by Manasevit and Simpson (1971) in which the gas flow is perpendicular to the substrate surface, and the horizontal version developed by Bass (1975) in which the gas flow is parallel to the substrate surface. Both types have been used successfully at low and atmospheric pressures using several different substrate heating methods involving radio-frequency induction, resistance, radiant, and laser heating.

The advantages of a low-pressure MOCVD process are

(i) the elimination of parasitic nucleations in the gas phase,
(ii) the reduction of out diffusion (i.e., the solid-state diffusion of impurities from the substrate through active layers or from one active layer to another),
(iii) the reduction of autodoping (i.e., the doping of an epitaxial layer by volatile impurities that originate from the substrate),
(iv) the improvement of the interface sharpness and impurity profiles,
(v) the lower growth temperature,
(vi) the thickness uniformity and compositional homogeneity and
(vii) the elimination of memory effect.

The presence of vortices (behind the susceptor) and the dead volumes (sharp corners in reactor inlet) will act as sources of unwanted materials, which cannot be removed easily. So during the growth of a sharp heterojunction or a steep doping profile, by switching a flow with another chemical composition, the original composition is still present in the vortices and trapped gases. The slow-out diffusion from these parts will smear out the doping or heterojunction profiles. This is called the memory effect. We can eliminate such trapped gases by rapid evacuation, i.e. by working at low pressure.

In relation to the reactor, the main advantage of the horizontal reactor is considered to be the uniform gas flow achieved over a slightly angled susceptor (7–15°C), which leads to uniform deposition as the reactants are depleted from the gas flow.

The most important feature in the growth of InP-based materials is the arrangement for mixing the gases in order to inhibit the prereaction between the constituents of the gas flow. The reactor design has concentrated on introducing the gases into the reactor separately. A good mixing between the metal alkyls and hydrides through a delivery tube near the heated substrate is vital. So, the use of N_2, which eliminates the prereaction, enables good mixing to take place and allows the efficient gas mixer designed for quaterneries growth to be utilized.

11. Growth Procedure

The LP-MOCVD growth of InP and related compounds can be explained by the rapid transport of reactive species from the pipeline to the deposition zone by forced flow (in this case, the control of flow dynamics, flow mixing, and the adjustment of the flow to geometrical effects and temperature changes is essential). Then, the transport of reactive species in the deposition zone to the hot substrates by diffusion (the knowledge about development of concentration profiles, the effect of thermal gradients on diffusion, and the effect of annihilation or creation of molecular species on diffusion is crucial).

Since the growth rate in MOCVD is limited by mass transport of the group-III growth component, flow dynamics coupled with diffusion govern the deposition rates and therewith, the gas phase depletion. A profound knowledge of flow patterns and concentration profiles in the reactor are essential for optimisation of the reactor construction. In this respect, two important factors include homogeneous growth on large surface areas and minimization of gas memory effects, which is essential in the growth of InP-based materials heterojunctions and superlattices with sharp interfaces. For that, the flow region must be laminar with no turbulence in order to achieve control over growth process, and it must develop its pattern in a controlled way when it enters the reactor. Also, when the gas is heated to the process temperature, no instabilities due to natural convection (buoyancy) in the boundary layer may occur. The term "boundary layer" is used here to define the regions of rapidly increasing compositional, thermal, or momentum gradients perpendicular to the substrate.

12. FLOW PATTERNS

One of the most important factors in the MOCVD reactor is the flow pattern. We can specify the macroscopic gas movements in the MOCVD reactor as follows: (1) laminar flow contrary to the turbulence, and (2) diffusion contrary to the free convection (buoyancy). In order to specify the circumstances in which the different types of flow occur, we need to introduce the concept of the Reynolds number for the first case, and the Rayleigh number for the second case.

The Reynolds number (R_e) (dimensionless) of reactor flow is (Tritton, 1982):

$$R_e \simeq \frac{\rho \bar{V} d}{\eta}$$

where d is the diameter of the tube (m), $\bar{V}$ is the average flowrate (m/sec), ρ the density (kg/m^3), and η the dynamic viscosity (kg/m.sec) of the gas. Laminar flows typically have low Reynolds numbers. At higher Reynolds numbers a transition takes place from laminar to turbulent.

When R_e is small (less than 100), the flow regime is laminar. In laminar flow, the velocity at a fixed position is always the same. Each element of reactive species travels smoothly along a simple well-defined path (Prandtl, 1934). Each element that starts at the same place follows the same path.

When R_e is high, the flow becomes turbulent, and none of these features is retained. The flow develops a highly random character with rapid irregular fluctuations of velocity in both space and time. In this case, an element of gas flow follows a highly irregular distorted path. Different elements starting at the same place follow different paths, since the pattern of irregularities is changing all the time (Shapiro, 1961).

The Rayleigh number (R_a) dimensionless of reactor flow is:

$$R_a \simeq \frac{g\alpha C p\, p^2 h^3\, \Delta T}{\eta K}$$

where α is the coefficient of thermal expansion in (1/T), g is gravity constant (9.81 m.sec^{-2}), Cp is specific heat (J.kg^{-1}.K^{-1}), ρ is density (Kg.m^{-3}), h is free height above susceptor (m), $\Delta T = T$(susceptor) $- T$(reactor wall), η is dynamic viscosity (Kg.m^{-1}.sec^{-1}), and K is thermal conductivity (J.m^{-1}.sec^{-1}.K^{-1}). When $R_a < 1700$, the gas is stable, for when $R_a > 1700$, free convection occurs.

With a high Rayleigh number, the free convection occurs, which affects the mass transfer, growth rate, and homogeneity. Convection usually occurs between hot substrate and reactor cold wall. The cause of convection is the action of the gravitational field on the density variations associated with temperature variations. The heavy cold gas is situated above the light hot gas. If the former moves downward and the latter move upward, there is a release of potential energy, which can provide kinetic energy for the motion. Thus, there is a possibility that the equilibrium will be unstable.

From the kinetic gas theory, it follows that both the dynamic viscosity, η, and the thermal conductivity, K, are independent of reactor pressure. But the density, ρ, is pressure dependent, which means that $R_e \propto \rho$ and $R_a \propto \rho^2$. So, the consequence is that lower pressures stabilize the convection behavior of the gas flows, and laminar flows are easily obtained.

In conclusion, for the growth of high-quality InP-based materials with sharp interfaces, the following remarks on reactor design are crucial:

(1) Laminar flows free of convection should exist (a) using horizontal reactor, (b) working at low pressure, and (c) decreasing the reactor diameter.

(2) No temperature gradient should be present across the susceptor.

(3) Eliminate the memory effect because (a) the geometry of the reactor is such that no vortices can develop, (b) no dead volumes are present inside the reactor, and (c) the elimination of sharp corners in reactor inlet, where the laminar flow can go by without having a strong interaction, are also behind the susceptor.

V. InP-InP Systems

Over the past few years, the low pressure metalorganic chemical vapor deposition (LP-MOCVD) growth technique has been used for the growth of InP and related compounds for many new device applications. The 1.3–1.5 μm lasers and the development of microwave devices provided strong stimulation to those studying the properties of InP and related compounds.

13. GROWTH AND CHARACTERIZATION OF InP USING TEIn

High-quality InP layers have been grown by LP-MOCVD. TEIn and PH_3 are used for In and P sources, respectively. The mixture of H_2 and N_2 are used as carrier gas, because H_2 is necessary in order to avoid the deposition of carbon and N_2 is necessary in order to avoid the parasitic reaction between TEIn and PH_3.

The InP layers can be grown at 76 Torr and low temperature, between 500°C and 650°C, by using TEI and phosphine (PH_3) in a $H_2 + N_2$ carrier gas. The growth rate depends linearly upon the TEI flow rate and is independent of the flow rate of PH_3 within 200 to 800 cm^3 min^{-1} (Fig. 11) of the substrate temperature and the substrate orientation, which suggests that

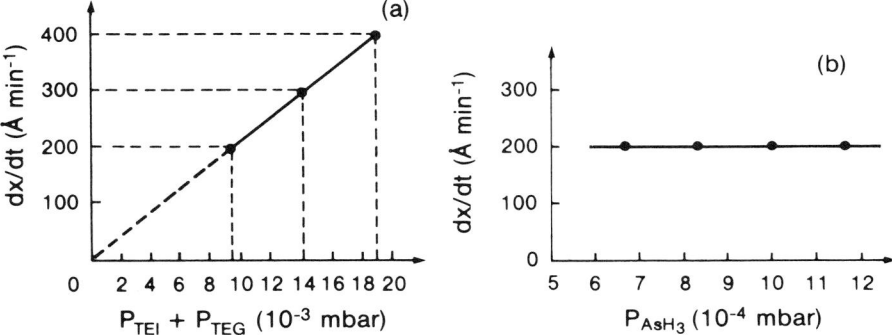

FIG. 11. Variation of the growth rate of InP with (a) TEI flow ($T = 550$ °C, $PH_3 = 300$ cm³ min⁻¹, total flow 6 l min⁻¹); (b) PH_3 flow ($T = 550$ °C, N_2 flow rate through TEI is 450 cm³ min⁻¹).

the epitaxial growth is controlled by the mass transport of the group-III species.

We have studied the growth of InP layers by using 100% H_2 and $H_2 + N_2$ mixtures as the carrier gas. The best morphology and the highest PL intensity were obtained by using 50% H_2 and 50% N_2.

Using Ar instead of N_2 gives InP layers with the same surface quality. Table IX lists the optimum growth conditions for LP-MOCVD growth of INP at 550°C and 650°C, which were used for this study. The InP layers grown by LP-MOCVD are less compensated at lower growth temperatures. Figure 12 shows that the carrier concentration of InP is $N_D - N_A = 6 \times 10^{14}$ cm⁻³ at 520°C and 6×10^{15} cm⁻³ at 650°C if all other growth conditions remain the same.

TABLE IX

OPTIMUM GROWTH CONDITIONS

Growth temp.	N_2-TEI bubbler flow	PH_3 (cm³ min⁻¹)	Total flow (1 min⁻¹)	Growth rate (Å min⁻¹)
550	450	260	6	200 ± 10
	225	200	6	100 ± 10
650	450	520	6	220 ± 10
	225	400	6	110 ± 10

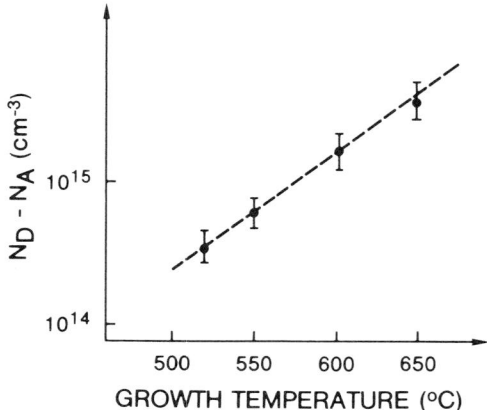

FIG. 12. Variation of residual carrier concentration of InP layers grown by LP-MOCVD as a function of growth temperature.

14. ORIENTATION EFFECTS

The growth of InP layers has been carried out on (100) substrates misoriented up to 4° toward (110). At the low-growth temperature (550°C), the surfaces of the grown layers are generally mirror smooth to the naked eye whether the substrates were accurately oriented or not, but at high-growth temperature, the best quality InP layers were obtained by using (100) substrates 2° off toward (110). The density of surface defects is higher on (100)-oriented substrates compared to 2° off material. Saxena *et al.*, (1981) reported similar results for GaAs.

Most epitaxial growth of III-V semiconductors has been performed on either (100) or (111) crystal faces. Shaw (1968) studied the chloride VPE growth of GaAs on (100), (111), (112), (113), and (115) GaAs substrates and concluded that the best morphologies in the 700-800°C temperature range are obtained on (113)A substrates. Olsen *et al.*, (1982) reported that the chloride VPE growth of InP at 700°C on (100), (111), (311), and (511) were both exact and 2° off toward (110)-orientation InP substrates. They found the best morphology and highest PL intensity for the (311)B 2° off substrate orientation.

We have performed a study of LP-MOCVD growth simultaneously on nine InP substrates with orientations of (100), (111), and (115), which were placed adjacent to each other within the reactor for a growth temperature of 650°C. Table X indicates relative PL intensity I, PL half-width $\Delta h\nu$, growth rate dx/dt, net carrier concentration evaluated by C-V measurements, and relative surface appearance of LP-MOCVD InP grown on InP substrates of

TABLE X

Relative PL Intensity I, Relative Surface Appearance, and Carrier Concentrations of InP Grown on InP Substrates by LP-MOCVD[a].

Orientation	$N_D - N_A$ (cm^{-3})	Surface quality[b]	I (a.u.)	$\Delta h\nu$ at 300 °K (meV)	dx/dt Å/min
InP(Sn)(100)2 °1$\bar{1}$0	3×10^{17}	E	16	60	220 ± 10
InP(Fe)(100)Exact	3×10^{17}	F	9	60	220 ± 10
InP(Fe)(100)2 °1$\bar{1}$0	3×10^{17}	G	16	60	220 ± 10
InP(Fe)(100)3 °1$\bar{1}$0	3×10^{17}	G	16	60	220 ± 10
InP(Sn)(100)4 °1$\bar{1}$0	3×10^{17}	G	16	60	220 ± 10
InP(S)(111)B2 °	2×10^{17}	G	45	60	220 ± 10
InP(S)(115)B2 °	2×10^{17}	E	180	95	220 ± 10
InP(Fe)(111)B Exact	3×10^{17}	F	10	60	220 ± 10
InP(Sn)(100)Exact	3×10^{17}	G	8	60	220 ± 10

[a]Growth rate, 220 ± 10 (Å min^{-1}); PL half-width $\Delta h\nu$, 60 meV at 300 °K.
[b]E, excellent; F, fair; G, good.

various orientations at 650°C. The relative PL intensity is a measure of the radiative recombination efficiency within the material, whereas the PL half-width is a measure of crystalline quality and impurity incorporation (a narrower half-width indicates purer material).

Table XI presents S, Si, Cr, Fe, Mg, and Mn distribution measured by SIMS in these layers. All of the layers were doped by H_2S, and C-V measurements give N_D-$N_A \simeq 2 \times 10^{17}$ cm^{-3} for all of them.

The secondary-ion mass spectroscopy (SIMS) analyses were performed by using a Cameca IMS 3F instrument (Huber et al., 1982). The surface of the sample was scanned with a focused oxygen primary-ion beam. The scanned area was 250×250 μm for the working conditions, and the analyzed region was 60 μm in diameter.

The precision of measurements is 50% for each element. Analysis of a large number of epilayers will be necessary for any given set of growth conditions in order to specify accurately the contribution of each impurity source and its relation with the substrate in MOCVD-layer growth.

15. Source-Purity Effects

The importance of source purity is illustrated in Table XII, where we have listed the range of 300-K mobilities and carrier concentrations with a variety of source materials under optimum growth conditions, as defined in Table IX.

TABLE XI

S, Si, Cr, Fe, Mg, and Mn Distribution Measured by SIMS in the InP Layers Grown by LP-MOCVD at 650 °C[a].

Reference layer	S	Si	Cr	Fe	Mg	Mn
InP(S)(111)	3×10^{17}	5×10^{16}	4×10^{14}	2×10^{14}	2×10^{14} lim	4×10^{14} lim
(115)	6×10^{17}	5×10^{16}	4×10^{14}	2×10^{15}	1×10^{14} lim	5×10^{14} lim
(100)2 °	3×10^{17}	5×10^{16}	4×10^{15}	1×10^{16}	2×10^{14} lim	5×10^{14} lim
InP(Fe)(100)2 °	6×10^{16}	3.5×10^{16}	3×10^{15}	2×10^{14}	4×10^{14}	7×10^{14} lim
(100)3 °	8×10^{16}	3×10^{16}	5×10^{14}	2×10^{15}	1×10^{15}	1×10^{15} lim
(100)4 °	1.5×10^{16}	5×10^{16}	5×10^{14}	2×10^{14}	2×10^{14}	5×10^{15} lim
InP(Sn)(100)2 °	2×10^{17}	7×10^{16}	5×10^{15}	2×10^{14}	2×10^{16}	5×10^{14}

[a] All layers doped by H_{2S} ($N_D - N_A \simeq 2 \times 10^{17}$ cm^{-3}).

The purity of undoped InP grown by LP-MOCVD depends very heavily on the particular source of TEI and PH$_3$ used in the growth. Dapkus et al., (1981) reported the same investigation for GaAs layers grown by MOCVD using TMG and AsH$_3$ and showed a similar dependence on source purity.

16. Characterization

a. Photoluminescence

Photoluminescence is the optical radiation emitted by a physical system resulting from excitation to a nonequilibrium state by irradiation with light. Three processes can be distinguished: creation of electron-hole pairs by

TABLE XII

Effect of Source Purity on Carrier Concentration and 300-K Mobility on InP Layers.

Sources		$N_D - N_A$ (cm^{-3})	$\mu(300\ K)$ (cm^2 V^{-1} sec^{-1})	$\mu(77\ K)$ (cm^2 V^{-1} sec^{-1})
TEI	PH$_3$			
α-Ventron(I)	Matheson(I)	6×10^{14}	5300	85000
α-Ventron(I)	Air Liquide(I)	3×10^{16}	3400	42000
Sidercom	Matheson(I)	1.3×10^{16}	3000	25000
α-Ventron(II)	Matheson(I)	7×10^{15}	3990	58000
α-Ventron(II)	Air Liquide(I)	7×10^{16}	3550	35000
Texas-Alkyle	Phoenix	3×10^{15}	4500	65000
α-Ventron(III)	Matheson(II)	5×10^{15}	5500	95000
SMI	Matheson(II)	10^{14}	5500	150000

absorption of the exciting light, radiative recombination of electron-hole pairs, and escape of the recombination radiation from the sample.

Photoluminescence (PL) is very useful for the assessment of semiconductor materials. Impurities and native defects, which are present in concentrations as low as 1×10^{15} cm^{-3} can be detected without destroying the sample, and surface irregularities are not important. By calibrating with known impurity concentrations, it is possible to calculate any unknown carrier concentration from the half width of the spectral line associated with a particular impurity. From the line shapes and half widths as a function of temperature, it is possible to distinguish between simple and complex centers and between simple donors and acceptors, if the effective masses of the electrons and holes are different. By simple donors and acceptors, we mean those centers that give rise to luminescence lines whose line shapes can be fitted by the effective mass theory for hydrogenic levels. A simple center is defined as an impurity that sits on the In or the P lattice site and that contributes only one additional carrier. It is analogous to hydrogen in that only s electrons take part in the binding. Consequently, the activation energy of the single carrier bound to the substitutional impurity is close to that which is calculated from the hydrogen model.

Figure 13 shows the PL spectra measured at 6.6 K of an undoped InP layer grown by LP-MOCVD using TEIn (from α-Ventron) and PH$_3$ (from Matheson). The PL was excited by using the 5145-Å line of an Ar$^+$ laser and was dispersed by a Jobin-Yvon monochromator. The PL was detected with a Ge photodiode.

Similar to GaAs, several elementary recombination mechanisms may occur and cause near-band-gap emission lines: free excitons (X: 1.4181 eV), excitons and shallow impurities (D°, X: 1.4169 eV), and (A°, X: 1.4147 eV). Silicon is the dominant donor and Zn the acceptor in the undoped InP samples grown by LP-MOCVD.

b. *Study of Impurity Redistribution in InP Layer by SIMS*

In SIMS analysis, the surface of sample under study is bombarded by a beam of energetic ions (O$_2^+$, Ar$^+$, or Cs$^+$ generally). A fraction of the sputtered species are ions. The secondary ions provide the information about the sample composition. A mass spectrometer is used to select the mass of the impurity ions. The secondary-ion intensity measurements are carried out by an electron multiplier.

The analyses at Thomson are performed by a CAMECA IMS3F (Lepareur, 1980; Huber, 1986), which has the essential characteristics required for semiconductor analysis, such as high trace sensitivity (PPb range), high mass resolution ($M/\Delta M = 10000$), high depth resolution (range nm), wide dynam-

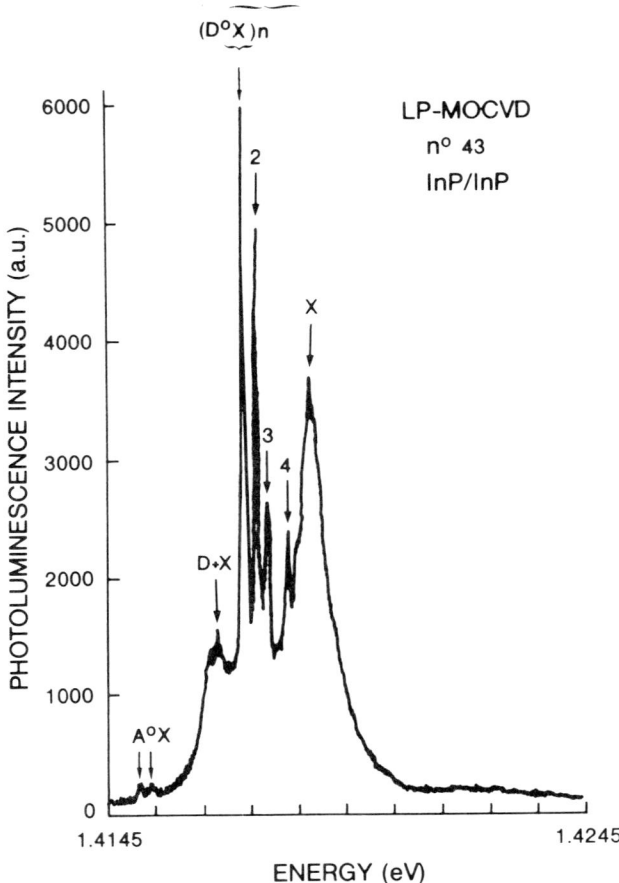

FIG. 13. Photoluminescence spectra of InP undoped layer grown by LP-MOCVD.

ic range in depth profiling (five in current use, seven under special conditions), mass separated primary ion beam, simultaneous depth profiling of several elements, and cesium gun equipment. The use of Cs^+ ion bombardment in conjunction with negative secondary ion detection in an ultra-high vacuum environment has dramatically increased the usefulness of SIMS, particularly for III–V compound technology (Magee, 1979). Therefore, SIMS, with its excellent sensitivity and depth resolution (30 to 100 Å), provides information on doping profiles and the quality of the interfaces.

The replacement in the primary and secondary ion optics of the stainless steel apertures by tantalum was one of the modifications. As a result, the background of Fe and Cr diminished by a factor of 100 and a detection limit of 5×10^{12} at.cm^{-3} was obtained in InP for these elements. This sensitivity is necessary for a good evaluation of InP, because Fe and Cr are deep acceptor level elements in InP. In an analysis of InP, the surface is scanned with a focused mass filtered oxygen ion beam ($Ip \sim 1.5$ μA at 10 keV). The scanned area is 250×250 μm, and the analysed region is 150 μm in diameter (Huber et al., 1984). An offset target voltage of -50 V was applied to avoid possible interference with ionized hydrocarbons. Ion-implanted samples were the standard for quantitative calibration of the instrument. The statistical results of various experiments show that the quantitative results of SIMs are given with an accuracy of $\pm 20\%$ at a concentration level of 1×10^{16} at.cm^{-3}. Below this level, results are less accurate, $\pm 50\%$ at 1×10^{14} at.cm^{-3}. The Talysurf measured depth precision is estimated at $\pm 10\%$. The detection limits of the impurities, which were measured, are Mg, Cr, Mn: 5×10^{12} at.cm^{-3}, Fe: 1×10^{13} at.cm^{-3}, and Si: 7×10^{13} at.cm^{-3}.

Figure 14 shows the typical SIMS depth profiles of Mg, Si, Cr, Fe, and Mn in a InP layer grown on a semi-insulating Fe-doped InP substrate by LP-MOCVD under standard growth conditions. Each sample was analysed in

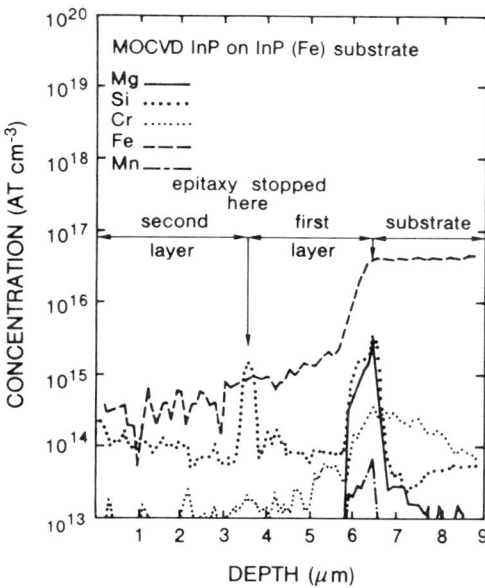

FIG. 14. Depth profiles of Mg, Si, Cr, Fe, and Mn in a MOCVD InP (Fe) substrate. Epitaxy was stopped, and the wafer was taken out of the reactor. A second layer was grown on the first layer.

two different areas about 10 mm apart. Generally, analysis of two clean areas gives reproducible and representative results for the material.

Accumulation of impurities such as Mg, Fe, Cr, Mn, and Si were detected by SIMS at the interface between the substrate and the epilayer. A similar phenomenon was observed in GaAs epilayers grown by MOCVD (Huber et al., 1982). The accumulation of these impurities may have different origins such as the following:

- Incorporation of an impurity from the ambient gas phase during preheating.
- Incorporation of impurities from the exposure to air.
- Contamination of the substrate by the chemical etching before the growth.
- Incorporation of impurities during in situ HCl etching.
- Out diffusion of impurities from the substrate to epilayer.
- The effect of substrate orientation on the impurity redistribution.

To determine the origin of these impurities, we tried to react as follows: an InP epilayer 3 μm thick was grown on the InP substrate under the conventional procedure. The wafer was removed from the reactor. After a 30-minute exposure to the air, the wafer was transferred again to the reactor, and a second epilayer, 3.5 μm thick, was regrown under the same growth conditions.

Figure 15 shows the "pile up" of impurities at the epilayer-substrate interface, where peaks in the concentration of Mg, Cr, Mn, Fe, and Si were found. At the epilayer-epilayer interface, the only concentration change was some increase in the Si level, which could be due to the heating of the susceptor. These experiments indicate that exposure to air and heating under $H_2 + N_2 + PH_3$ pressure under conventional growth conditions does not account for the accumulation of impurities.

In order to examine the possibility of contamination of the substrate by the pregrowth chemical treatment, we grew an InP epilayer 10 μm thick under conventional growth conditions. We separated the sample into two parts. On one part, pregrowth chemical etching was used to remove 2 μm of this epilayer. This sample was put into the reactor, and 2 μm InP was grown under the conventional growth conditions. Figure 16 shows the impurity profile of this sample after regrowth. One can observe similar accumulations of impurities at the epilayer-substrate interface and epilayer-epilayer interfaces. These results clearly showed that the majority of impurities results from the chemical etching, even though very pure reagents were used (e.g., "suprapure" Merck H_2SO_4, Fe (10^{-6}%), Cr (10^{-7}%) and Mg (10^{-6}%)).

In order to evaluate the effect of some growth parameters "such as in situ HCl etching prior to layer growth" on impurity redistribution, the second

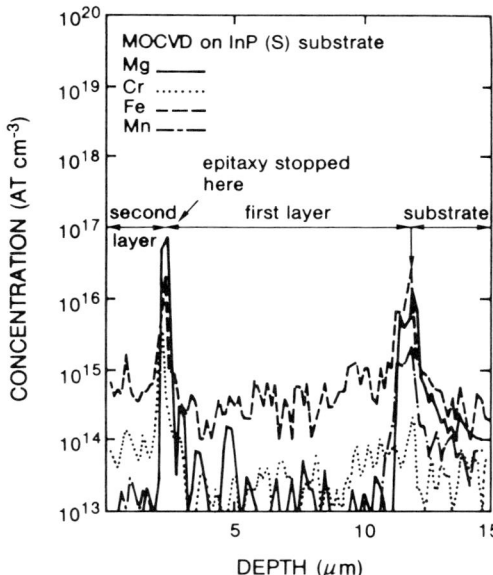

FIG. 15. Depth profiles of Mg, Cr, Fe, and Mn in a MOCVD InP layer grown on InP (S) substrate. The first layer was partially removed by controlled chemical etching. A second layer was grown on the chemically treated first epilayer.

part of the 10 μm thick InP epilayer was again placed in the reactor and etched in situ with HCl for a minute at 550°C. After that, 2 μm InP well grown on this sample. Figure 17 shows the SIMS impurity profiles of this sample. The large peaks of Si, Fe, Mg, Mn, and Cr at the interface between epilayer-epilayer show that impurities were incorporated from the HCl gas phase.

In situ HCl etching without PH_3 flow created the same interface contamination with an enhanced impurity redistribution in the layer for Fe and Cr, which occurred as shown in Fig. 18. It is possible that in this case a surface with a large degree of nonstoichiometry and a large density of defects (phosphorus vacancies) accelerates impurity diffusion.

To evaluate the effect of doping elements from the substrates, a study was made of LP-MOCVD growth simultaneously on three InP substrates with Fe, Sn, and S doped placed adjacent to each other within the reactor for growth under conventional growth conditions.

Figure 19 shows Fe profiles. One can observe that the Fe concentration and the Fe accumulation at the Fe-doped substrate epilayer interface and in the layer are about 10 times lower than in the case of Sn- or S-doped substrates. Results are similar for Mg, Mn, and Si. The diverse levels of the

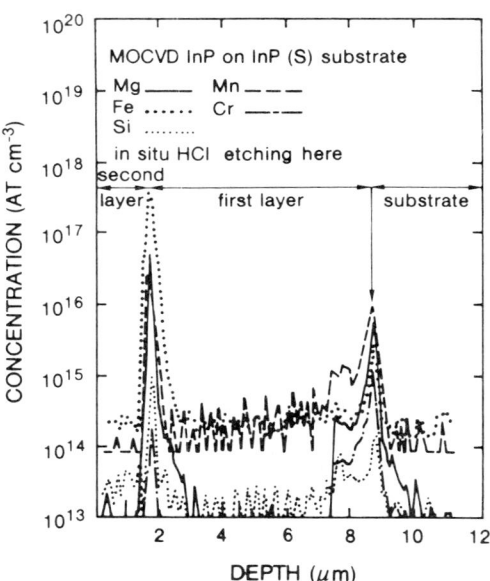

FIG. 16. Depth profiles of Mg, Si, Cr, Fe, and Mn in a MOCVD InP layer ground on InP (S) substrate. The first layer was partially removed by in situ HCl etching in PH_3/H_2 flow. A second layer was grown on the first layer.

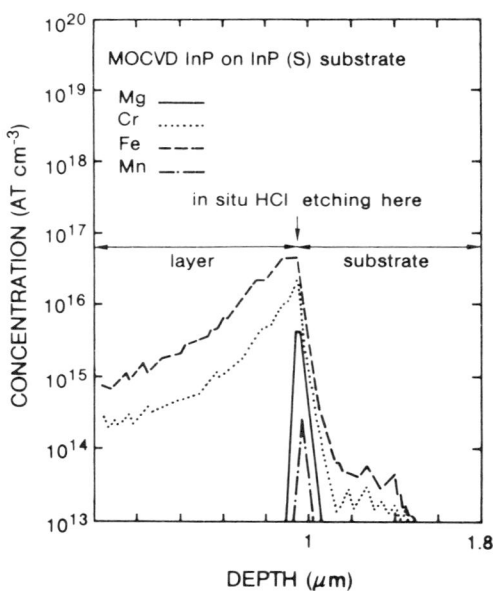

FIG. 17. Depth profiles of Mg, Cr, Fe, and Mn in a MOCVD InP layer grown on InP (S) substrate. The substrate was in situ HCl etched without PH_3 in H_2.

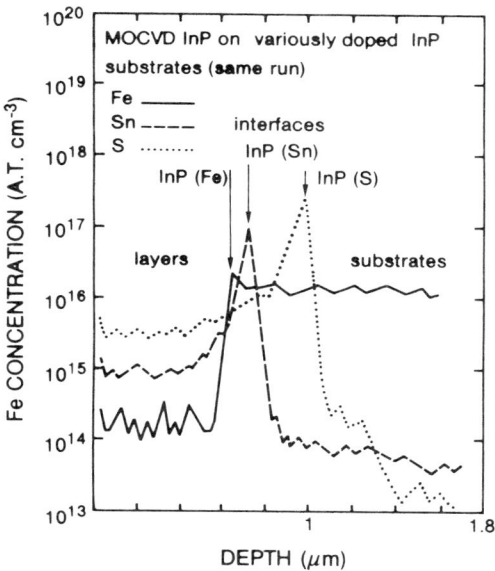

FIG. 18. Depth profiles of Fe on InP (Fe) and InP (Sn) substrates.

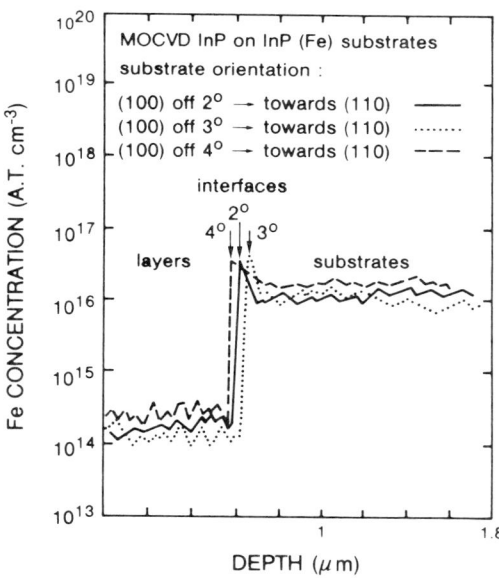

FIG. 19. Depth profiles of Fe semi-insulating substrates oriented (100) off 2°, 3°, and 4° toward (110).

same elements in the layers grown in the same run are probably due to the substrate quality. The degree of nonstoichiometry on the surface and in the substrates could influence the chemisorption, the crystal quality of layers, the formation of vacancies, and, consequently, the impurity redistribution. Layers grown on Fe-doped substrates often present lower impurity peaks at the interface and lower levels in epitaxial layers. Nevertheless, similar results were sometimes obtained on S- and Sn-doped substrates coming from different growths and different suppliers.

In order to estimate the effect of substrate orientation on the impurity redistribution, growth of InP layers was carried out on Fe-doped substrates oriented (100) off 2°, 3°, and 4° toward (110) (from the same InP ingot). The three samples were prepared simultaneously, and layers were grown in the same run under conventional growth conditions. Figure 15 shows the Fe depth profiles of layers. No significant difference in Fe redistribution due to misorientation of InP substrate can be observed. Similar results were found concerning Mg, Cr, Si, and Mn. It was shown that the adsorption of atoms during chemical etching before epitaxy represents a major source of impurities, which accumulate at the epilayer-substrate interface.

These results showed that during in situ HCl etching, the same phenomenon occurs. However, impurity peaks at the substrate and epitaxial layer interfaces, in some cases $\sim 10^{17}$ at.cm^{-3}, could have dire consequences on device quality (Huber et al., 1984). Whiteley et al., (1983), observe a fall in the electron mobility in the vicinity of the GaInAs–INP interface grown by MOCVD. They propose using in situ HCl etchng of InP prior to layer growth. Razeghi (1983) showed that there is no improvement of the quality of InP layer with in situ HCl etching. SIMS analysis clearly showed the effect of HCl etching before growth. It is now necessary to find a technique for elimination of impurities from the substrate surface before growth.

c. Carrier Concentration Measurements

When a metal is brought into intimate contact with a semiconductor, the conduction and valence bands of the semiconductor are brought into a definite energy relationship with the Fermi level in the metal. Once this relationship is known, it serves as a boundary condition on the solution of the Poisson equation in the semiconductor, which proceeds in exactly the same manner as in *p-n* junctions. Energy-band diagrams of a Schottky barrier on an *n*-type semiconductor under different biasing conditions, such as (a) thermal equilibrium, (b) forward bias, and (c) reverse bias, are shown in Fig. 20.

In *p*-type material, the Fermi level is located near the valence bands (V_B). A depletion region is established by the combined effect of the built-in voltage

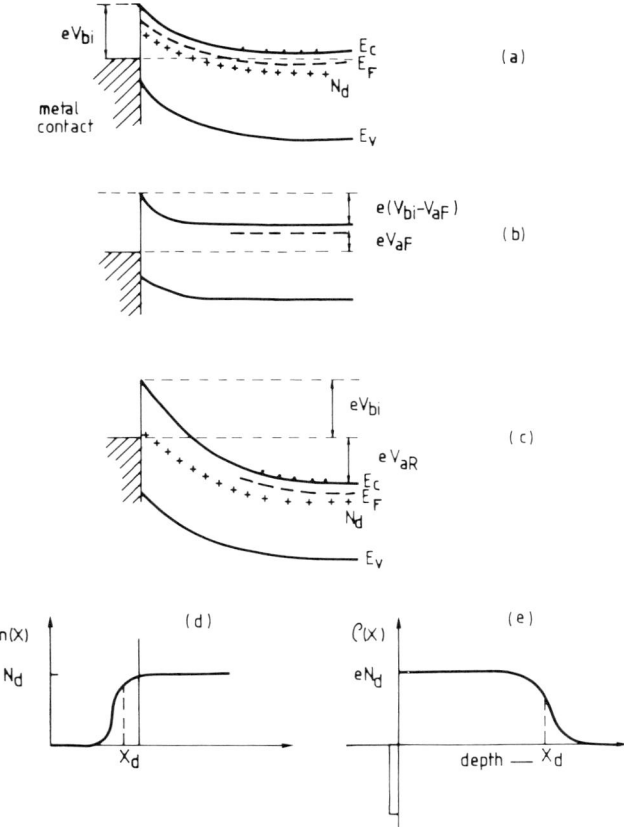

FIG. 20. Energy band diagram of a Schottky barrier on an n-type semiconductor under different biasing conditions. (a) thermal equilibrium; (b) forward bias; (c) reverse bias; and (d) electron density distribution $n(\alpha)$ at the edge of the depletion region.

(V_{bi}) and applied reverse bias (V_{aR}) or applied forward bias (V_{aF}), and the electron density distribution $n(X)$ at the edge of the depletion region as is illustrated in Fig. 20d. The plot of space-charge density ($\rho(x)$), as a function of depth for uniform material, illustrates the presence of positive charge in the semiconductor and an accumulation of electrons in the metal (Fig. 20e). Under the abrupt approximation that $\rho \simeq qN_d$ for $X \leqslant x_d$ and $\rho \simeq 0$, and $dV/dX \simeq 0$ for $X > X_d$, where X_d is the depletion width N_d can be given by:

$$N_d = \frac{2}{q\varepsilon_s} - \left[\frac{1}{d\left(\frac{1}{C^2}\right)/dV} \right].$$

If N_d is constant throughout the depletion region, one should obtain a straight line by plotting $1/C^2$ versus V.

For determining the carrier concentration in InP and related compounds, the conventional nondestructive, capacitance-voltage (C-V) method is usually used. The main disadvantages of the C-V method is that the maximum depth that can be profiled is limited by electrical breakdown at high reverse bias, and this can be restrictive in highly doped materials, where the depletion depths are small. Originally, this was overcome by alternate chemical etching and profiling with a temporary mercury barrier, but this process is time consuming and tedious. Ambridge *et al.*, (1973; 1974; and 1975) used an electrolyte to make the barrier and to remove material electrolytically, so both processes can be carried out in the same electrochemical cell and controlled electronically using automatic equipment to perform the repetitive etch/measure cycles to generate a profile plot. The etch depth can be measured continuously by integrating the etch current and applying Faraday's law. The basic principles of electrochemical C-V profiling are documented in a number of original papers by Ambridge *et al.*, (1975), and a review paper by Blood (1985). This method is destructive, but the profile can, in principle, be measured to unlimited depth. The requirements for the electrolyte are rather demanding, calling for satisfactory barrier and dissolution properties on *n*- and *p*-type material.

Actually, a versatile instrument manufactured by Polaron Equipment Ltd., Watford, UK (sometimes called the "Post Office Plotter" or POP) is available commercially.

Figure 21 shows the schematic diagram of the electrochemical cell used in the profiles. The semiconductor sample is held against a sealing ring, which defines the contact area by means of spring-loaded back contacts. The etching and measuring conditions are controlled by the potential across the cell, and this is established by passing a DC current between the semiconductor and the carbon electrode to maintain the required overpotential measured potentiometrically with reference to the saturated calomel electrode (SCE). The AC signals are measured with respect to a Pt electrode located near the semiconductor surface to reduce the series resistance due to the electrolyte. The capacity associated with the accumulation of carriers at the interface between electrolyte and semiconductor, due to the charged depletion zone in the material, is then determined. When the contact is illuminated with photons of energy greater than the bandgap of the semiconductor, the reverse current is increased, because of a flow of holes or electrons from the semiconductor into the electrolyte in *n*- or *p*-type material, respectively. This causes a change in the voltage for zero current, which is of the opposite sign for *n*- and *p*-type material, and can, therefore, be used to indicate the material type. Material is dissolved when an anodic current is drawn by a flow of holes

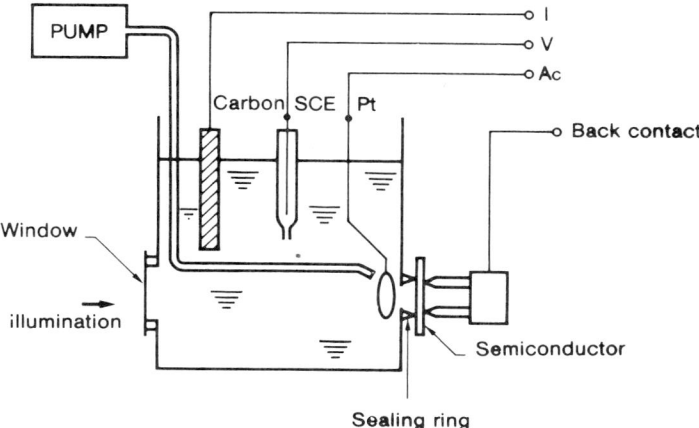

FIG. 21. Schematic diagram of the electrochemical cell used in the profiles.

from the InP, where as a cathodic curret causes deposition of material from the electrolyte (which is HCl for InP) onto the semiconductor surface. With p-type material, the holes required for the dissolution reaction are obtained from the valence band by simply applying an anodic potential, which drives the barrier into forward bias, whereas, in order to etch n-type material, the holes are generated by illumination under reverse bias. Smooth removal of n-material is achieved when the anodic current depends upon the illumination intensity but not upon the potential. For a reverse bias C-V measurement on p-type material, the potential is switched from an anodic to a cathodic value, and to avoid contamination of the sample surface by the cathodic reaction promoted by electrons from the conduction band, it is important that the cathodic potential is such that the reverse current during the measurement is very small.

To calculate the depth etched (X_e), one can use Faraday's law of electrolysis (Blood, 1985) from the total charge transferred by integrating the etch current I:

$$X_e = \frac{M}{qFDS}\int I \, dt,$$

where M is molecular weight, and D is the density of the semiconductor, F is the Faraday constant, S is the dissolution surface, and q is the charge transferred per molecule dissolved. For InP, $q = 6$. The depletion depth (x_d) can be obtained from:

$$X_d = \frac{\varepsilon\varepsilon_o S}{C}$$

by measurement of C at 3 KHz (≈ 0.14 V peak-to-peak), and $N(X_d)$ is derived from:

$$N(X_d) = -\frac{C^3}{e\varepsilon\varepsilon_o S^2}\left(\frac{\Delta C}{\Delta V}\right)^{-1}$$

by measuring $\Delta C/\Delta V$ by modulation at 30 Hz ($\simeq 0.28$ V peak-to-peak) at a low fixed reverse bias.

Fig. 22(a) shows the electrochemical (Polaron) profile for a InP over a Sn-doped InP substrate ($N_D - NA = 10^{18}$ cm^{-3}) establishing uniform doping along the layer thickness and an abrupt change in carrier concentration.

Fig. 22(b) shows the carrier-concentration profile evaluated from C-V measurements for InP over a semi-insulating InP substrate exhibiting uniform doping along the layer thickness and an abrupt change in carrier concentration. The C-V measurements and Polaron profile give the concentration of shallow dopants ($N_D - N_A$), which agrees with the carrier concentration deduced from Hall effect measurements.

d. Hall Mobility Measurements

All of the devices using InP-based materials require an exact specification in terms of carrier concentration and thickness, and top performance often depends on material quality. The use of the Hall effect is extremely common to extract carrier density and mobility between room temperature and 2 °K. In many cases, mobility is used as a simple figure of merit for quality of starting materials such as hybride (PH_3 or AsH_3) and alkyle (TEIn or TMIn). Hall data can yield excellent quantitative information regarding the electrically active impurities in a semiconductor.

The total mobility in InP-based material results from the combination of the different scattering processes. However, the lattice is in constant motion even at absolute zero, and the quantized normal modes of its vibration are known as phonons. The localized contractions and dilatations of the lattice in the presence of phonons create potential fluctuations of various origins, which interact with and scatter the electrons. This interaction may be through local band structure perturbations, known as deformation potential scattering, which either involve acoustic phonons or nonpolar optic phonons.

The most important scattering mechanisms are defect scattering, which includes alloy scattering, impurity (neutral or ionized) scattering, and crystal defects; lattice scattering, which includes: intervalley scattering, intravalley optic (polar or nonpolar) scattering, and intravalley acoustic (piezoelectric or deformation potential) scattering, and carrier-carrier scattering.

The lack of a center of symmetry in InP-based materials leads to piezoelectric behavior, and through this the phonons can scatter the electrons. In ternary and quaternary III–V alloys, an extra scattering process

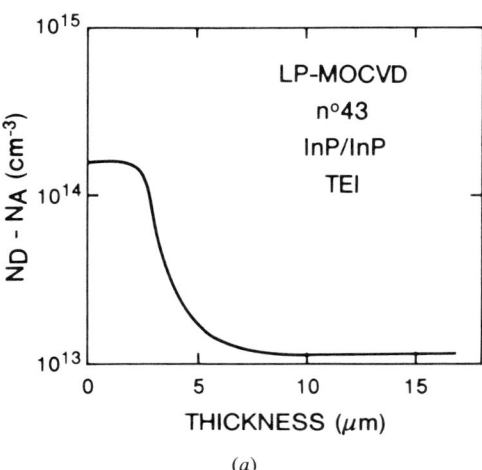

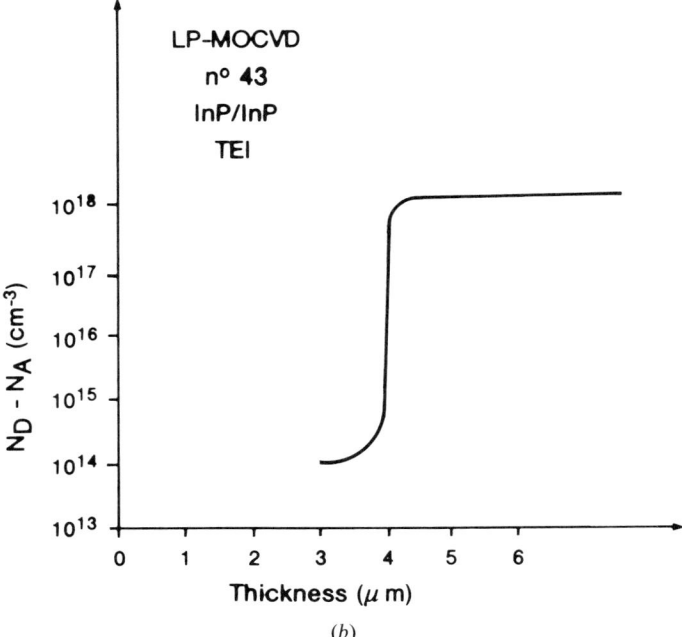

FIG. 22. (a) Carrier concentration profiles from C-V measurements for InP layers on semi-insulating InP substrate. (b) Electrochemical polaron profile.

exists—alloy scattering—which arises from the fact that there is a random distribution of the group III or the group V atoms on their respective sites. The presence of impurities, lattice defects, and complexes is most important when the impurity is charged, whereupon the electron scatters off the long-range coulomb potential, ionized impurity scattering. Neutral centers can also interact in a complex manner, but this is generally only important in the highest purity samples at low temperatures. Naturally, all of these processes proceed together and result in carrier scattering, which significantly affects the mobility. The mobility from ionized impurities (μ_i) and from acoustic phonon due to deformation potential interaction (μ_A) can be given by (Sze, 1981)

$$\mu_i \simeq (m^*)^{-1/2} N_I^{-1} T^{3/2},$$

where N_I is the ionized impurity density

$$\mu_A \simeq (m^*)^{-5/2} T^{-3/2}.$$

For polar semiconductors such as InP, optical-phonon scattering is significant.

For making Hall measurements, the Van der Pauw clover leaf (Van der Pauw, 1958) geometry has been used. This pattern is ideal for making Hall measurements on epitaxial layers in which the layer thickness is orders of magnitude less than the lateral dimension of the sample.

Figure 23 shows the typical Hall mobility as function of temperature for InP layers grown by LP-MOCVD. From these data one can usually estimate the carrier density $N_D - N_A$, the compensation ratio $K = N_A/N_D$, and the donor ionization energy E_d.

If we assume that all common donor species in InP and related compounds are hydrogenic with binding energy, E_d, given by

$$E_d = 13.6 \times \frac{m^*}{m^o} \frac{1}{\varepsilon_r^2} \quad (eV)$$

with $m_e^* = 0.08\, m_o$, $m_p = 0.5\, m_o$, $\varepsilon_r = 12.4$ (ε_r is the relative dielectric constant). One obtains for donor impurities E_d (donor) $\simeq 7$ meV and for an acceptor impurities E_d (acceptor) $\simeq 40$ meV. This shows that for InP and related compounds, the donors are all shallow and the acceptors are sufficiently deep. This means that for temperature between, 4K and 400K, the Fermi energy movement is not sufficient to change the charge state of the deeper compensating acceptors. Note that an acceptor in this context is any

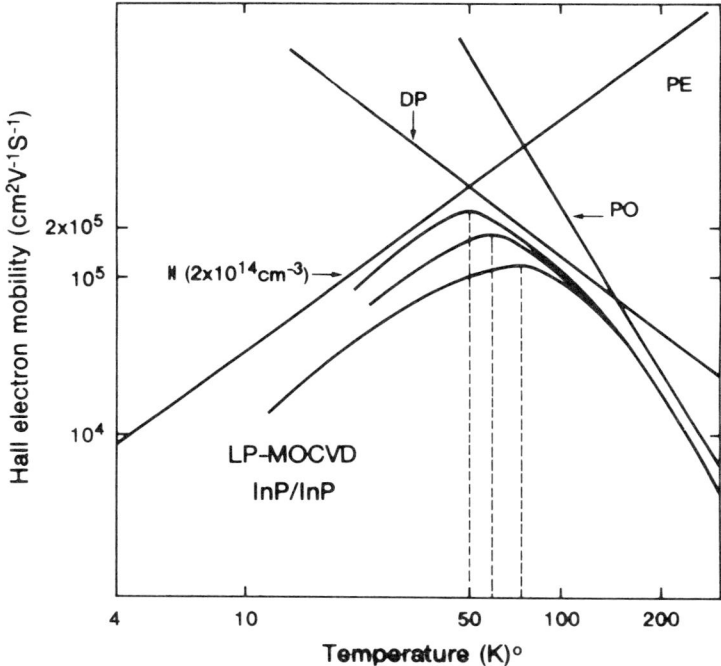

FIG. 23. Electron Hall mobility as a function of temperature.

center that is negative when occupied, regardless of depth (Anderson and Aspslay, 1986).

Hall mobilities of epitaxial InP layers grown on semi-insulating substrates were measured in a magnetic field of 4000 G by a conventional Van der Pauw technique. In Table XIII, we give the measured mobility at 300 K and 77 K in undoped InP, using pure starting materials (TEI from α-Ventron and PH_3 from Matheson). Figure 23 shows the electron Hall mobility as a function of temperature of sample 43.

e. *Deep-Level Transient Spectroscopy*

Deep-level transient spectroscopy (DLTS) is a method for obtaining information about defects giving rise to electrically active deep energy levels. Conventional DLTS can only, in a Schottky-barrier junction, detect majority-carrier traps. It is easy, however, to resolve different traps and to derive their concentrations. The DLTS measurements show that there are two

TABLE XIII

AVERAGE MEASURED VALUES OF MOBILITY FOR InP EPILAYERS.

Sample n°	Thickness (μm)	300 K		77 K	
		$N_D - N_A$ (cm^{-3})	(cm^{2}V$^{\mu-1}$ sec^{-1})	$N_D - N_A$ (cm^{-3})	(cm^{2}V$^{\mu-1}$ sec^{-1})
119	2.5	2×10^{15}	$5350 \pm 2\%$	1.5×10^{15}	59,800
151	2.5	5×10^{15}	$5240 \pm 2\%$	3.6×10^{15}	56,700
127	3.3	5.7×10^{15}	$4950 \pm 2\%$	5.7×10^{15}	53,320
43	4	10^{14}	$5500 \pm 2\%$	10^{14}	150,000

electron traps in InP layers grown by LP-MOCVD (see the following tabulation) (Lim et al., 1982):

Activation energy E(meV)	Capture cross section S (cm^{-2})
E_5, 433	3×10^{14}
E_6, 661	8×10^{14}

17. INTERFACES

The surface of InP layers and the epitaxial-layer-substrate interface have been studied by Auger electron spectroscopy. Figure 24 shows the Auger

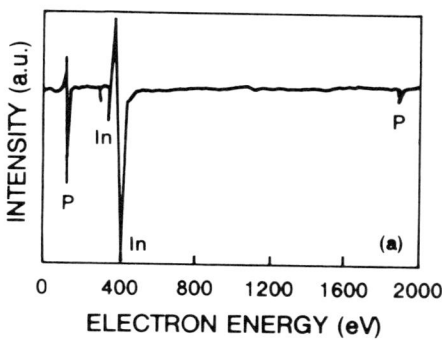

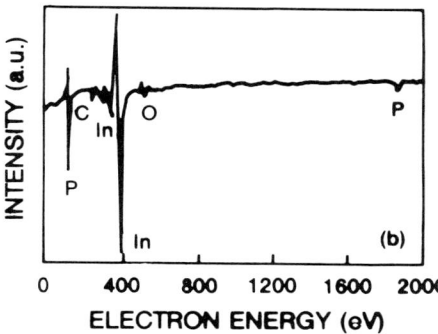

FIG. 24. Auger spectrum for (a) the InP epilayer-substrate interface; (b) the surface of the epilayer. (From Razeghi, 1985, in "Lightwave technology for communication," (W. T. Tsang, ed.) Academic Press, New York.)

spectrum for such an interface eroded by sputtering, and it can be seen that there are no impurities at the interface. Figure 25 indicates the Auger spectrum for the surface of the InP epilayer and shows the presence of O and C as well as P and In. The number of dislocations in the epilayer is the same as in the substrate, and it is not possible to see the interface of the substrate and the epilayer after normal chemical etching.

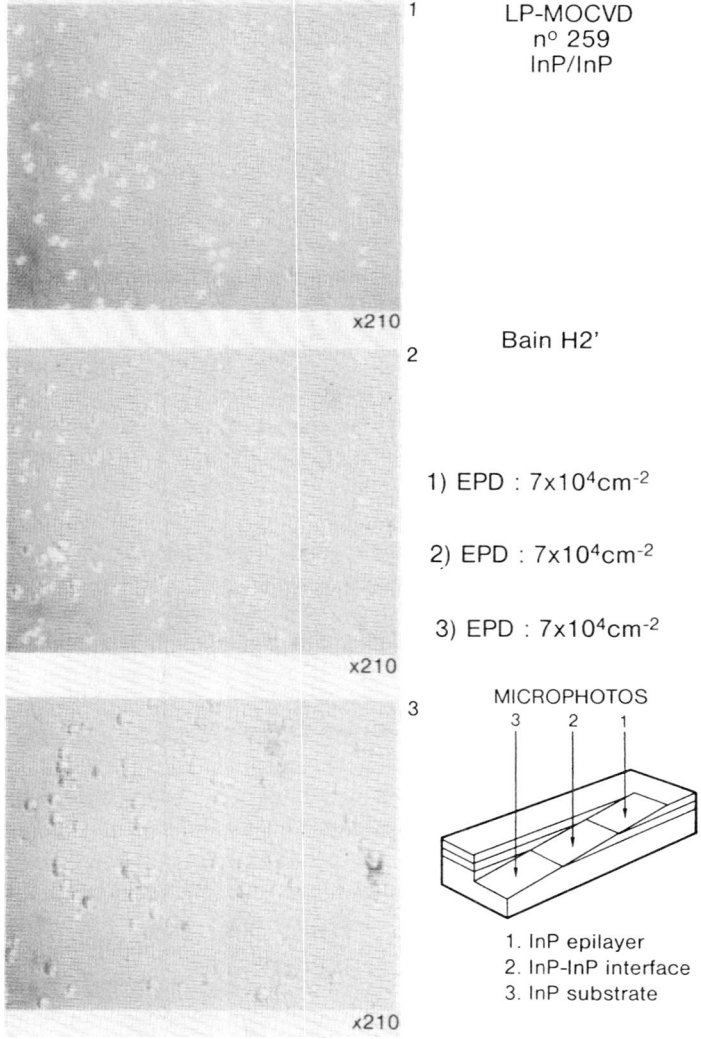

FIG. 25. Transition region after chemical etching of a InP–InP epilayer grown LP-MOCVD.

VI. Growth and Characterization of InP Using TMIn

High-quality InP epilayers have been grown by LP-MOCVD using trimethyl indium (TMIn) (Alfa product) and PH_3 (Matheson) for In and P sources, respectively. Pure hydrogen is used as a carrier gas. The growth temperature is 550 °C, and the growth pressure is 100 m bar. The details on the growth optimization conditions, which were determined during these investigations, are indicated in Table XIV.

All the InP layers have been grown on (100) oriented Sn- and Fe-doped InP substrates. The main characteristics of the different samples are summarized in Table XV.

Figure 26 shows the transition region after chemical etching of an InP/InP epilayer grown by LP-MOCVD using TMIn. The etch-pit density (EPD) was the same in the substrate as in the confinement InP layers, and the interfaces are defect free.

Figure 27 indicates the electrochemical polaron profile of a InP layer using TMIn. A carrier concentration as low as 3×10^{13} cm^{-3} has been measured, which is the purest InP epilayer that has been grown by any growth technique.

Electrical measurements were performed using the classical Van der Pauw technique with ohmic contacts defined by evaporation and annealing of Au–Ge. The maximum mobility was generally found between 50 K and 55 K

TABLE XIV

OPTIMIZED GROWTH CONDITIONS OF InP USING TMIn.

Growth temperature	°C	550
Total H_2 flow	l/min	6
H_2 through TMIn bubbler	CC/min	50
PH_3 flow	CC/min	100
Growth rate	Å/min	300

TABLE XV

CHARACTERISTICS OF THE DIFFERENT InP EPILAYERS.

no	66	68	70	75	76
width	5	8	10	5	3.5
Nd-Na	7.10^{13}	2.10^{14}	3.10^{13}	6.10^{13}	10^{14}
μ(300K)(cm^{2}V^{-1}s^{-1})	5500	4500	6000	5500	5000
μ(70K)(cm^{2}V^{-1}s^{-1})	150.000	100.000	200.000	145.000	110.000

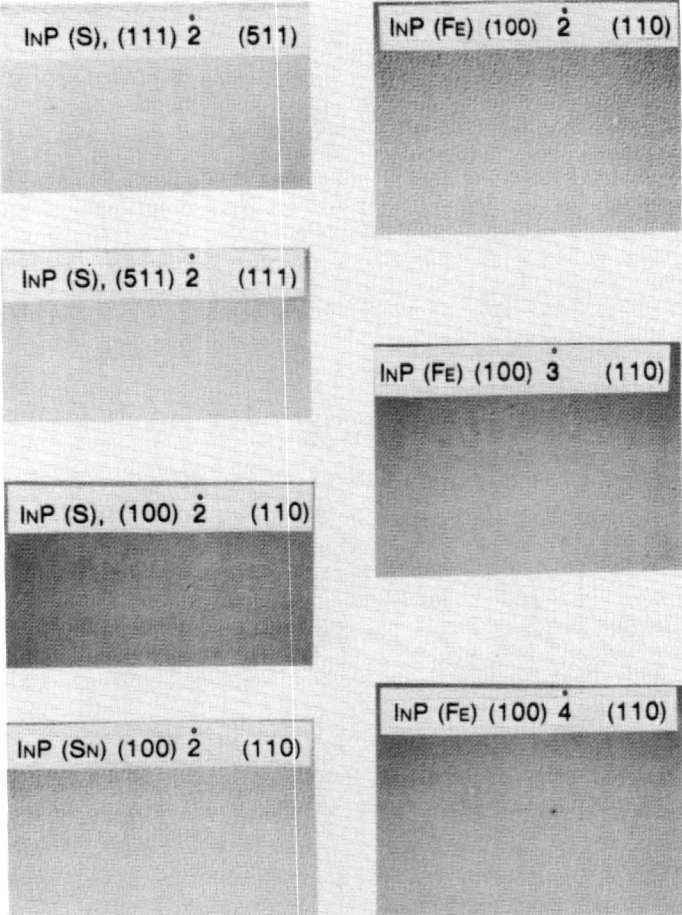

FIG. 26. Chemical etching of a InP-InP epilayer.

and exceeded 100.000 cm^2 V^{-1} s^{-1}, proving the high quality of the samples under study. At the same time, the residual doping level was found to be very low, generally less than 10^{14} cm^{-3}.

Photoluminescence measurements were performed at 2 K using a dye laser as a source of excitation. The energy of excitation was tuned just above the band gap of InP, at about 1430 meV, with a slitwidth of 50 μm.

Figures 28–30 show the photoluminescence spectra of samples 66, 68, and 70. All of the spectrum exhibit different lines, which were identified as free exciton (X), exciton bound to neutral donor $(Do\ X)$, charged donor $(D^+\ X)$,

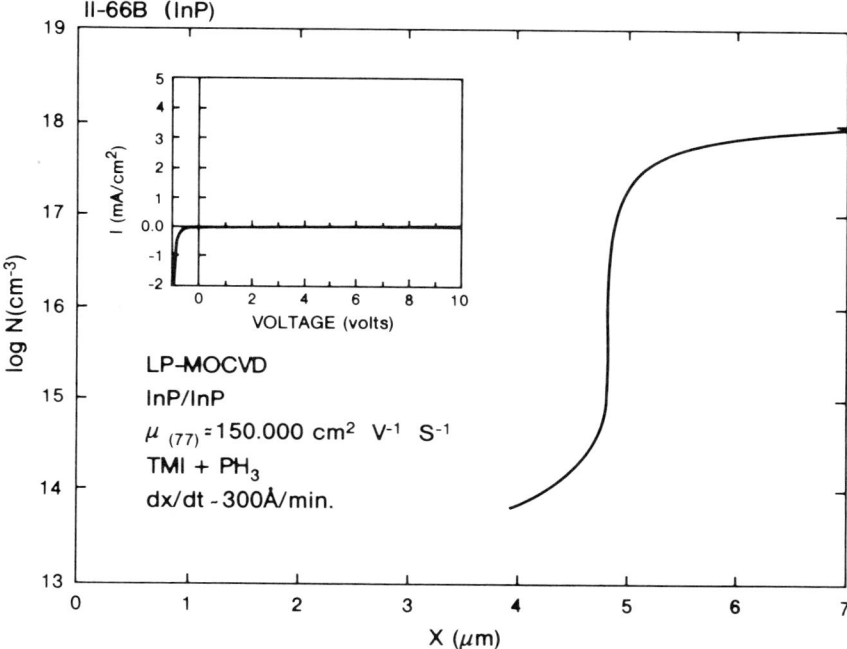

FIG. 27. Typical electrochemical polaron profile of an InP epilayer grown by LP-MOCVD.

donor-valence band recombination (Do-h), and exciton bound to acceptor (Ao X).

The energy position of all of the peaks and their identification are summarized in Table XVI.

The free exciton line is located at about 1418.8 meV, which is in good agreement with previous work on the subject. Various components of the bound exciton (Do X) recombination, identified as $(Do\ X)_n$ with $n = 1,...,5$, have been observed. They are more complicated, since an additional electron is present, which makes the donor neutral before binding the exciton. The different components $(Do\ X)_n$ result from different angular momentum states of the $j = 3/2$ hole. The states of the complex are generally described by those of the hole in a central potential, with the pair of electrons in the singulet state. These lines are located between 1417.3 meV ($n = 1$) and 1418.5 meV ($n = 5$). Their linewidths are less than 0.1 meV, thus necessitating a very sensitive optical system.

D^+X and Do-h lines appear at lower energies (1416.6 and 1416.8 meV, respectively), since those systems are more energetically bound. All of these features appear in the three samples under study.

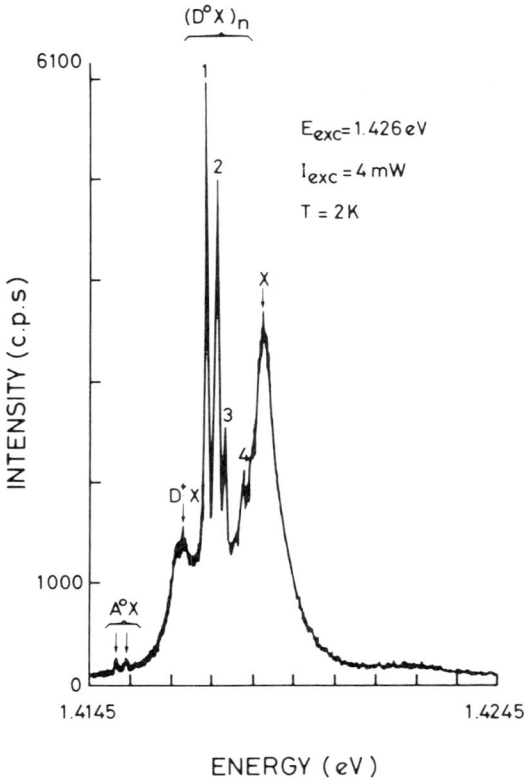

FIG. 28. Near gap photoluminescence (NGPL, at 2 K of a InP epilayer grown by LP-MOCVD (sample no. 66).

In two of the samples (66 and 68), two very weak exciton bound to acceptor (A^oX) lines located in 1414.2 meV and 1414.4 meV appear. We have su... narized in Table XVII the ratio of intensity of X/DoX and AoX/DoX lines for the three samples.

In all cases, the acceptor-bound-exciton lines appear very weak when compared to the other recombination processes. The sample, therefore, should have very little compensation.

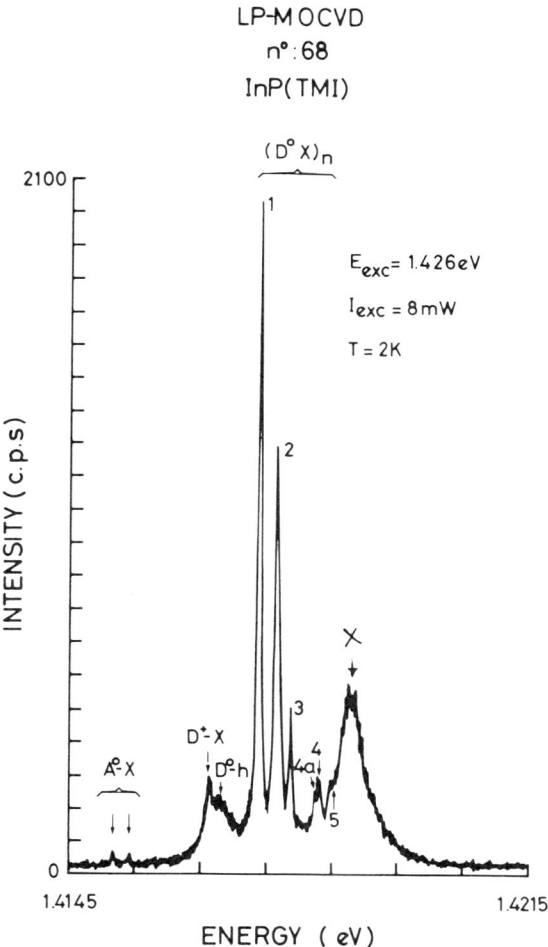

FIG. 29. Low temperature NGPL spectrum at 2 K of a InP epilayer (sample no. 68).

For the sample no. 70 (Fig. 30), the sensitivity of detection was multiplied by 100 in the region of the (A_0X) peak, with no trace of an acceptor. This further demonstrates the purity of the sample and explains the exceptionally high mobility measured ($\mu(_{50K}) = 200.000$ cm^2 V^{-1} s^{-1}).

On the other hand, the free exciton peak (X) appears more intensive than the donor bound like (D_0X). This is the first time that such a fact has been observed in any InP crystal, which confirms the very low concentration of donors (3 10^{13} cm^{-3}) determined by a polaron profile.

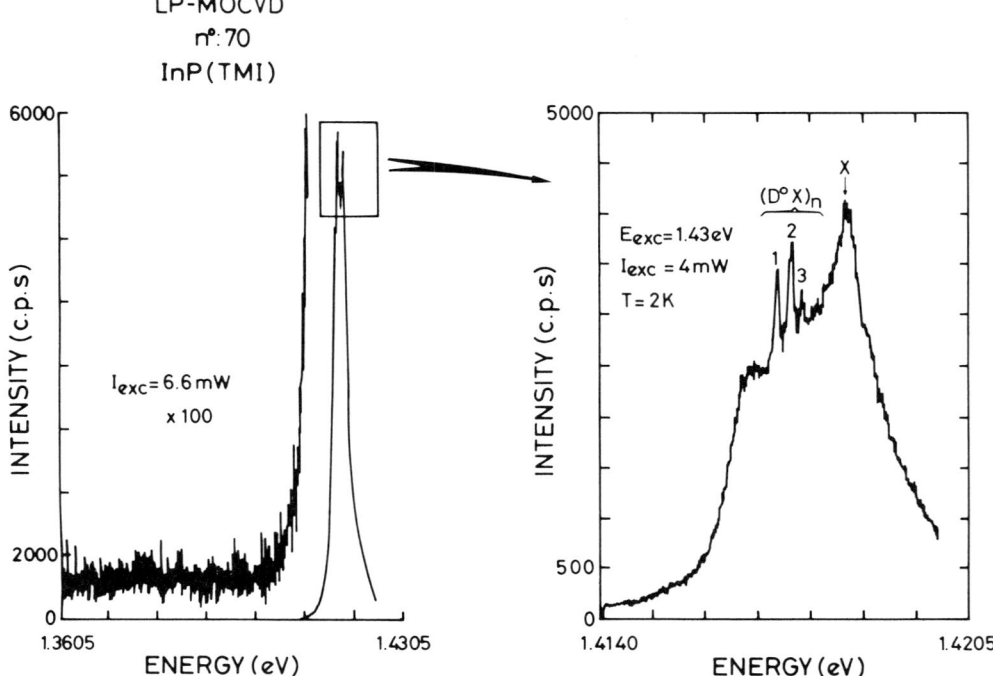

FIG. 30. Low temperature photoluminescence spectrum at 2 K of a InP epilayer (sample no. 70). Inset: Detail of the NGPL spectrum.

TABLE XVI

Position in Energy of PL Recombination in InP Epilayer.

Type of recombination		Energy (meV)
X		1418.8
Do X	n = 1	1417.3
Do X	n = 2	1417.6
Do X	n = 3	1417.8
Do X	n = 4	1418.3
Do X	n = 5	1418.5
D^+ X		1416.6
Do h		1416.8
Ao X		1414.2
Ao X		1414.4

TABLE XVII

THE RATIO OF INTENSITY OF X/DoX AND AoX/DoX FOR InP EPILAYER USING TMI.

	66	68	70
X/DoX	$5.8\ 10^{-1}$	$2.8\ 10^{-1}$	1.25
AoX/DoX	$2.0\ 10^{-2}$	$3.0\ 10^{-2}$	no acceptor

To our knowledge, this is the purest InP crystal ever grown, with the highest low temperature mobility reported in the literature. This can be related to the very low residual doping level of these layers.

VII. Incorporation of Dopants

a. p-Type

InP layers grown by LP-MOCVD can be doped p-type by using diethylzinc (DEZ). Figure 31(a) shows the variation of net carrier concentration as a function of H_2 flow through the DEZ bubbler ($-15\ °C$) (growth temperature was 650 °C). When the flow rate of DEZ is kept constant, the free carrier concentration varies exponentially with $1/T$, as shown in Fig. 31(b), where T is growth temperature.

The incorporation of dopants during the growth on InP using the doping species such as DEZ can be explained by using a model described by Duchemin (1977a). If we suppose that all the DEZ arriving at the hot surface is decomposed, then the Zn concentration becomes limited by the diffusion of the DEZ through the boundary layer to the hot surface. After decomposition, there are two possible limiting cases.

(1) In the simplest case, all of the decomposed material is incorporated into the growing layer. Thus, the impurity concentration is independent of temperature and inversely proportional to the growth rate. This behavior is observed for the doping of silicon by germanium using germane (GeH_4) (Duchemin, 1977b).

(2) The second case is one in which only a small fraction of the secondary form (Zn) of the dopant is incorporated into the growing layer. Here, the major part of the dopant is vaporized and is then lost by diffusion away from the substrate. As the temperature is raised, more of the dopant is vaporized, and the doping concentration decreases. Here, then, the doping concentration decreases as the temperature increases but is independent of the growth rate. This behavior is typical of the doping of the InP by zinc (Fig. 32).

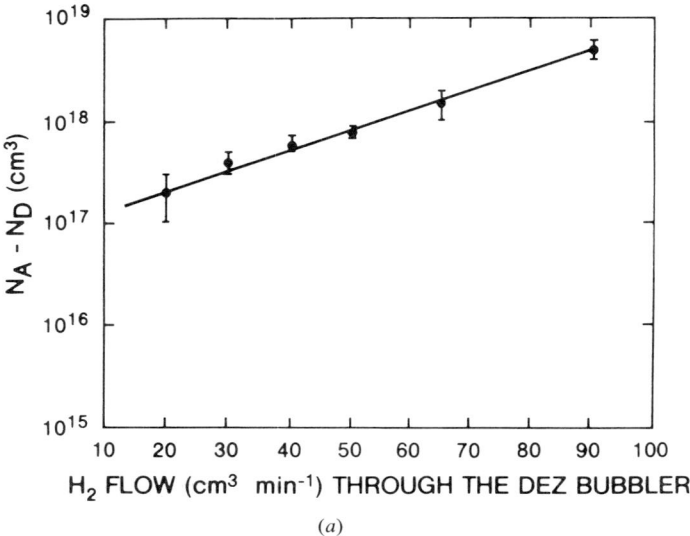

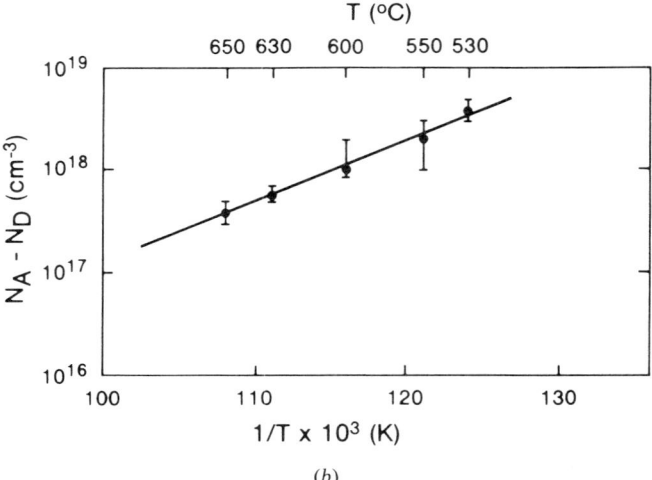

FIG. 31. (a) Variation of acceptor level $N_A - N_D$ in InP with DEZ flow rate ($T_G = 650$ °C; total flow, 7 liters min^{-1}). (b) Variation of acceptor level in InP with growth temperature: $N_A - N_D \propto \exp(-E_i/kT)$; $E_i \simeq 10$ meV; 30 cm^3 H$_2$ min^{-1} through the bubbler of DEZ.

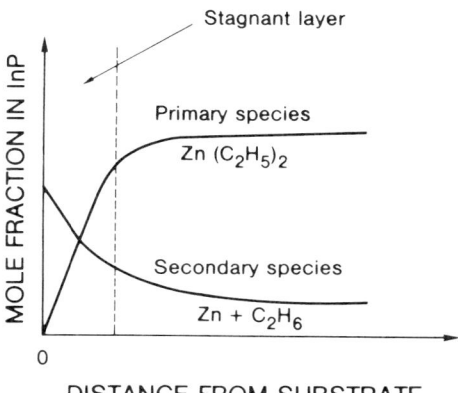

FIG. 32. Total decomposition of DEZ (primary species) with partial incorporation of ZN (secondary species).

b. *n-Type*

InP and related compound epilayers grown by LP-MOCVD can be doped *n*-type using H_2S or SiH_4. When the flow rate of H_2S is kept constant, the free-carrier concentration varies exponentially with $1/T$, as in the case of DEZ. The free-carrier concentration in the epilayer decreases when the growth temperature increases (Fig. 33).

The decomposition of the H_2S (primary species) is the rate-limiting step. After decomposition, most of the secondary species (S) is incorporated into the growing layer (Fig. 34).

A typical post-office-plotter electrochemical profile (using a 0.1 cm² area and 0.5 M HCl) of a InP Gunn-diode structure ($n^+ - n - n^+$) grown by LP-MOCVD is shown in Fig. 35. Using H_2S for n^+ doping, the interfaces between the layers are sharp, thanks to LP-MOCVD growth. Figure 36 presents a SIMS profile for the Gunn-diode structure of InP growth by LP-MOCVD using sulfur (H_2S) as an *n*-type dopant.

Sulfur has been widely used as an *n*-type dopant for DH layers grown by LP-MOCVD for laser applications. There have been indications in the literature (Benz *et al.*, 1982; Giles *et al.*, 1984; Stormer *et al.*, 1979a) that at growth temperature, significant sulfur diffusion can occur, which will seriously degrade doping interfaces.

In the case of SiH_4, when the flow rate is kept constant, the free-carrier concentration increases with increasing growth temperature. At higher temperatures the decomposition of the SiH_4 is more efficient. We found that for high doping level $\simeq 10^{18}$ cm^{-3} (such as for lasers), it is better to use H_2S.

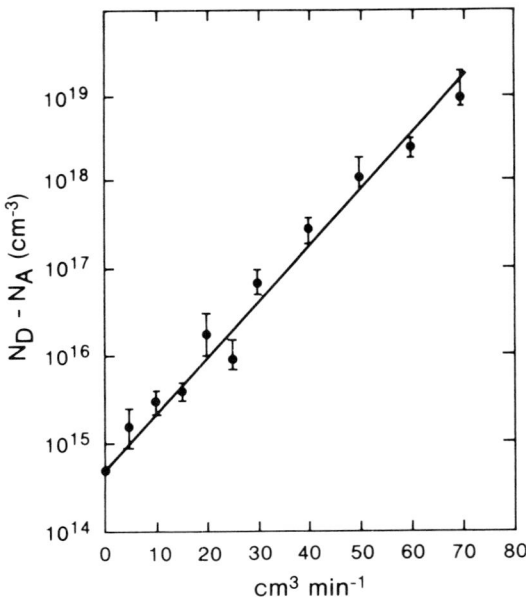

FIG. 33. Variation of donor level $N_D - N_A$ in InP layers with H_2S flow rate: 1000 ppm H_S in H_2; $T_G = 550\ °C$; $dx/dt \simeq 200$ Å min^{-1}.

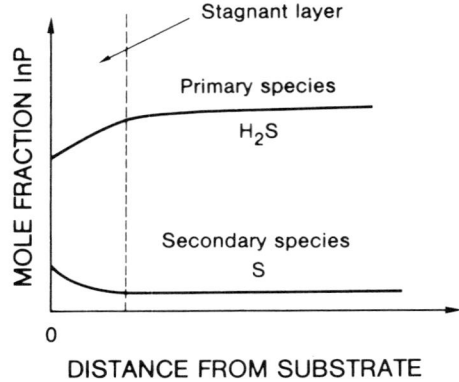

FIG. 34. Partial decomposition of H_2S (primary species) with incorporation of S (secondary species).

5. LP-MOCVD OF $Ga_xIn_{1-x}As_{1-y}$ ALLOYS

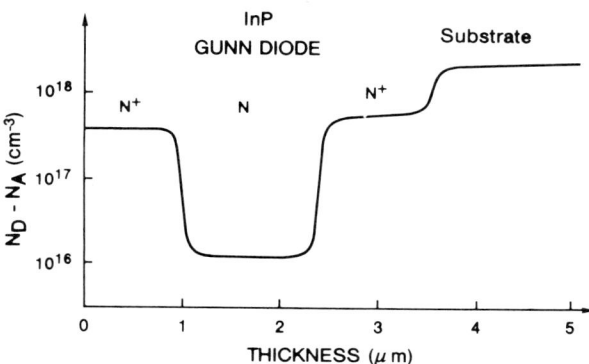

FIG. 35. Electrochemical profile of an InP Gunn-diode structure.

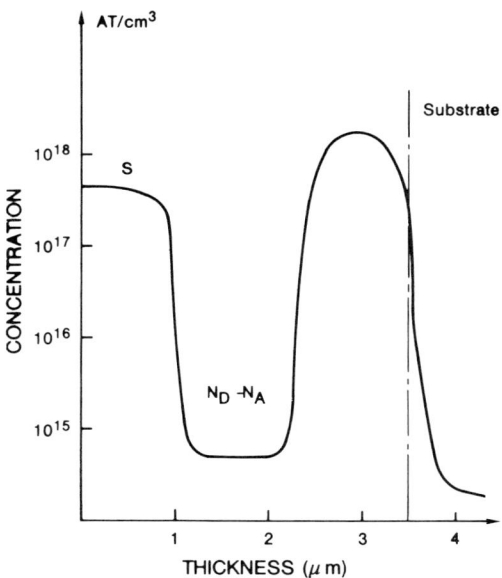

FIG. 36. SIMS profile for Gunn-diode structure of InP grown by LP-MOCVD.

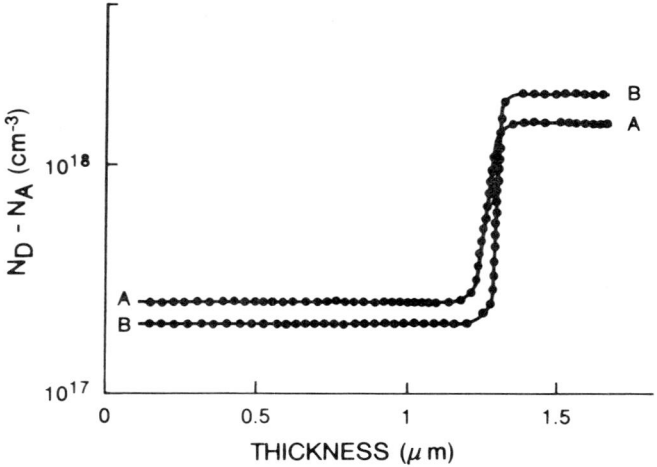

FIG. 37. Electrochemical profiles of typical doping interfaces for InP doped with $S(A)$ and $Si(B)$: LP-MOCVD; dots indicate values with light, solid curves indicate values without light.

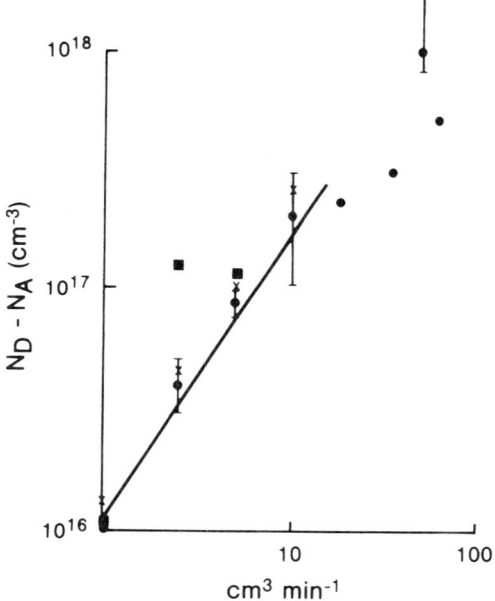

FIG. 38. Variation of donor level in InP layer with SiH_4 flow rate: LP-MOCVD InP; $(X)C - V$; (●) electrochemical profile; (■) Hall mobility (250 ppm SiH_4 in H_2); $T_G = 550\ °C$; $dx/dt = 200\ Å\ min^{-1}$; $N_D - N_A = 66, 57,$ and 47% for $\mu = 2600, 1400,$ and $2700\ cm^2\ V^{-1}\ sec^{-1}$, respectively).

By using H_2S, the epitaxial layers are less compensated. Silicon in III–V compounds is amphoteric, the incorporation of Si in InP depends on the ratio of III–V elements. But the diffusion coefficient of *Si* is less than that for *S*, so for modulation doping it is better to use Si.

In order to examine these topics and to identify a satisfactory *n*-type dopant for Gunn-diode InP and modulation-doped GaInAs–InP layers grown by LP-MOCVD, we examined the electrical properties of epitaxial layers doped with SiH_4. Figure 37 shows electrochemical profiles of typical doping interfaces for InP doped with S and Si, indicating that the interfaces of sulfur-doped layers do not differ significantly from those doped with Si.

For the same carrier concentration, the measured Hall mobility is lower in Si-doped InP layers than in the S-doped ones (Fig. 38). The results show that the Si-doped InP layers are compensated. Autocompensation was not present in the S-doped samples. Also, with S, it is generally simpler to obtain highly doped InP layers. There is a satisfying agreement between our results and those obtained by Giles *et al.*, (1984) using chloride-process VPE growth of InP layers.

VIII. Microwave Applications

InP is an excellent candidate for microwave devices. Microwave devices can be made with operating frequencies covering the range from about 0.1 GHz to 1000 GHz with corresponding wavelengths from 300 cm to 0.3 mm.

The applications in this frequency range are numerous types of radar for both military and civil applications. For example, communication via satellite is made at 12 GHz for broadcasting and at 20/30 GHz for communication. But communication flow is increasing so rapidly that new channels are necessary, and the use of frequencies up to 100 GHz is anticipated in the near future.

Until recently, all of these applications were in the professional field, which implied that only a limited number of devices were needed (typically a world market of $\sim 10^6$ devices per year), but in the near future there could be an opening for a large consumer market with satellite television broadcasting systems working at 12 GHz. Each individual receiver will utilize several solid state microwave devices, which can be advantageously made from InP and related compounds.

18. GUNN DIODES

The Gunn diode discovered by J. Gunn and C. Hilsum is a transferred electron device. They found that coherent microwave output was generated when a d.c. electric field in excess of a critical threshold value of several thousand volts per centimeter was applied across an *n*-type sample of GaAs or InP. Hilsum (1962) and Ridley (1961) proposed that all the observed properties of the microwave oscillation were consistant with a theory of negative differential resistance, which is due to a field-induced transfer of conduction band electrons from a low-energy, high-mobility valley to higher energy low-mobility satellite valleys.

To understand how this effect leads to negative differential resistivity, consider the energy-momentum diagrams for GaAs and InP in Fig. 39(a), the two most important semiconductors for transferred-electron devices. Figure 39(a) shows that the band structures of GaAs and InP are similar. The energy separation between the two valleys is ΔE, which is about 0.31 eV for GaAs and 0.53 eV for InP.

As a consequence, the electrons are slowed down and accumulate in domains that are propagated along the sample. The propagation time, which is proportional to the sample length, determines the frequency.

Figure 40 shows the measured room-temperature velocity field characteristics for GaAs and InP. The threshold field ε_T defining the onset of negative differential resistivity is 3.2 KV/cm for GaAs and 10.5 KV/cm for InP. The peak velocity V_p is about 2.2×10^7 cm/S for high-purity GaAs and 2.5×10^7 cm/S for high-purity InP. In order to have the electron transfer mechanism, the following conditions are necessary:

1) In the absence of a bias electric field, most electrons are in the lower conduction-band minimum, or $kT < \Delta E$.

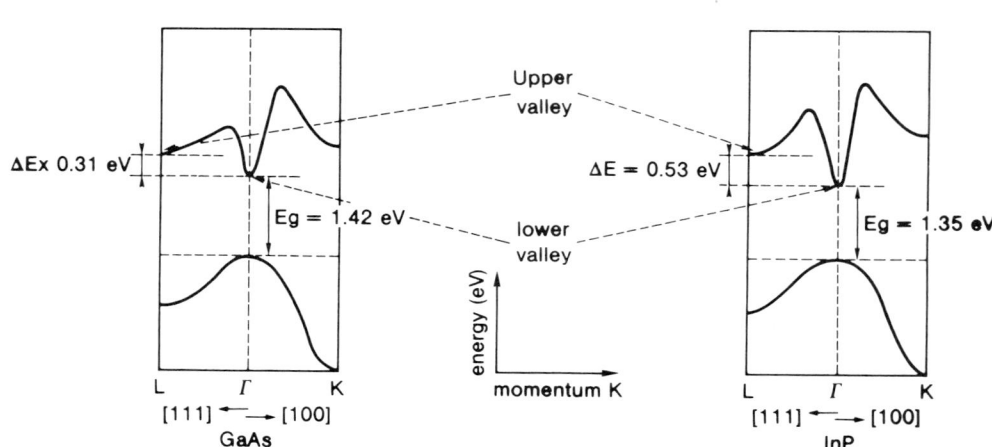

FIG. 39. Band structures of InP and GaAs.

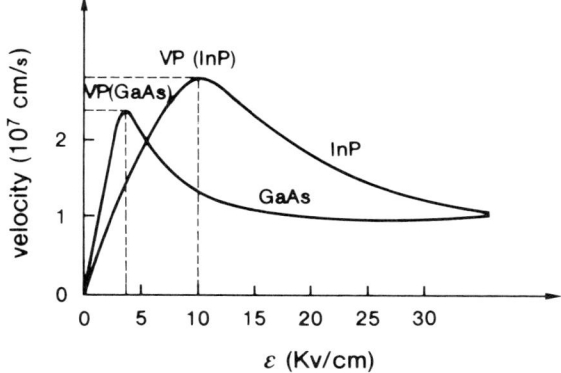

FIG. 40. Room temperature velocity field characteristics for GaAs and InP.

2) In the lower conduction-band minimum, the electrons must have high mobility, small effective mass, and low density of state. In the upper satellite valleys, the electrons must have low mobility, large effective mass, and a high density of states.

3) The energy separation between the two valleys must be smaller than the semiconductor bandgap so that avalanche breakdown does not set in before electrons are transferred into the upper valleys.

The n-type InP is the semiconductor satisfying these conditions and is widely studied and used.

The transferred electron effect has been referred to as the Hilsum-Ridley effect or as the Gunn effect.

The important characteristics of a Gunn diode are the power efficiency (3–10%), which depends on the material perfection, and the maximum dissipated power, which depends on the heat removal (mainly a technological problem). The application of Gunn diodes in the field of small Doppler radars and solid state sweepers tends to be supplanted by other devices, mainly the FETs.

High power and high efficiency InP Gunn diodes, which were made from layers grown by LP-MOCVD, have been developed in the millimeter-wave range. The Gunn diodes, processed using the integral heat sink technique, have delivered up to 100 mW CW output power with 2.5% efficiency at 94 GHz, while average power levels in excess of 90 mW were obtained at 94 GHz (Poisson, 1985).

Gunn diodes, which are well known as low noise devices, are better adapted for communication receivers than avalanche diodes. Until recently, gallium arsenide had been the only material for such applications, but since its introduction a few years ago, indium phosphide has been recognized as a very attractive material for making millimeter wave oscillators (Corlett *et al.*, 1975; Cromley, 1980; Hamilton, 1976).

The indium-phosphide Gunn device has rivaled the GaAs Gunn device in performance and system application. With the exception of mobility, which is higher in GaAs, the other characteristics (cutoff frequency, acceleration-deceleration time for electrons to gain or lose energy in the central valley T_{ad}, intervalley relaxation time T_e, peak/valley ratio V_P/V_V) favor InP over GaAs in terms of superior Gunn device performance.

The cutoff frequency is approximately 200 GHz at 3 db for InP compared with 100 GHz for GaAs. The key characteristic, which is necessary to obtain high efficiency, is a high peak-to-valley ratio V_P/V_V of the electron velocity as a function of applied electric field. The peak-to-valley ratio is significantly higher in InP than in GaAs (4 for InP; 2.4 for GaAs).

Other InP properties compared to those in GaAs that aid in obtaining higher power output are higher threshold field, higher electric breakdown field, and higher thermal conductivity. For these reasons, indium phosphide offers significant advantages over gallium arsenide for high power and efficiency at millimeter wavelengths.

Until recently, the output powers and efficiency obtained using InP represented only modest improvement over GaAs. But recent progress in 94 GHz InP Gunn device development at Thomson's laboratory showed that higher output power and efficiency can be obtained from InP in the millimeter range.

The ability to grow high-purity indium phosphide material, as required for Gunn effect devices, has led to some significant improvements in the performance of these devices.

Thomson developed a microwave diode structure whose performance is claimed to outclass any other structure reliable enough to be used in professional equipment. In the CW mode, output power up to 103 mW was measured at 94 GHz with 2.5% efficiency. This device is based on a three-layer $N+/N/N+$ structure, where Si (from silane) was used as the dopant (Razeghi, 1985). This dopant allows a precise control of doping profiles. High quality nonintentionally doped and Si-doped epitaxial InP material was reproducibly obtained by low-pressure metalorganic chemical vapor deposition (Poisson *et al.*, 1985, and Razeghi *et al.*, 1984).

Nonintentionally and Si-doped InP layers were evaluated by capacitance/voltage (C/V), Van der Pauw, and photoluminescence measurements. Nonintentionally doped layers exhibited net residual donors as low as 8.10^{13} cm^{-3} with Hall mobility as high as 5,300 cm^2 V^{-1} s^{-1} at room temperature, 185,000 cm^2 V^{-1} s^{-1} at liquid nitrogen temperature, and 200,000 cm^2 V^{-1} s^{-1} at 55 °K. This result was found to be in very good agreement with the observed PL spectra at 1.7 °K of these layers. The ability to grow InP by LP-

MOCVD so pure that its mobility surpasses some estimates of the theoretical limit for the material is one of the keys to its success in the millimeter range.

The device structure consists of a $N+$ buffer layer with carrier densities of 10^{18} cm^{-3} followed by an active layer doped in the 1.0×10^{16} to 1.5×10^{16} cm^{-3} range, and a contact layer in the 10^{18} cm^{-3} range.

Excellent results have been obtained from wafers ranging in active layer thickness from 0.9 μm to 1 μm. A typical 94 GHz InP Gunn device doping profile is shown in Fig. 20.

The grown wafers are 2.5 cm $\times$ 4 cm and are processed using the integral heatsink (IHS) technique. This technique consists of various epitaxial layers, alloyed AuGe, Ti, Pt, and Au contacts, and an electroplated 40 μm gold heatsink to reduce thermal resistance and facilitate chip bonding and packaging. On the episide of the as-grown wafers, channels are etched 10–20 μm deep and spaced 3 mm apart to provide a depth gauge when the substrate side is thinned. The entire episide is then metallized with AuGe, Ti, Pt, and Au. The Au heat sink is electroplated 40 μm thick. The wafer is then turned over and uniformly thinned to 10 μm with a lap and chemical polish technique until the channels on the other side are exposed.

A total substrate thickness less than 10 μm is required at 94 GHz in order to minimize the significant resistive loss created by the substrate.

AuGe, Ti, Pt, and Au contacts are formed on this new thinned substrate surface. Mesas of 30–60 μm diameter are defined using a light-sensitive FeCl$_3$ etch (Fig. 41).

The thickness for a 94 GHz InP Gunn diode is about 10 μm. Diode chips are mounted in W5 packages. Different bonding ribbon geometries were used to minimize parasitic self connection.

To obtain a reduced bonding-wire inductance with no capacitance increase, new V5 packages, in which the ceramic ring was thinner and lower than before, were used in conjunction with several bonding-wire geometries.

The integral heat-sink (cathode) side of the diode was bonded to the bottom of the package using a Au-Sn preform (80%–20% by weight) at 300 °C in a nitrogen-hydrogen atmosphere to prevent oxidation. The top (anode) bond was then thermocompressed at 250 °C.

A low-inductance bond was obtained by using two 100 μm gold ribbons stitch bonded from the metallized ceramic across the mesa. The best result, however, was obtained using a star-pattern top bond. In both cases, the reduced parasitic inductance allowed the diode to operate at a higher frequency with no power loss.

The packaged devices were tested in a cavity using a tuning "cap." The cavity is tuned by adjusting the cap diameter. The impedance match of the device is given by the diode position in the cavity to get the optimal power.

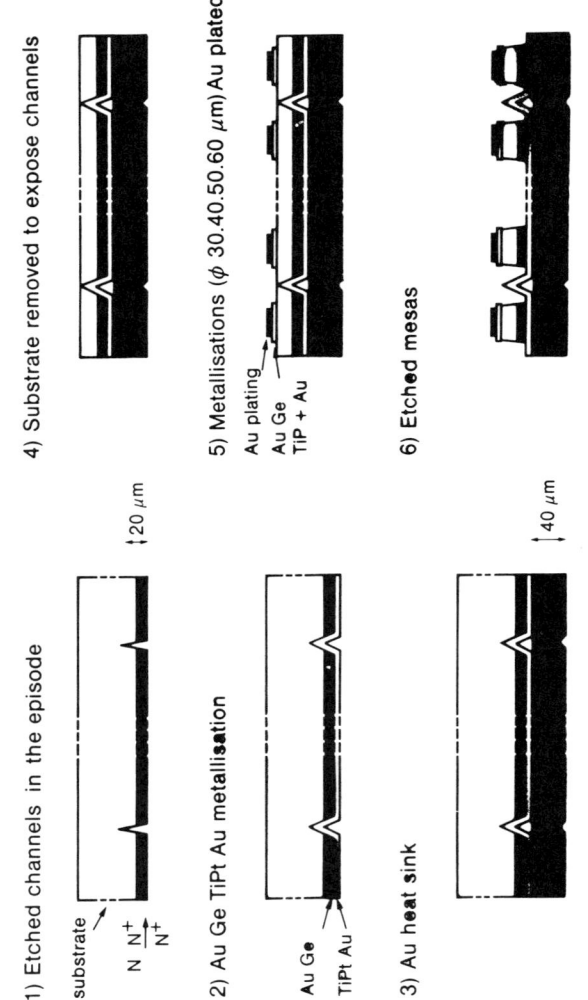

FIG. 41. The various sequences of Gunn-diode technology.

This type of circuit, which was conceived and built at Thomson, facilitates the testing of Gunn diodes at frequencies in the 94 GHz range. The auxiliary test equipment consists of a variable attenuator, an absorption-type frequency meter, and a power meter with a thermistor sensor. The power losses due to the measurement apparatus have been corrected for use in the previously mentioned power levels.

The major RF characteristics of importance for oscillators are efficiency, microwave output power, noise level, and stability with temperature. A cap cavity was used to evaluate InP Gunn diodes in CW mode operation, and the bias voltage is provided by the cap. A sliding short circuit positioned behind the resonator provides impedance matching and optimum coupling, which results in maximum power output. The frequency of oscillation is principally determined by the cap diameter, and the position of the sliding short circuit allows mechanical tuning of the Gunn diode.

In the 94 GHz region, the most powerful device (60 μm diameter mesa) produced a CW output power of 103 mW with 2.5% efficiency. Good I/V characteristics were measured on the wafer in pulse mode, and no variations in these I/V characteristics were observed on the diode after all of the operations of packaging.

Thermal resistance for these 94 GHz InP Gunn devices with 60 μm diameter mesa ranges between 40–45 °C/W. The working temperature of the active layer is less than 200 °C. Both the CW power level and the efficiency of these devices were studied over a temperature range of $-40°$ to $+60°$C. A power shift of 0.005 db/°C and a frequency shift of 3.5 MHz/°C have been observed.

AM noise (-140 dbc/Hz SSB at 10 KHz from carrier) and FM noise (-60 dbc/Hz SSB at 10 KHz from carrier) at 94 GHz were found to be similar to those obtained with gallium arsenide TEOs (Poisson *et al.*, 1985).

IX. LP-MOCVD Growth of InP on Alternative Substrates

We have shown that the LP-MOCVD is well adapted for the growth of a variety of III–V semiconductor binary, ternary, and quaternary heterojunctions, multiquantum wells (MQW), and superlattices on lattice matched substrates for optoelectronic or microwave device applications.

In heterostructures, it is certainly desirable to select a pair of materials closely lattice matched in order to minimize defect formation or stress. However, heterostructures lattice mismatched to a limited extent can be grown with essentially no misfit dislocations, if the layers are sufficiently thin, because the mismatch is accommodated by a uniform lattice strain. Without the requirement of lattice matching, the number of available pairs for device applications and integrated circuits can be greatly augmented.

High-quality InP and related compound-strained heterostructures have been grown on alternative substrates by the low pressure metalorganic chemical vapor deposition growth technique. Photoluminescence, SIMS, and Auger measurements showed the high quality optical and electrical properties of these layers. Table XVIII shows the different strained heterostructure grown by LP-MOCVD.

19. Growth Procedure

The growth apparatus has been described in Part IV, with growth carried out at 76 torr. The optimum conditions for the low-pressure growth of these layers, as determined during these investigations, are presented in Table XIX.

Smooth, single-crystal films exhibiting mirrorlike surfaces have been obtained in the presence of a large layer-substrate lattice parameter mismatch. These layers tend to be heavily dislocated, however, and their electrical and optical characteristics, especially those related to minority carrier properties such as diffusion length and lifetime, are generally inferior to the typical lattice-match system. These effects tend to be especially severe for thin layers, but we found that they can be partly eliminated by the use of thick buffer layers or special grading or superlattice techniques.

Featureless mirrorlike surfaces have been grown over a wide temperature range of 500–650 °C. The x-ray diffraction rocking curve about the (400) $K\alpha$ reflection from InP epilayer on GaAs on InP substrate is shown in Fig. 42, and the x-ray diffraction rocking curve of heterostructure of InAs–GaAs on InP substrate is shown in Fig. 43. We have performed a simultaneous study of LP-MOCVD growth of InP on InP, GaAs, and InAs substrates with

TABLE XVIII

Strained Heterostructure Grown by LP-MOCVD.

Substrate	First epilayer	Second epilayer
InP	GaAs	
InP	InAs	
GaAs	InP	
GaAs	InAs	
InAs	InP	
InAs	GaAs	
InP	GaAs	InAs
InP	$Ga_{0.47}In_{0.53}As$	$Ga_{0.49}In_{0.51}P$
GaAs	$Ga_xIn_{1-x}As$	
GaAs	InP	$Ga_xIn_{1-x}As_yP_{1-y}$

TABLE XIX
OPTIMIZED GROWTH PARAMETERS.

(1.3 μm)		InP	GaAs	InAs	GaP	GaInAs	GaInP	GaInAsP
Growth temperature	°C	550	550	550	550	550	550	630
Total flow rate ($N_2 + H_2$)	l/min	6	6	6	6	6	6	74
N_2/TEI bubbler flow	cm³/min	200	—	200	—	200	200	350
H_2/TEG bubbler flow	cm³/min	—	120	—	120	120	120	120
PH_3 flow	cm³/min	300	—	—	300	—	300	530
AsH_3 flow	cm³/min	—	90	90	—	90	—	21
Growth rate	Å/min	100	100	100	100	200	200	150

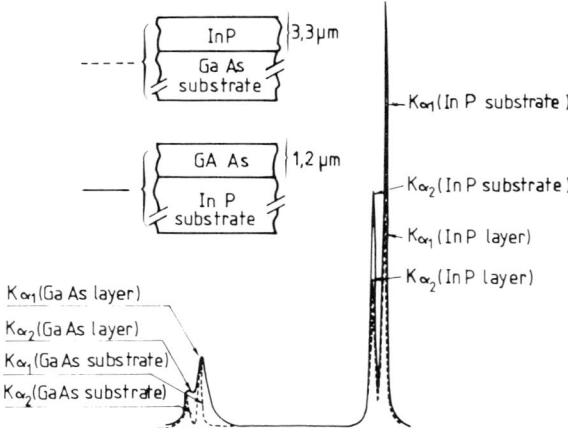

FIG. 42. X-ray diffraction rocking curve of (400) $CuK\alpha$ reflection from InP epilayer on GaAs substrate and GaAs epilayer on InP substrate.

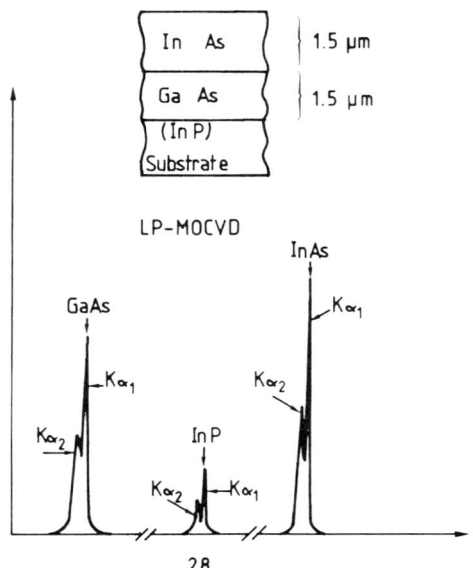

FIG. 43. X-ray diffraction rocking curve of (400) $CuK\alpha$ reflection from InAs–GaAs–InP heterostructure grown by LP-MOCVD.

orientations of (100) placed adjacent to one another within the reactor for a growth temperature of 550 °C.

Figure 44 shows the photoluminescence spectra of these layers at 5 K, using a Helium gas-flow variable-temperature cryostat. Luminescence was excited using a He–Ne laser and was analyzed in a 60 cm grating spectrometer and detected with a high-sensitivity N-cooled Ge photodiode. A series of luminescence transitions, which can be attributed to recombination mechanisms such as free excitons, exciton bound to shallow impurities (such as Zn), and donor-acceptor recombination were observed on these spectra. The exciton recombination energy of InP on InP, InP on GaAs, and InP on InAs substrates are 1.419 eV, 1.423 eV, and 1.421 eV, respectively. Considering that the lattice parameters of InP, GaAs, and InAs are 5.869 Å, 5.633 Å, and 6.057 Å, respectively, one expects the InP layer on the GaAs substrate to be compressed, and the InP layer on the InAs substrate to be expanded. Usually these layers are pseudomorphic (i.e., the elastic straining of the deposited lattice produces a zero misfit with the substrate). The accommodation of the mismatch by elastic strains induces changes in the magnitude of the gap. Thus, in the case of InP epilayer on InAs substrate, one expects lower energy for exciton recombination than InP epilayer on InP substrate.

a. Secondary Ion Mass Spectrometry (SIMS)

SIMS analysis has been described in Section 14. Figure 45 shows depth profiles of Mg, Si, Cr, Fe, Si, As, and P in InP layers grown on InP, InAs, and

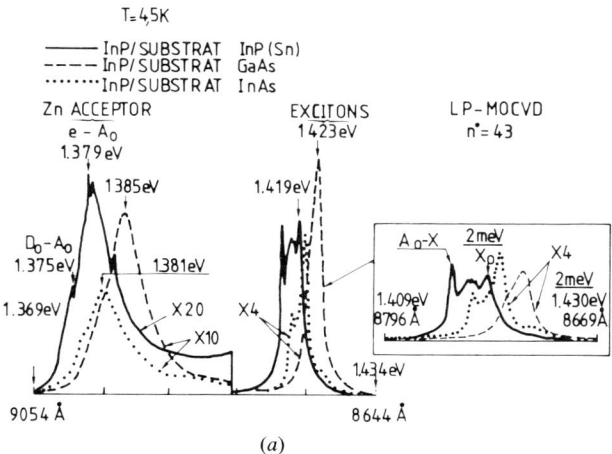

FIG. 44. (a) Photoluminescence spectrum of epilayer on InP, GaAs and InAs substrates at 5 °K using an He–Ne laser. (b) Transition region after chemical etching of InP/GaAs and (c) InP/InAs epilayers grown by LP-MOCVD.

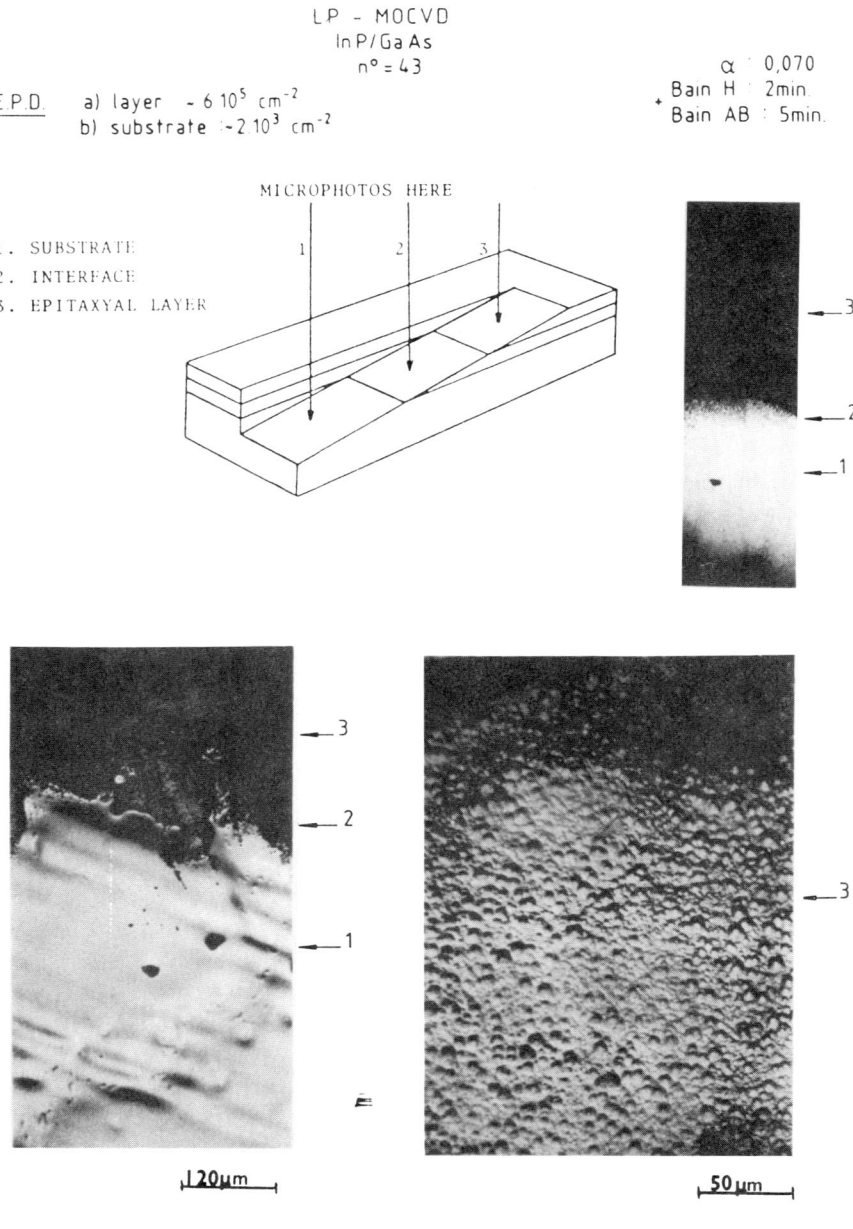

FIG. 44. (*Continued*)

6. LP-MOCVD GROWTH

E.P.D. a) layer : ~ 5 × 10^5 cm^{-2}
b) substrate : ~ 10^2 cm^{-2}

LP – MOCVD
InP/InAs
n° = 43

α : 0,0650
Bain H : 2min
Bain AB : 5min

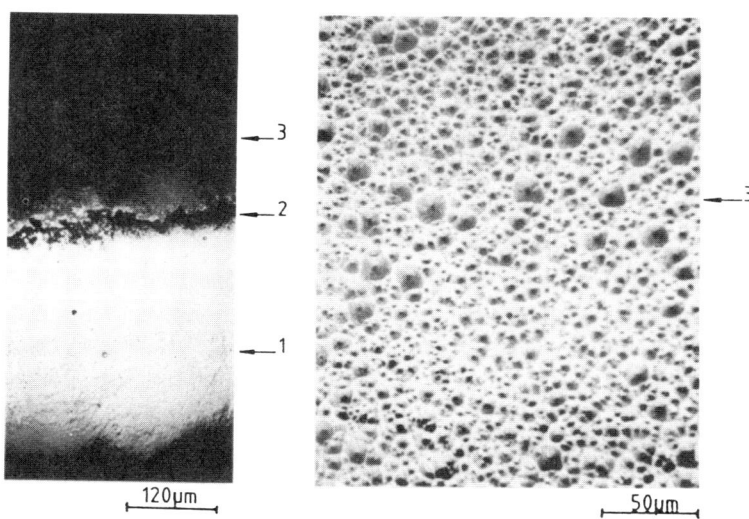

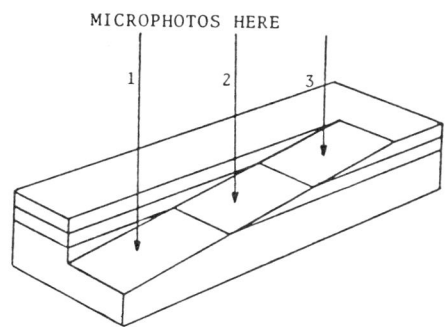

1. SUBSTRATE
2. INTERFACE

(c)

FIG. 44. (*Continued*)

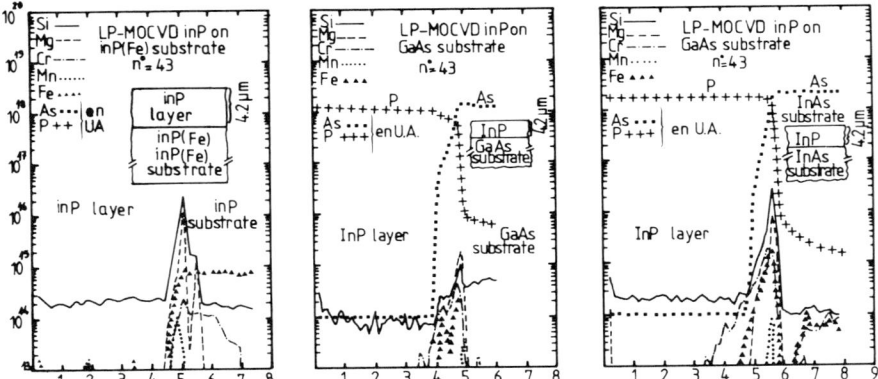

FIG. 45. Depth profile of Mg, Si, Cr, Fe, Mn, P, and As in a LP-MOCVD growth of: (a) InP on InP substrate; (b) InP on GaAs substrate, and (c) InP on InAs substrate.

GaAs substrates by LP-MOCVD. Each sample was analyzed in two different areas about 10 mm apart. Generally, the analysis of two clean areas gives reproducible and representative results for the material. We have already shown that the major source of impurities at the interfaces is the adsorption of atoms on the substrate surface during chemical etching prior to epitaxy. The pretreatment of the InP and InAs substrates are similar, so their SIMS profiles are identical. But the chemical etch of the GaAs substrate prior to epitaxy is $SiH_4 + H_2O + H_2O_2$, and the concentration of impurities at the interface of InP–GaAs is lower than InP/InP or InP/InAs.

These results show that the quality of InP layers far from the interfaces is independent of substrate origin. Figure 46 shows an automatic electrochemical profile through an InP layer grown on a GaAs substrate with and without light (under illumination, for n-type semiconductors in order to generate holes required for the reaction). The result shows that near the interface there are some perturbations, but far from the interface of epilayer-substrate, the quality of the epilayer and the carrier concentration become similar to InP epilayer grown on InP substrate under the same conditions.

b. Auger Analysis

The constituent concentration gradients at an InP/GaAs interface were determined by Auger analysis. The sample was chemically etched using a methanolbromine solution (15% Br) in order to obtain a level with a mean amplification coefficient (M) of 2100 (measured with a Talysurf). This means that a change of one micron along the surface corresponds to a change of 4.5 Å in depth (z-direction). Figure 47 shows a schematic representation of the

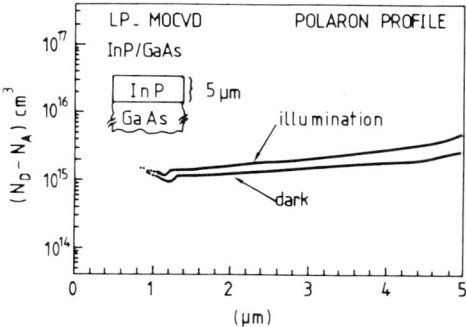

FIG. 46. Electrochemical polaron profile of an InP epilayer on GaAs substrate.

bevel. By scanning the incident electron beam four times along the bevel, the successive Auger profiles of the four elements P, As, In, and Ga have been obtained. All of the four profiles, shown on the Fig. 48, were obtained in 12 minutes, but additional profiles following the inverse sequence (Ga, In, As, P) were obtained subsequently in order to verify that there were no changes either in the intensity scale or in the position scale. All of the four profiles followed the same line (same starting point) along the bevel. It can be observed that the In and P profiles are rather smooth, but that the Ga and As profiles show rather large fluctuations. This was in spite of the use of a larger modulation bevel and a larger time for the acquisition of a scan, and is probably due to the poor Auger sensitivities of Ga and As relative to In and P. Due to the large magnification coefficient obtained by the chemical beveling, the spatial broadening of the profiles related to the incident spot size and back scattering effects can be neglected because the incident beam diameter corresponds to an error in the depth of $\Delta Z = d_o/M < 2$ Å. For the signal intensity for a given element, the change of the backscattering contribution when the incident electron beam is scanned along the interface can also be neglected due to the fact that the mean atomic number $\bar{Z}$ of the sample does not change when going from GaAs $(Z(Ga)31 + Z(As)33 = 64)$

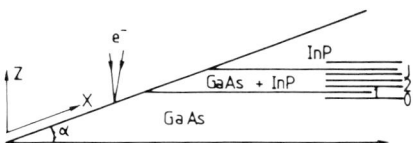

FIG. 47. Schematic representation of the bevel of the InP/GaAs structure showing how a surface analysis along the bevel is converted into depth analysis.

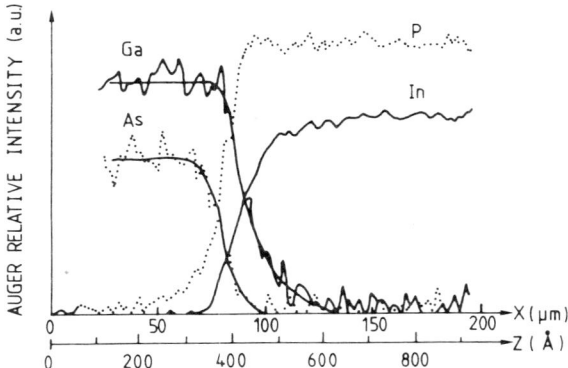

FIG. 48. Corresponding Auger profiles relative to the four components: P, In, As, and Ga.

to InP ($Z(In)49 + Z(P)15 = 64$). Under the previous simplifications, the relation between Auger intensity I^A and concentration C^A of an element A can be easily obtained by assuming that the interface region consists of n slices each of thickness "a," where the slices are numbered starting from the deepest one (at the end of the homogeneous GaAs concentration, here) (Gazaux et al., 1986), see Fig. 47.

The intensity due to the element A in the Jth slice corresponds to the intensity flowing from outside of the slice, I_J^A, minus the intensity coming from all of the other deeper slices, I_{J-1}^A, which is attenuated by the factor K by traveling through the Jth slice of thickness a. The interface composition (transition region) of InP-GaAs is $Ga_xIn_{1-x}As_yP_{1-y}$, where $0 \leqslant x, y \leqslant 1$.

After Auger analysis, the transition region constituting the interface can be subdivided into three parts (starting from the GaAs substrate).
1) A region (thickness = 150 Å), where In is absent, its chemical composition is $GaAs_yP_{1-y}$ with $0.87 < y < 1$, which corresponds to the heating of the GaAs substrate under PH_3 before growth, so that it is possible to have the adsorption of As and absorption of P. This can be remedied by heating the GaAs substrate under AsH_3 pressure, before introducing PH_3 into the reactor.
2) The mid region (thickness = 120 Å), where all of the four components are present. Its chemical composition is $Ga_xIn_{1-x}As_yP_{1-y}$ with $0 \leqslant y \leqslant 0.87$, $0.24 \leqslant x \leqslant 1$.
3) The region (thickness = 115 − 130 Å), where As is absent, and its chemical composition is PGa_xIn_{1-x} with $x < 0.24$.

Such analyses can be developed on any hetero-epitaxial structure if a good method of the chemical etching is found (Gazaux et al.)

c. *Etch-Pit Density (EPD)*

Figure 44 shows photomicrographs of these layers after forming a chemical bevel with a very low angle and selective etching. The EPD of epitaxial layers and substrates are indicated in Table XX. These results show that the EPD in InP/GaAs and InP/InAs interfaces is independent of the EPD of the InAs or GaAs substrates.

20. GROWTH OF GaInAs–InP MULTIQUANTUM WELLS ON GARNET ($GGG = Gd_3Ga_5O_{12}$)

The composition of the garnets is given by the formula $R_3A_2B_3O_{12}$, in which the trivalent R ions are rare earths such as lutetium, yttrium, gadolinium, or other large ions, and the A and B ions are smaller ions, such as Ga^{3+} or Al^{3+}. The R ions are surrounded by a decahedron (not the regular one, but a distorted cube of eight oxygen ions). The B ions are surrounded by a tetrahedron of four oxygen ions and the A ions by an octahedron of six oxygen ions. The basic structure of GGG is cubic, its lattice parameter is 12.383 Å, and GGG is an insulator, which can be used as a substrate.

$Y_3Fe_5O_{12}$ (YIG) is a magnetic garnet material having the same lattice parameter as GGG, and YIG can be epitaxially grown on GGG (see Fig. 49). YIG is a potentially useful material for application to integrated optical devices, nonreciprocal elements such as isolators, and circulators in planar integrated optics.

The $Ga_xIn_{1-x}As_yP_{1-y}$ – InP double heterostructure lasers emitting in the 1.0- to 1.67- µm wavelength region are promising candidates as light sources for optical fiber communication systems because of the low-loss, low-dispersion optical fibers available in this spectral region.

The growth of these III–V semiconductor compounds on GGG substrate can be used for
1) optical telecommunication such as integrated magneto-optical isolators and the integration of a laser diode with an isolator (YIG);

TABLE XX

EPD IN EPITAXIAL LAYERS OF InP ON InP, InP ON GaAs, AND InP ON InAs SUBSTRATES.

Epilayer/substrate	EPD (epilayer) cm^{-2}	EPD (substrate) cm^{-2}
InP/InP	8×10^3	8×10^3
InP/GaAs	6×10^5	2×10^3
InP/InAs	5×10^5	1×10^2

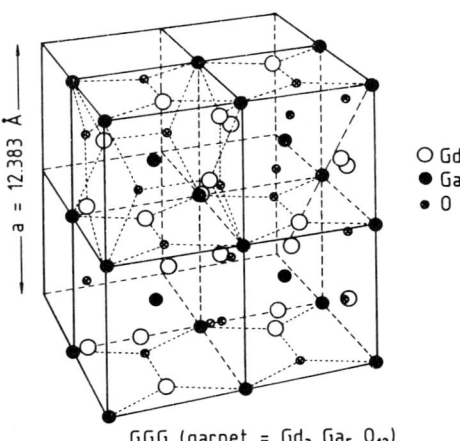

FIG. 49. Crystal structure of GGG.

2) integrated magneto-optical circulators with possibilities for integrated emitters, detectors, isolators, circulators, nonreciprocal phase shifters, and modulators;
3) magnetic recording with integrated magnetic read heads; and
4) magnetometry.

The growth processes have been presented in Table XVIII. And the GGG ($Gd_3Ga_5O_{12}$) substrates were supplied by Crismatec.

The wafers had an orientation of (100), (110), or (111) and were free from defects, precipitates, or second-phase materials. The inclusions were less than $10/cm^3$. The surface defect density was determined by etching a polished substrate in $1:1::H_2SO_4/H_3PO_4$ mixture at 200 °C for two minutes. The density of crystalline defects was < 1 defect/cm^2.

A GGG (100) orientation and an InP substrate were placed on the susceptor, which was then placed inside the reactor. The substrates were heated to 550 °C, and the growth pressure was 76 Torr. (The growth conditions for InP and GaInAs are indicated in Table XVIII.)

The epitaxial structure is as follows. One 5 μm InP buffer layer, five quantum wells of GaInAs (5200 Å thick) with InP barrier (200 Å thick), and one 5-μm InP top layer.

The growth process for the GaInAs-InP multiquantum wells is the same as indicated in Part III.

X-ray diffraction was used to examine the quality of the deposited layers. Figure 50 presents the x-ray diffraction pattern of GaInAs-InP multilayers grown on GGG substrate, which are indicated in Table XXI.

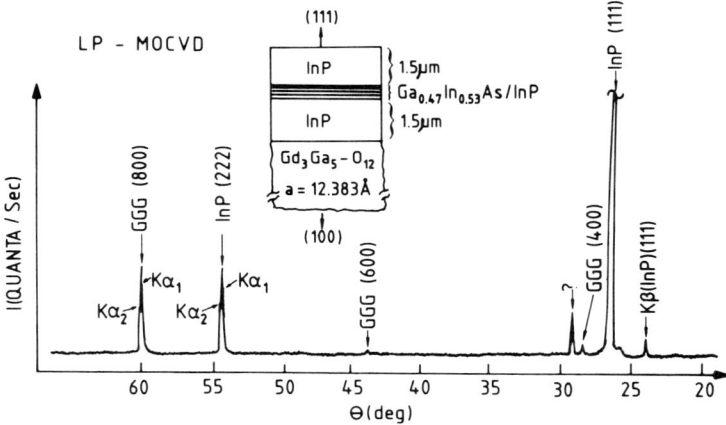

FIG. 50. X-ray diffraction pattern of the 3 μm-thick InP/GaInAs/InP epitaxial layer grown by LP-MOCVD on a (100) GGG substrate. The inset shows a schematic view of the sample configuration.

This data shows that the orientation of the epitaxial layer is (111), while the orientation of the substrate is (100). The lattice mismatch between the substrate and epilayer is $\Delta a/a = 0.526$. However, there are many important cases where the epitaxial layer has either a different orientation than the substrate example CdTe (111) on GaAs (100) ($a_{GaAs} = 5.653$ Å and $a_{CdTe} = 6.481$ Å (Asuka et al., 1985), or it has a totally different crystal structure (for example: silicon on sapphire) (Cullen, 1985).

TABLE XXI

COMPARISON BETWEEN THE EXPERIMENTAL AND CALCULATED VALUE OF 2θ (X-RAY DIFFRACTION DATA) DUE TO THE GaInAs–InP MULTILAYERS GROWN ON GGG SUBSTRATE.

Layers	hkl	2θ(exp)	2θ(cal)
InP	111	26.60	26.30
InP	222	54.30	54.10
InP	311	51.85	51.63
GGG	800	60.08	59.70
GGG	600	43.85	43.83
GGG	400	28.60	28.82

In such cases, the criterion of lattice match, namely, comparing the lattice parameters, is no longer applicable, and a new criterion should be used. Zur and McGill (1984) defined the concept of lattice match for any pair of crystal lattices in any given crystal direction, allowing for a periodic reconstruction of the interface.

Instead of comparing the bulk lattice parameters, they compared the interface translational symmetry with that of the bulk materials on both sides of the interface.

A cut in any crystal direction through a three-dimensional lattice results in a surface that has a two-dimensional (2-D) translational symmetry. Any additional reconstruction of the atoms near the interface during the interface formation may, in general, reduce the symmetry of the reconstructed surface. Thus, the symmetry group of the reconstructed surface will be a subgroup of the symmetry group of the unreconstructed surface. Therefore, the problem of lattice match is reduced to scanning the 2-D cuts in a given pair of lattices, and then comparing the two resulting 2-D lattices and looking for a common superlattice. The theory of geometrical lattice matching is applicable to any pair of crystal structures and for any direction.

In general, it is not possible to achieve a perfect match. We think, however, using a thick epitaxial growth layer while growing superlattices makes the growth of monocrystalline III–V semiconducting compounds on alternative substrates possible.

Figure 51 shows the photoluminescence (PL) spectrum of this sample at 77 K. It is evident that the PL peak in Fig. 51 can be deduced from a comparison of the PL spectra of GaInAs–InP multiquantum wells grown on InP and GGG substrates in the same run. The difference of the PL wavelength may be due to misfit dislocations at the interface between the epilayer and the GGG substrate.

We have grown thick layers (7 μm) of InAs, InP, GaAs, and GaInAs on GGG substrate, and in these cases, all of the epilayers were polycrystalline. After growing the MQW structure, we succeeded in growing single crystals of III–V compounds on GGG substrates. These results indicated that a wide range of potentially useful new phenomena based on multilayer epitaxy on alternative substrates is thus becoming available and, with imaginative implementation, may be expected to lead to further novel device structures.

We present for the first time the InP/GaInAs/InP layers grown by LP-MOCVD on the GGG-Garnet ($Fe_3Ga_5O_{12}$) substrate. The lattice parameter of GGG is 12.383 Å, and the crystal symmetry is cubic. The lattice mismatch between epitaxial layers and the substrate is 52%. The x-ray diffraction pattern shows that the orientation of the epitaxial layer is (111), and the substrate orientation is (100). The PL peak is due to the GaInAs layers. The main interest of these results is to show that by using MQW epilayers, it is possible to grow III–V compounds on alternative substrates.

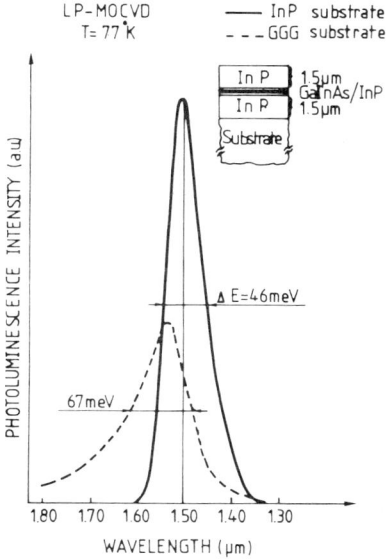

FIG. 51. Photoluminescence spectra of InP/GaInAs/InP epitaxial layer on a GGG (100) substrate (dotted line), InP (100) substrate (solid line). The inset to the PL spectra shows a schematic view of the sample configuration.

21. GaInAsP/InP Heterostructures Grown by LP-MOCVD on Si Substrates

High-quality bulk InP and double heterostructure InP/GaInAsP/InP have been grown on silicon substrates by low pressure chemical vapor deposition. X-ray diffraction patterns, as well as structure characterizations, indicate that the layers have very good crystalline quality, and an intense photoluminescence signal from quaternay alloy $Ga_xIn_{1-x}As_yP_{1-y}$ has been observed at room temperature, at the expected value of 1.3 microns.

Recently, there has been a great deal of interest in the growth of GaAs on silicon substrate (Christou et al., 1985; Nishi et al., 1985; Sheldon et al., 1984; Soga et al., 1985; Akijama et al., 1984; Wang, 1985). Devices such as GaAs FET transistors (Aksum et al., 1986) or GaAs lasers (Van der Ziel, 1987) have been realized on silicon substrates. On the other hand, by virtue to their large application for long wavelength components, especially 1.3 μm and 1.5 μm lasers, quaternary compounds $Ga_xIn_{1-x}As_yP_{1-y}$ lattice matched to InP have become very important materials for the optoelectronic industry (Razeghi et al., 1984; 1985), and it would be of great interest to combine these structures with silicon technology. But, due to the large lattice mismatch (8%) between InP and Si, only a few attempts have been made (Lee et al., 1987) to grow InP on silicon substrates. Now, we report the successful growth of bulk InP, as well as a double heterostructure $InP/Ga_xIn_{1-x}As_yP_{1-y}$ (1.3 μm)/InP on silicon substrate, by the low pressure metalorganic chemical vapor deposition (LP-MOCVD) technique.

a. Growth and Characterizations

For the LP-MOCVD technique, the growth temperature is 550 °C, and the growth pressure is 100 mbars (the growth details are the same as previously described).

Figures 52a and 52b exhibit the simple diffraction pattern of sample no. 166, using $CuK\alpha_1$ and $K\alpha_2$ of wavelengths 1.54 Å and 1.544 Å, as x-ray

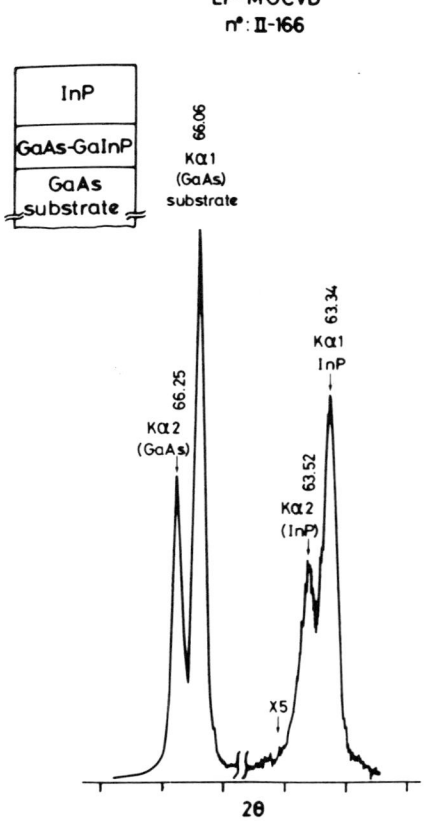

FIG. 52(a). Simple x-ray diffraction pattern of the $InP/GaAs-Ga_{0.49}In_{0.51}P$ structure grown on GaAs substrate.

sources. The structure has been grown on GaAs and Si substrates, using a GaAs/Ga$_{0.49}$In$_{0.51}$P superlattice as a buffer layer in both cases. The 2 μm-thick InP layer is deposited on top.

The silicon and GaAs substrates were (100) oriented. The InP epilayer keeps the same orientation, and no trace of disorientation has been found. In both cases, $K\alpha_1$ and $K\alpha_2$ peaks, relative to InP signal, appear well separated, which proves the good crystalline quality of the layer under study.

Structural properties have also been investigated by a method of ball polishing. Figure 53 compares the results obtained on the alternate substrates. It is worth noting the abruptness of the interfaces in the case of the growth on the silicon substrate, further proving that the epitaxy is high-quality InP. The heating was done under AsH$_3$ flow, thus inducing a strong

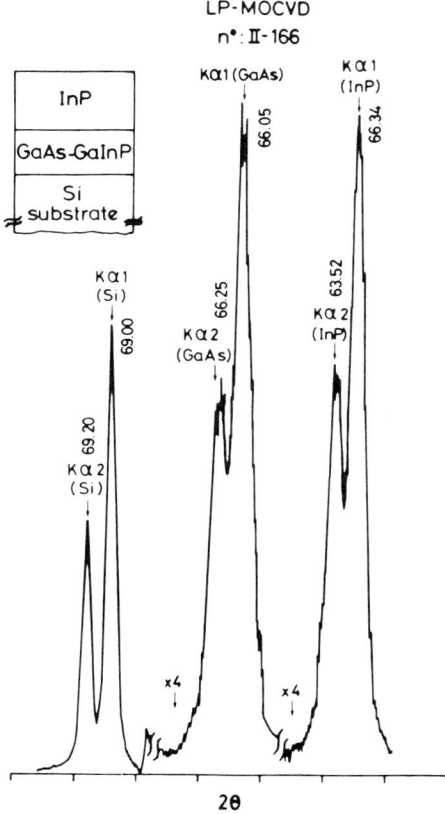

FIG. 52(b). Simple x-ray diffraction pattern of the InP/GaAs-Ga$_{0.49}$In$_{0.51}$P structure grown on Si substrate.

LP-MOCVD
n° : II-166

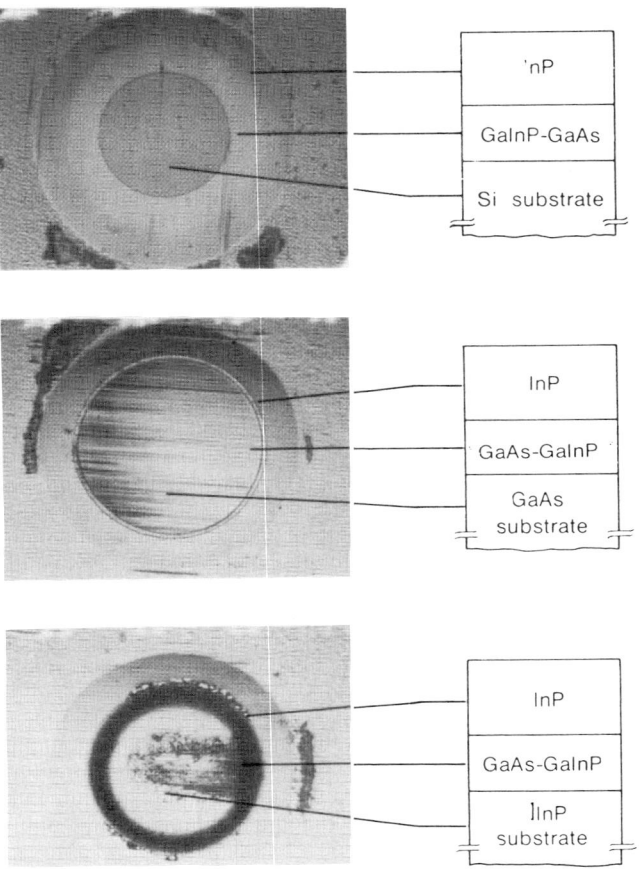

FIG. 53. Ball polishing photograph of the InP/GaAs–$Ga_{0.49}In_{0.51}P$ structure grown on the Si, GaAs, and InP substrates.

desorption of volatile element phosphorus on InP substrate, which is why the interface appears more perturbed in that case.

Figure 54 compares the three SIMS profiles. It was performed using Cs^+ as etching ions. The signals relative to the majority species Ga, In, As, P, and Si in the silicon substrate have been analyzed. The final InP layer, approximately 2 μm thick, is clearly shown. It must be added that SIMS is more sensitive

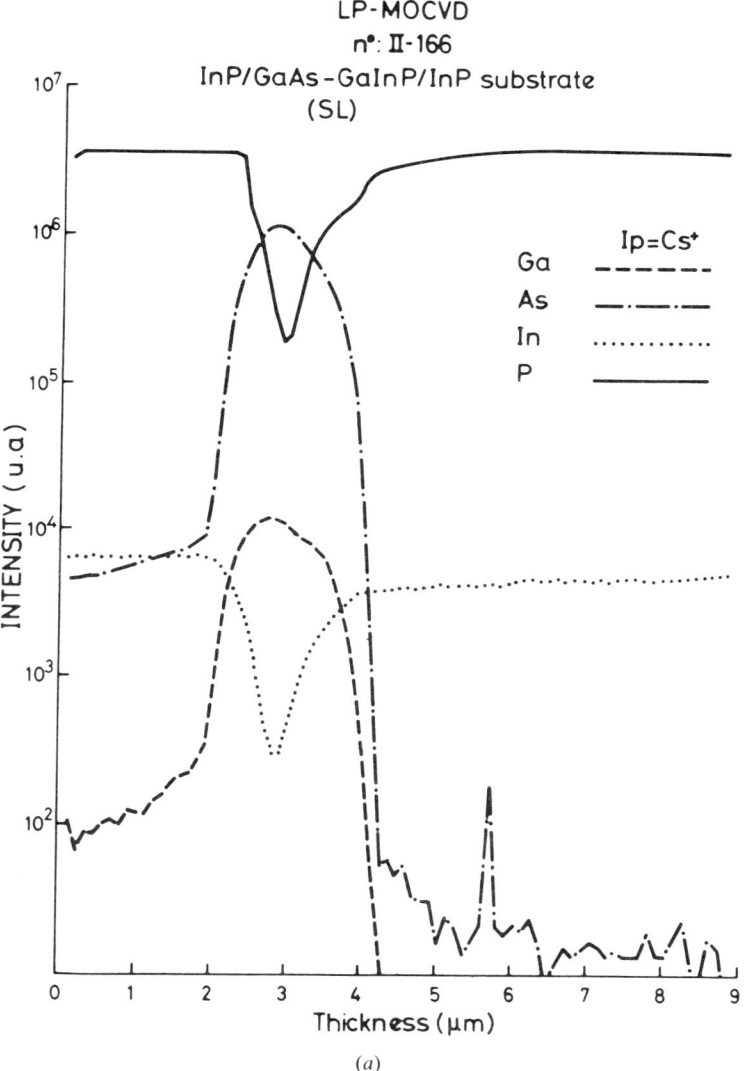

FIG. 54. SIMS profile of the $InP/GaAs\text{-}Ga_{0.49}In_{0.51}P$ structure grown on the (a) InP, (b) GaAs, and (c) Si substrates.

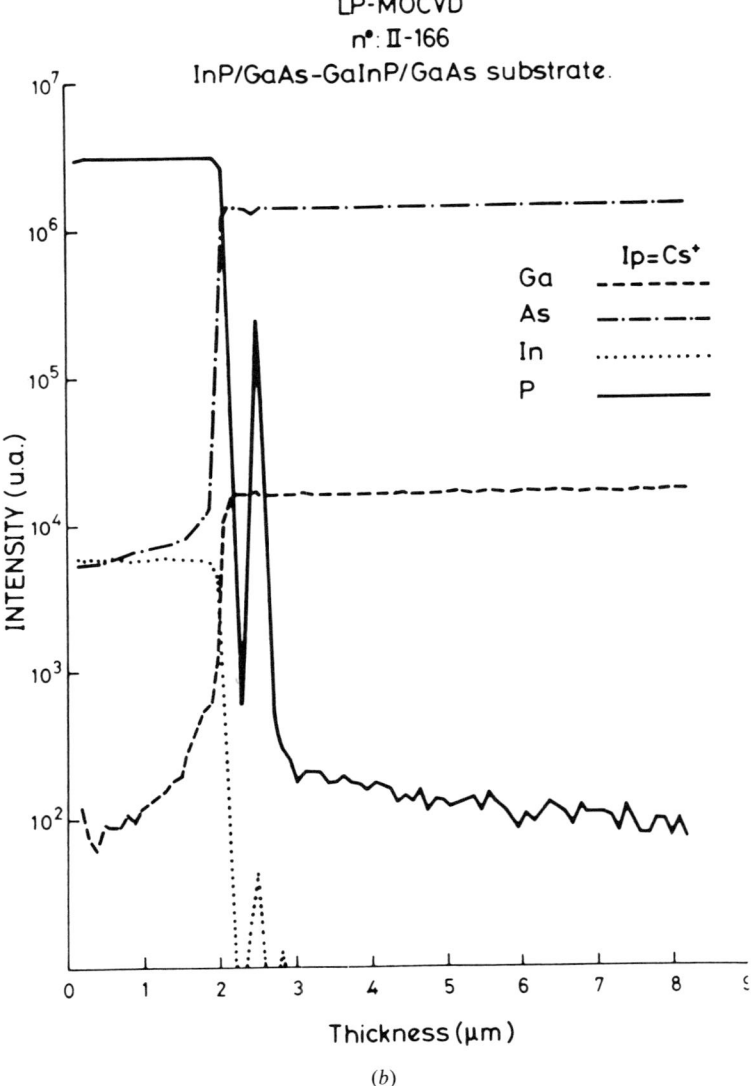

(b)

FIG. 54. (*Continued*)

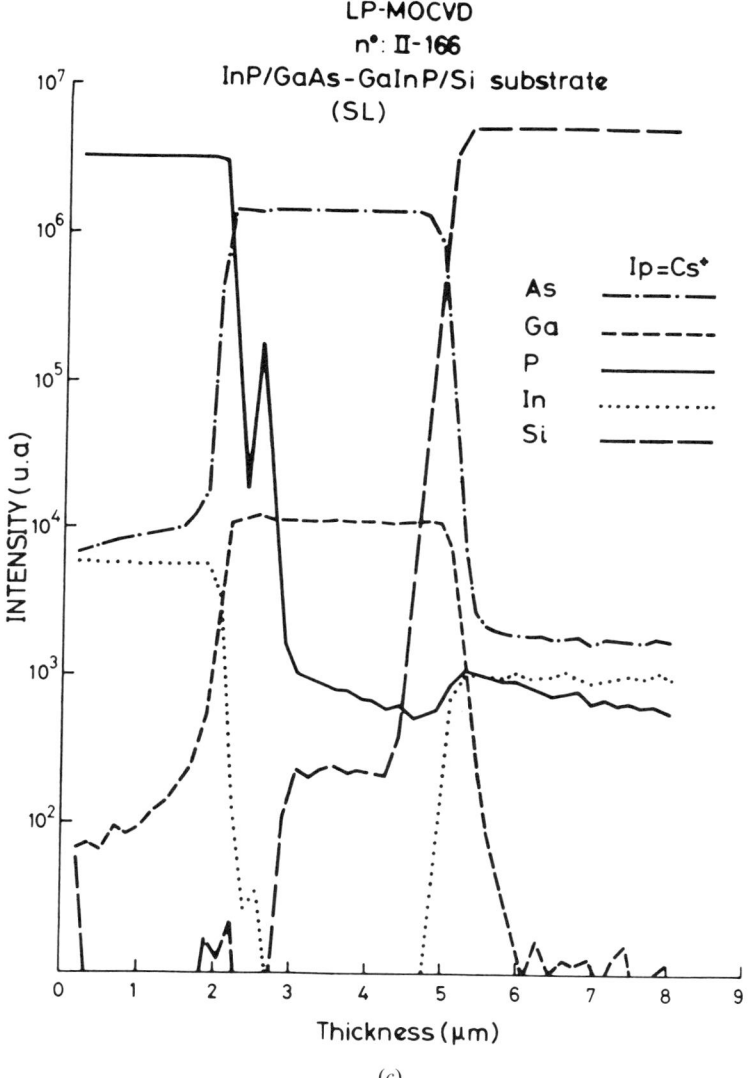

(c)

FIG. 54. (*Continued*)

to the group V elements *As* and *P* than to the group III elements Ga and In. That is why As and In signals appear nearly equivalent in the InP layer. The presence of arsenic can also be partly explained by its diffusion during the growth.

Between the substrate and the InP layer, the mean presence of the four elements in the GaAs/Ga$_{0.49}$In$_{0.51}$P superlattice has been detected.

InP was intentionally *n*-doped with sulphur during the growth to about: $N_d - N_a = 10^{18}$ cm^{-3}. The Polaron profile is reported in Figs. 55a and 55b (GaAs and Si substrates), and the doping level is in good agreement with the expected value. The figure confirms, as well, a thickness of about 2 μm for the InP layer.

Optical properties have been investigated by obtaining the $T = 20$ K photoluminescence spectrum for the structures on the alternate substrates (Fig. 56). Recombination energies are situated between 1410 meV (InP on silicon) and 1425 meV (InP on InP). The intensity of photoluminescence and the full width at half maximum ($\delta E = 28$ meV) are comparable in the three cases, thus confirming the high optical quality of the InP/Si structure.

A double InP/Ga$_x$In$_{1-x}$As$_y$P$_{1-y}$ heterostructure has been grown on silicon substrate as well. The global structure of the layer is shown in Fig. 57. The

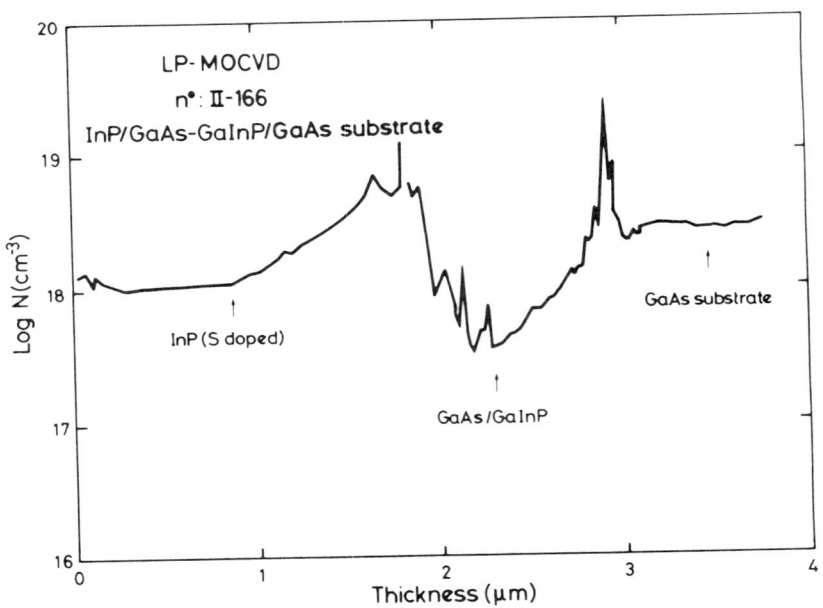

FIG. 55(a). Doping profile of the InP/GaAs–Ga$_{0.49}$In$_{0.51}$P structure grown on GaAs substrate.

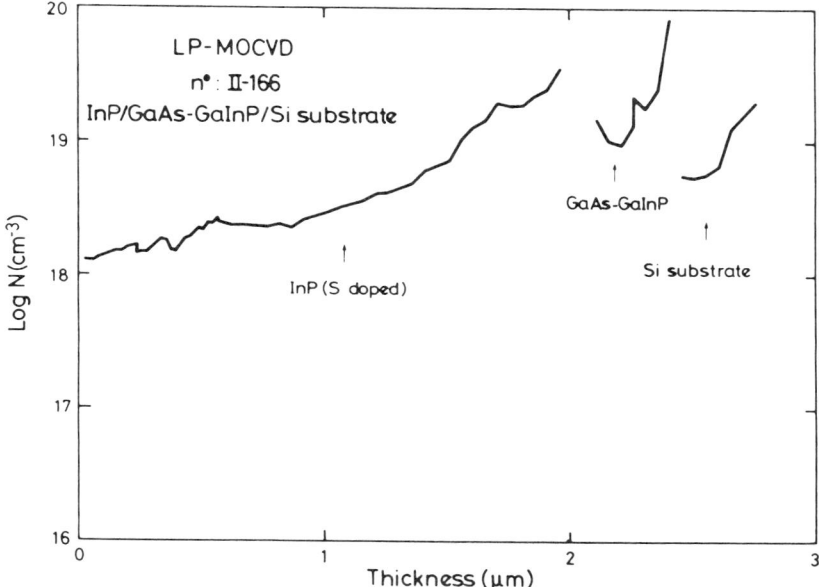

FIG. 55(b). Doping profile of the InP/GaAs–Ga$_{0.49}$In$_{0.51}$P structure grown on Si substrate.

growth has been initiated with a 1000 Å GaAs buffer layer and followed by 3 μm InP. The Ga$_x$In$_{1-x}$As$_y$P$_{1-y}$ active layer is about 2000 Å thick. As in Fig. 42, we can deduce from the single diffraction pattern the good crystallographic quality of the InP/GaInAsP/InP structure. $K\alpha_1$ and $K\alpha_2$ peaks appear to be very well resolved, and the GaAs buffer layer remains very thin, so its related signal is hardly detectable. Nevertheless, $K\alpha_1$ and $K\alpha_2$ can still be separated, even for this epilayer situated near the silicon substrate.

The quaternary compound has a wavelength of 1.3 μm. Figure 58 shows the 300 K photoluminescence spectrum, which is evidence of its good optical properties. The recombination peak is situated at the expected energy, with a full width at half maximum of 55 meV, which can be favorably compared with those of the same structure grown on InP substrate.

22. Growth of GaAs on Si Substrate, Using GaAs/Ga$_{0.49}$In$_{0.51}$P Superlattices as a Buffer Layer

GaAs/Ga$_{0.49}$In$_{0.51}$P superlattices have been used as a buffer layer, for growing GaAs on silicon substrates. The substrate was initially heated up to 1000 °C, after being introduced in the reactor, in order to remove the oxide present on its surface. It was then cooled down to 550 °C, and the growth was

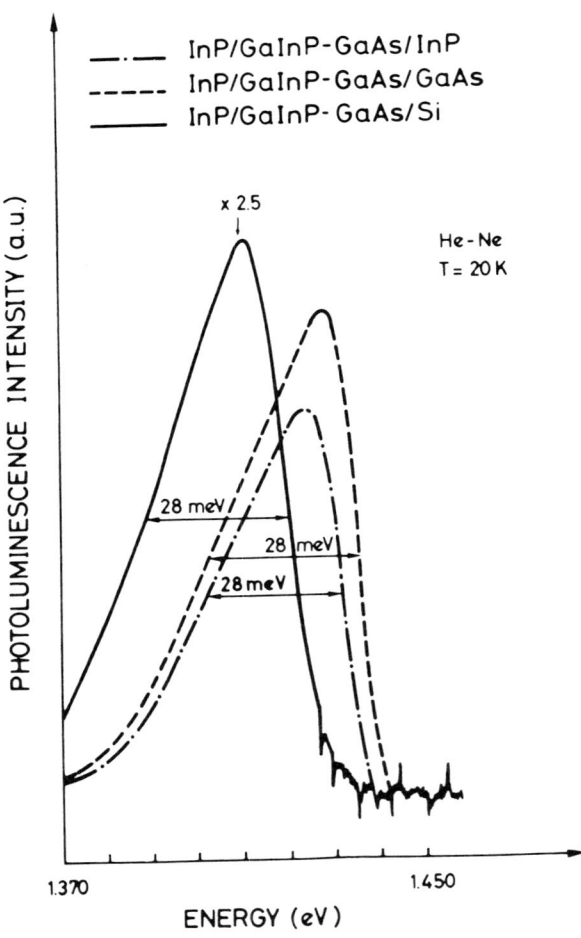

FIG. 56. Photoluminescence spectra at $T = 20\ °K$ of the InP/GaAs–$Ga_{0.49}In_{0.51}P$ structure grown on the InP, GaAs, and Si substrates.

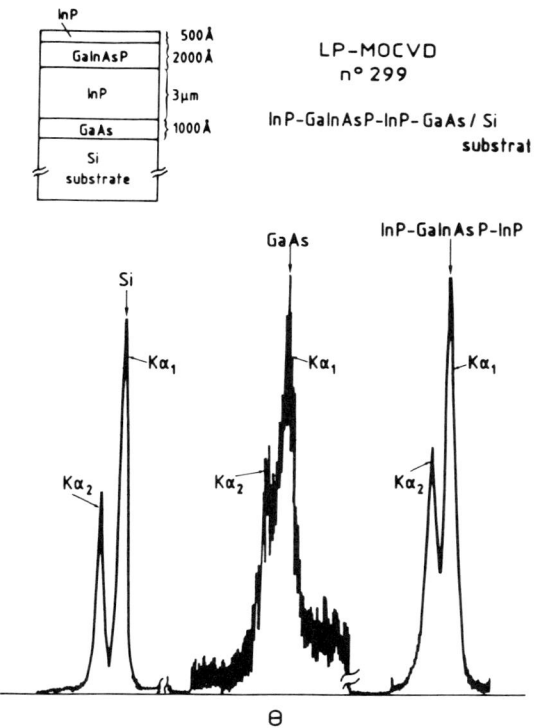

FIG. 57. X-ray simple diffraction pattern of the double heterostructure $InP/Ga_xIn_{1-x}As_yP_{1-y}/InP$ grown on Si substrate.

initiated with the conditions of Table XVIII by simply introducing gallium and arsenic.

The structure of the sample is shown in Fig. 59. Two thin layers (500 Å) were grown before the thick top layer of 0.6 μm. They were separated by two $GaAs/Ga_{0.49}In_{0.51}P$ superlattices, which were introduced to prevent the propagation of misfit dislocations due to the lattice mismatch between GaAs and Si. An intense $T = 77$ K photoluminescence signal has been obtained, which shows the good optical properties of the GaAs epilayer. A full width at half maximum of 20 meV is determined.

X-ray simple diffraction pattern of the structure is shown in Fig. 60. The epilayer has the same (111) orientation as that of the substrate. No trace of disorientation has been found. The good cristallographic quality of the GaAs is evidenced by clear separation between $K\alpha_1$ and $K\alpha_2$ related signals.

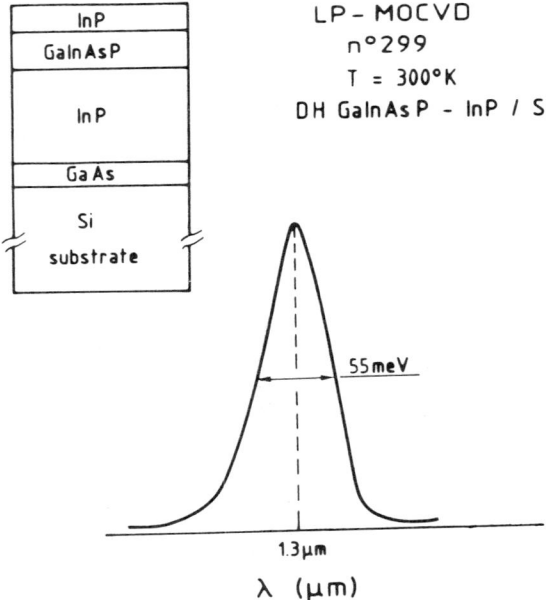

FIG. 58. 300 °K photoluminescence spectrum of the double heterostructure $InP/Ga_xIn_{1-x}As_yP_{1-y}/InP$ grown on Si substrate.

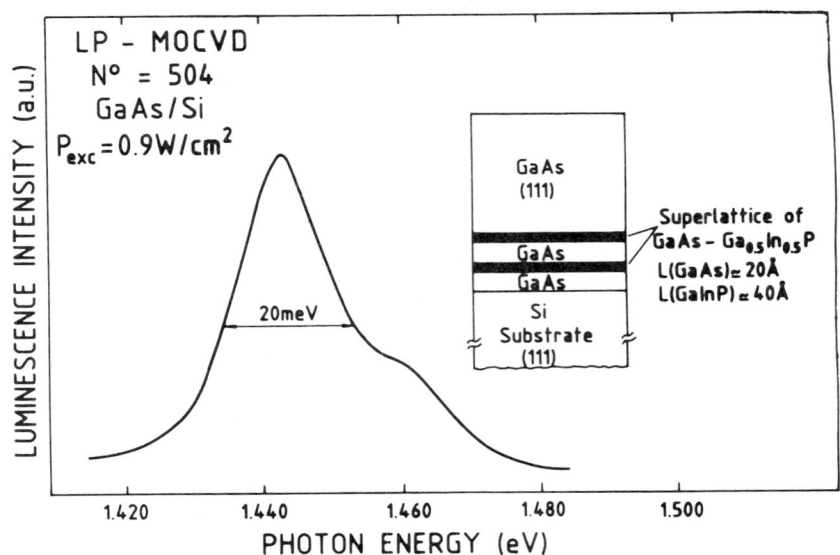

FIG. 59. Photoluminescence spectrum at $T = 77$ °K of bulk GaAs grown on a silicon substrate, using $GaAs/Ga_{0.49}In_{0.51}P$ superlattices as buffer layers.

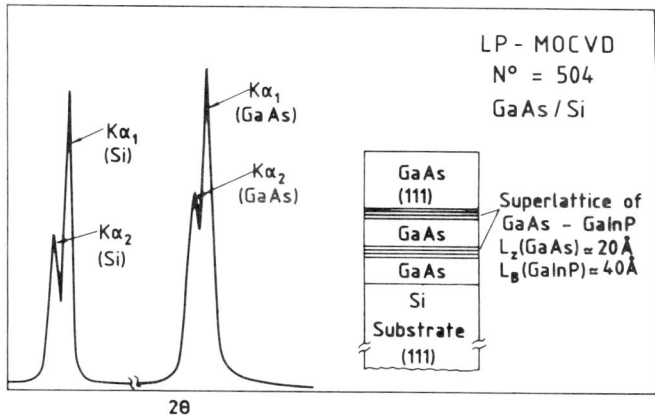

FIG. 60. X-ray simple diffraction profile of bulk GaAs grown on a silicon substrate, using GaAs/Ga$_{0.49}$In$_{0.51}$P superlattices as buffer layers.

X. Optoelectronic Applications

Long distance optical links use lasers emitting at 1.3 μm fabricated by material grown on InP substrate. Unfortunately, the technology of ICs (integrated circuits) for signal treatment is a lot more difficult on this material than on a GaAs substrate. A solution would be to combine the advantages of these two materials, but this goes against nature, because of the large lattice mismatch.

However, we have shown that it is possible to accomplish this with the LP-MOCVD growth technique. Ga$_{0.25}$In$_{0.75}$As$_{0.5}$ — InP buried ridge structure (BRS) lasers emitting at 1.3 μm have been fabricated on GaAs substrates using the LP-MOCVD growth technique. The BRS laser structure was manufactured as follows; First, the following layers were successively grown by LP-MOCVD on a Si-doped GaAs substrate oriented (100) off 2°: a 2 μm InP confinement layer, sulphur doped with $N_D - N_A \simeq 10^{18}$ cm^{-3}; a 0.2 μm thick undoped GaInAs (composition 1.3 μm) active layer; and a 0.2 μm thick Zn-doped ($N_A - N_D \simeq 2 \times 10^{17}$ cm^{-3}) InP layer in order to avoid the formation of defects near the active layer during the etching.

The morphology of these layers was excellent, and the photoluminescence intensity and PL half width were *the same* as for material grown on a InP substrate. The details of the growth conditions are given in Table XVIII. Next, a ridge of about a 2 μm width was etched in the InP (P) and GaInAsP active layers through a photolithographic resist mask. With the hope of having a good control over the etching, we used a selective etchant composed of H$_2$SO$_4$, H$_2$O$_2$, H$_2$O (1:8:40).

After removing the resist mask, the ridges were then covered with 1 μm of Zn-doped InP confinement layer and 0.5 μm Zn-doped GaInAs (with $N_A - N_D \simeq 10^{19}$ cm^{-3}) cap layer, grown by LP-MOCVD. In order to localize the injection current in the buried-ridge active region only, a deep proton implantation was performed through a 5 μm-wide photoresist mask after the metallization of the contacts. Further localization of the current in the buried ridge is achieved by the built-in potential difference between the $P - N$ InP homojunction on each side of the active region and the $N - P -$ InP $-$ GaInAsP heterojunction of the active region.

Figure 61 shows schematically the resulting GaInAsP–InP–BRS laser. The devices were cleaved and sawed, producing chips of 350 μm in width with

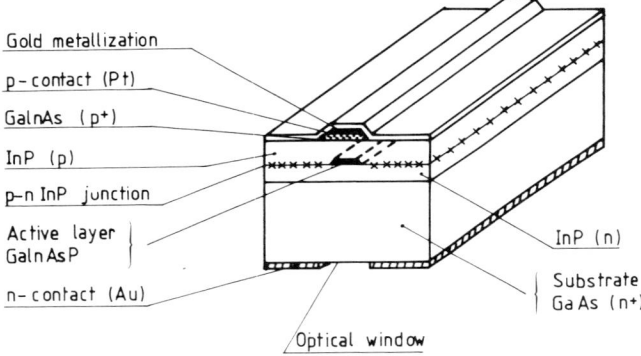

FIG. 61. Schematic diagram of the cross section of a GaInAsP/InP BRS laser emitting at 1.3 μm grown by LP-MOCVD on a GaAs substrate.

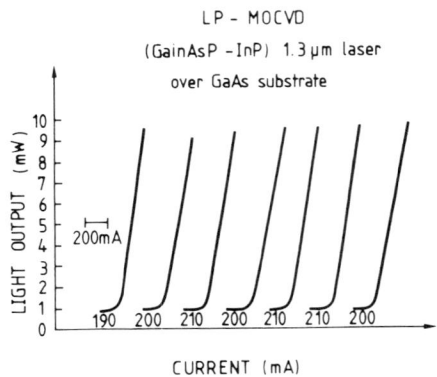

FIG. 62. Light-current characteristics of laser diodes emitting at 1.3 μm grown by LP-MOCVD on a GaAs substrate.

cavity lengths of 300 μm. The laser chips were tested while unmounted under pulsed conditions at a repetition rate of 10^4 Hz with a pulse length of 100 nsec.

Figure 62 shows the light-current characteristics of seven LP-MOCVD laser diodes obtained from the same wafer. A pulse threshold current of 190 mA at room temperature was measured with an output power up to 10 mW.

23. GaInP/GaInAs/InP MESFETs

The $Ga_{0.47}In_{0.53}As$ lattice matched to InP is a potentially important material for field-effect transistors (FET) with high peak electron velocity and high electron mobility for application in optoelectronic integration.

The metal-Schottky barrier heights on GaInAs are too low to be used as MESFET gates. FETs in lattice matched $Al_{0.48}In_{0.52}As$/InP, which exploit the increased Schottky barrier height provided by AlInAs (Barnard et al., 1980; Scott et al., 1984), and lattice-mismatched GaAs gate/GaInAs structures have also been prepared (Chen et al., 1985).

The preparation of a $Ga_{0.49}In_{0.5}lP/Ga_{0.47}In_{0.53}As$/InP MESFET fabricated from material grown by LP-MOCVD has also been reported by M. Razeghi et al., 1987. The energy gap of $Ga_{0.49}In_{0.51}P$ is 1.9 eV (Hilsum, 1964) and lattice parameter is 5.65 Å.

The FET device structures consist of n-type $Ga_{0.47}In_{0.53}As$ of 1500 Å thick, doped to 3×10^{17} cm^{-3} with sulphur and undoped $Ga_{0.49}In_{0.51}P$ of 800 Å thick with a carrier concentration $N_D - N_A = 10^{16}$ cm^{-3} grown at 550 °C onto (100) oriented Fe-doped semi-insulating InP substrates.

Large geometry FETs with 2 μm gate lengths, 150 μm gate widths and 5 μm source drain spacing have been fabricated. Source-drain current-voltage characteristics of a GaInP-GaInAs-InP FET exhibit a gate bias step of 0.5 V. The transconductance of this device is $g_m \simeq 50$ ms/mm.

24. Planar Monolithic Integrated Photoreceiver for 1.3–1.55 μm Wavelength Applications using GaInAs–GaAs Heteroepitaxies

Planar monolithic integrated photoreceivers are desirable devices for fiber-optic communication systems. In particular, the planar structure simplifies the problem of interconnecting the components, and, for high-speed applications, such a technique reduces the parasitic capacitances. The feasibility of such devices has recently been extensively demonstrated for 0.8 μm wavelength applications (Wada et al., 1983; Kollas et al., 1983; Verriele et al., 1985) and to a lesser extent for the 1.3–1.55 μm wavelength applications (Hata et al., 1984). A planar monolithic integrated photoreceiver suitable for long-wavelength optical communication systems ($\lambda \simeq 1.3-1.55\ \mu m$) using $Ga_{0.47}In_{0.53}As$ — GaAs strained heteroepitaxies was fabricated. It consists of a planar $Ga_{0.47}In_{0.53}As$ photoconductive detector associated with a GaAs field-effect transistor (FET). The strained heteroepitaxy consists of an undoped $Ga_{0.47}In_{0.53}As$ layer grown on a classical GaAs FET epitaxy (Fig. 63).

A scanning electron microscope photograph of the integrated circuit and its electrical circuit is shown in Fig. 64. The photoconductive detector has been formed by etching (ion) the $Ga_{0.47}In_{0.53}As$ up to the n^+ GaAs layer, except for the area corresponding to the photoconductor. The detector consists of two ohmic contacts (AuGeNi 2000 Å) separated from each other by 20 μm, leading to a $20\mu m \times 80\ \mu m$ photosensitive area. The FET has been fabricated on the GaAs layers. To obtain good amplification at frequencies up to several gigahertz, the FET has a $2\ \mu m \times 900\ \mu m$ gate (Ti 500 Å; Pt 300 Å; Ti 300 Å; Au 2000 Å). To produce a compact design integrated circuit, the FET surrounds the photoconductor. The interconnection between one of the photoconductor ohmic contacts and the FET source (AuGeNi 2000 Å) is made using a polymide bridge. The same technique allows an interconnection to be made between the other photoconductor ohmic contact and the FET gate. It can be observed that the n^+ GaAs layer reduces the dark resistance of the $Ga_{0.47}In_{0.53}As$ photoconductor; this effect will be discussed in the next. A photoconductor was also processed on the same ship but without the electronics, in order to perform test experiments on the photodetector only. All experimental results presented on the photoconductor were performed on this device.

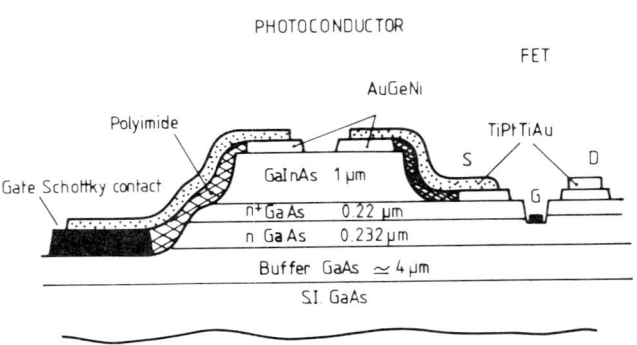

FIG. 63. Schematic cross section (CC) of an integrated photoreceiver. The gate metallization has been deposited on the n GaAs layer to fabricate the FET and on a part of the buffer GaAs (called GaAs Schottky contact of the figure) to realize an insulated interconnection between the gate and the photoconductor.

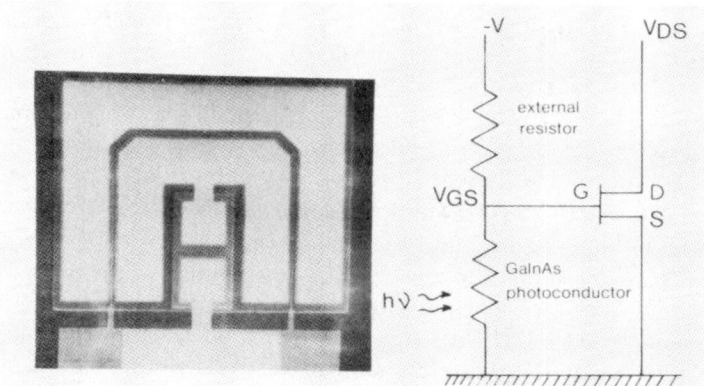

FIG. 64. Photograph and electrical circuit of monolithic integrated photoreceiver.

In Fig. 65, a typical steady-state gain (defined as the number of charges collected in the external circuit per incident photon) obtained for the photoconductor versus bias voltage is shown. These measurements were made using a cw 1.06 μm wavelength yttrium aluminum garnet (YAG) laser and a cw 1.3 μm wavelength laser diode. Observe that the gain value increases when the bias voltage increases; this result is due to the reduction of the transit time τ_t when the electron drift velocity increases corresponding to the expression of the gain: $G = \tau_v/\tau_t$ (Gammel et al., 1981; Vilcot et al., 1985), where τ_v is the electron-hole pair lifetime, whose value could be connected to

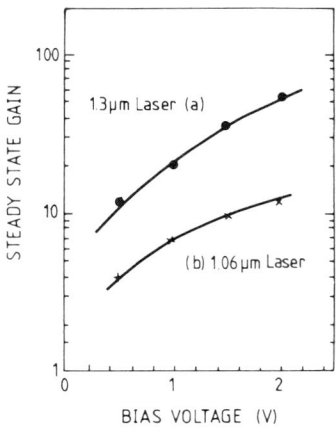

FIG. 65. Steady-state gain G of the GaInAs–GaAs photoconductor vs. its bias voltage V using (a) a cw 1.3 μm wavelength laser diode, and (b) a cw 1.06 μm wavelength YAG laser $P_L = 0.2$ μm.

trapping effects at the interface $Ga_{0.47}In_{0.53}As$-GaAs. The effect of the light intensity upon the gain, observed in Fig. 65, is due to a modification of τ_v probably related to a saturation phenomenon of the interface states under illumination. The low value of the gain is explained by a short lifetime τ_v, as can be observed on the picosecond response (Fig. 66). This result indicates that this photoconductor could be useful for gigabit photoreceiver applications. Typical dynamic gains $G(\omega)$ are given in Fig. 67. Our experimental results can be described by the expression $G(\omega) = \tau_v/\tau_t\sqrt{1 + \omega^2\tau_v^2}$ corresponding to an assumption of a generation-recombination process governed by a Poisson law; ω is the pulsation of the signal. The gain-band-width product is close to 1 GHz, a value that is related to the 20 μm electrode spacing. Obviously, this performance can easily be improved by using a shorter electrode spacing. The FET has been characterized separately. Its transconductance is close to 90 mS corresponding to the gate size (2 μm × 900 μm). A current amplification of two has been achieved on the picosecond response (Fig. 66) using 0.580 μm wavelength modelocked dye laser with a 2 V drain source and a 2 V gate source bias voltage. The same typical amplification factors are obtained for the low-frequency response (Fig. 67) and are equal to the transductance of the FET multiplied by the photocon-

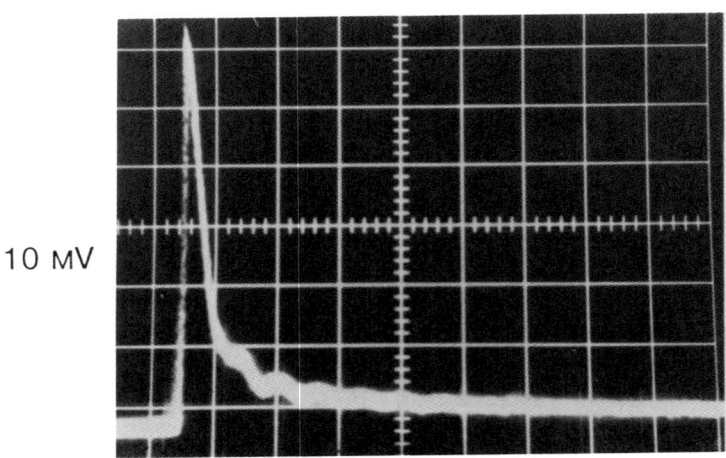

10 MV

TIME BASE 1 ns/DIV

FIG. 66. Picosecond response using a mode-locked dye laser; $\lambda = 0.580$ μm, $P_L = 0.25$ mW. The light has been focused on the photosensitive area of the photoconductor using a microscope objective, to be sure that the light impinging on the device created electron-holes pairs only in the $Ga_{0.47}In_{0.53}As$ of the photoconductor and not in the GaAs of the FET. The voltage measurements have been performed using a 10 dB attenuator.

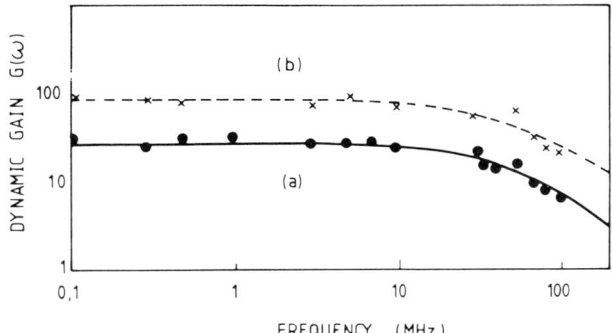

FIG. 67. Dynamical gains vs. frequency using a sinusoidally modulated 1.3 μm laser diode. (a) Photoconductor; bias voltage 2 V. (b) Integrated circuit; $V_{BS} = -2$ V; $V_{DS} = 2$ V.

ductor ($\simeq 50\ \Omega$) parallel with the bias resistor. A better amplification factor can be obtained by growing the $Ga_{0.47}In_{0.53}As$ photosensitive area on a GaAs buffer layer to realize photoconductors with higher dark resistance.

The noise properties of the photoconductor and of the integrated circuit have also been investigated in the 10 MHz–1.5 GHz frequency range using a HP 8970A noise figure meter. Our experiments (Fig. 68) show a high $1/f$ photoconductor noise for frequencies lower than 100 MHz. For frequencies higher than 100 MHz, this $1/f$ noise is greatly reduced as is commonly observed in III–V photoconductors (Chen et al., 1985; Vilcot et al., 1985). The noise of the integrated circuit is higher, and its value can be explained by

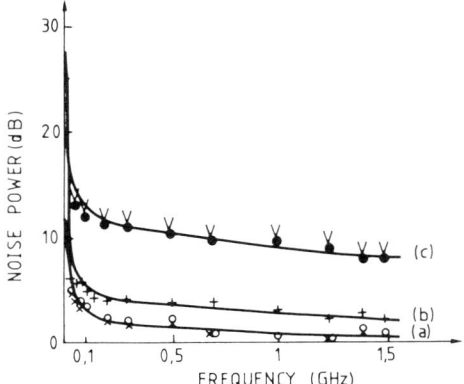

FIG. 68. Noise power vs. frequency. cw 1.06 μm YAG laser illumination. P_L 0.2 mW. (a) Photoconductor; bias voltage 0.5 V: (X) without illumination and (O) with illumination. (b) Field-effect transistor; $V_{GS} = -0.5$ V; $V_{DS} = 2$ V; (c) Integrated circuit $V_{GS} = -0.5$ V; $V_{DS} = 2$ V; (●) without illumination and (V) with illumination.

the FET amplification plus the noise of the FET itself (Fig. 68). It can be noted that the noise under illumination is very close to the noise in darkness. This result can be explained by the low value of the gain of the photoconductor, according to the expression for the noise due to illumination of such a device (Vilcot et al., 1985).

In conclusion, our results show that it is possible to realize a planar monolithic integrated photoreceiver for 1.3–1.55 μm wavelength applications using strained heteroepitaxies. In particular, our results show that the use of the strained epitaxies does not degrade the noise performance of the photoconductor in the gigahertz frequency range. Moreover, several improvements can be proposed to obtain better performances: (1) An interdigitated photoconductive detector would lead to a larger photosensitive area with a shorter electrode spacing, and obviously a higher gain-bandwidth product, and (2) An increase in the dark resistance of the photoconductor by growing the $Ga_{0.47}In_{0.53}As$ photosensitive area directly on the GaAs buffer would constitute the main improvement of this device.

More work is currently in progress and results are encouraging.

25. Monolithic Integration of a GaInAs/GaAs Photoconductor with a GaAs FET for 1.3–1.55 μm Wavelength Applications

Here we present a monolithic integrated photoreceiver for long wavelength optical systems that associates a GaInAs/GaAs strained photoconductor with a GaAs FET. In this realization, in order to reduce the dark current of the photoconductive detector, the GaInAs photosensitive layer has been deposited, by LP-MOCVD, on the undoped GaAs layer. Static, dynamic, and noise properties of the photoconductor and the integrated circuit are presented and discussed, and the special structure of the material and integrated circuit are taken into account.

The purpose of this work is to present an optoelectronic integrated circuit (OEIC) suitable for long wavelength optical communication systems. This OEIC associates a GaInAs planar photoconductor with a GaAs field effect transistor (Fig. 69). The photoconductor is realized on a $Ga_{0.47}In_{0.53}As$/GaAs strained heteroepitaxy (Razeghi et al., 1987), which allows long-wavelength (1.3–1.55 μm) photodetection. Several theoretical and experimental studies have shown that the sensitivity in the high frequency range of such a photoreceiver increases when the dark current of the photoconductor is reduced (Forrest, 1985). This OEIC represents an improvement compared to our first similar device (Razeghi et al., 1986), since the GaInAs photoconductor has been realized on an undoped GaAs layer in order to increase the dark resistance of the photoconductor. Therefore, the GaInAs heteroepitaxy has been grown by LP-MOCVD in boxes previously

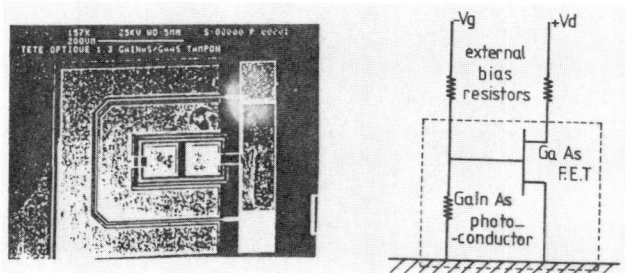

FIG. 69. SEM. photograph and electrical equivalent circuit of the photoreceiver.

etched in a classical GaAs FET epitaxy (Fig. 70a). The photoconductor is delineated by an ion milling of the GaInAs layer of about 1 μm. A schematic cross section of the device is shown in Fig. 70a. The ohmic contacts of the photoconductor are separated by 20 μm leading to a photosensitive area of 80 μm × 20 μm. The FET gate has a 2 μm length and 900 μm width. The interconnection between the photoconductor, and the FET is achieved using a polyimide bridge. Details showing this interconnection are given in Fig. 70b. On the same chip, a photoconductor alone is processed to perform experimental photoconduction evaluation.

Static, dynamic, and noise properties of the photoconductor and IC have been extensively studied. As an example, we present typical steady-state gain, versus bias voltage, obtained for the photoconductor (Fig. 71). Observe that the gain value, which lies between 10 and 100, increases when the bias voltage increases and decreases when the light intensity increased. This result can probably be explained by trapping effects at the GaInAs/GaAs interface.

Typical dynamical gains for the photoconductor and the IC are given in Fig. 72. These measurements have been performed using a sinusoïdally modulated 1.55 μm laser diode.

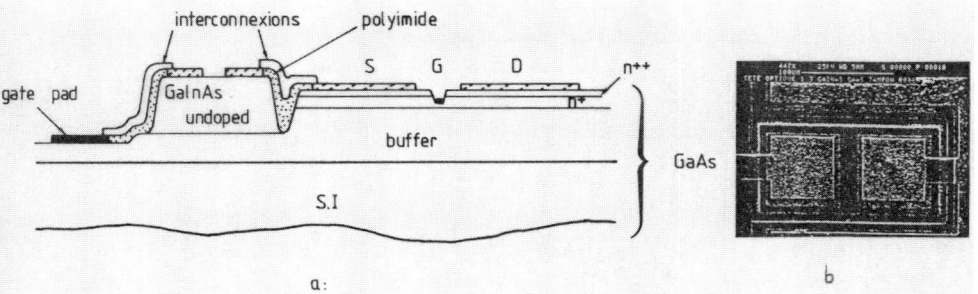

FIG. 70. (a) Schematic cross of the IC and (b) SEM photograph of interconnection between FET and photoconductor.

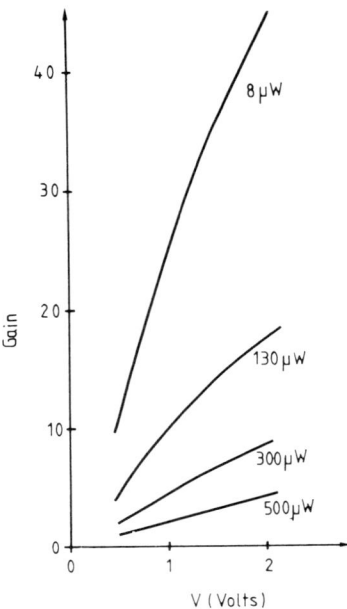

FIG. 71. Photoconductor steady-state gain vs. bias voltage (measurements, performed with a 1.55 μm wavelength laser diode).

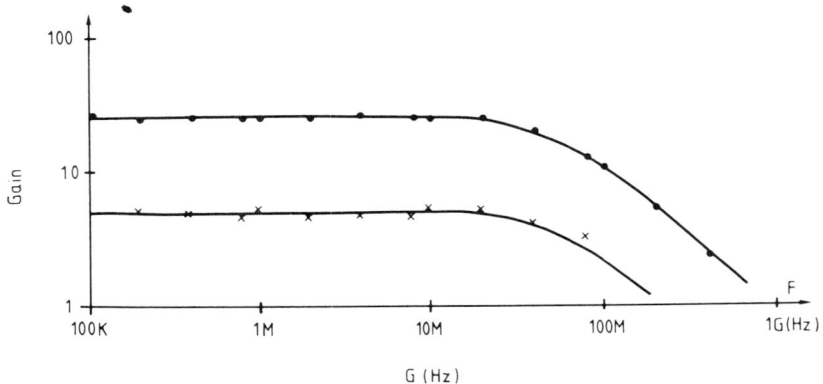

FIG. 72. Dynamical gains XXX Photoconductor ......I.C.

6. LP-MOCVD GROWTH   349

An amplification factor closes to five connected to the FET transconductance of about 50 mS has been achieved for a 200 Ω bias resistor. The gain-bandwidth product of the IC is close to 1 GHz and is related to the electrode spacing of the photoconductor.

26. MONOLITHIC INTEGRATION OF A SCHOTTKY PHOTODIODE AND A FET USING A $Ga_{0.49}In_{0.51}P/Ga_{0.47}In_{0.53}As$ STRAINED MATERIAL

For the realization of optoelectronic integrated circuits (OEICs) suitable for long wavelengths optical communications systems (1.3–1.55 μm), the $Ga_{0.47}In_{0.53}As$, latticematched to InP, is a potentially important material. Metal Schottky barrier heights on GaInAs are too low to be used as MESFET gates, for example. In this work, we demonstrate for the first time the realization of an OEIC, constituted of a Schottky photodiode associated with a FET using a $Ga_{0.49}In_{0.51}P/Ga_{0.47}In_{0.53}As$ strained heteroepitaxy. The energy bandgap of $Ga_{0.49}In_{0.51}P$ is 1.9 eV, and this superficial epilayer should improve the quality of Schottky barriers.

The different epilayers have been grown by LP-MOCVD on a S.I. InP substrate. The $Ga_{0.47}In_{0.53}As$ epilayer ($N_D \simeq 2.10^{17}$ At/cm^3: 2000 Å thick) is suitable for long wavelengths (1.3—1.55 μm) photodetection and FET

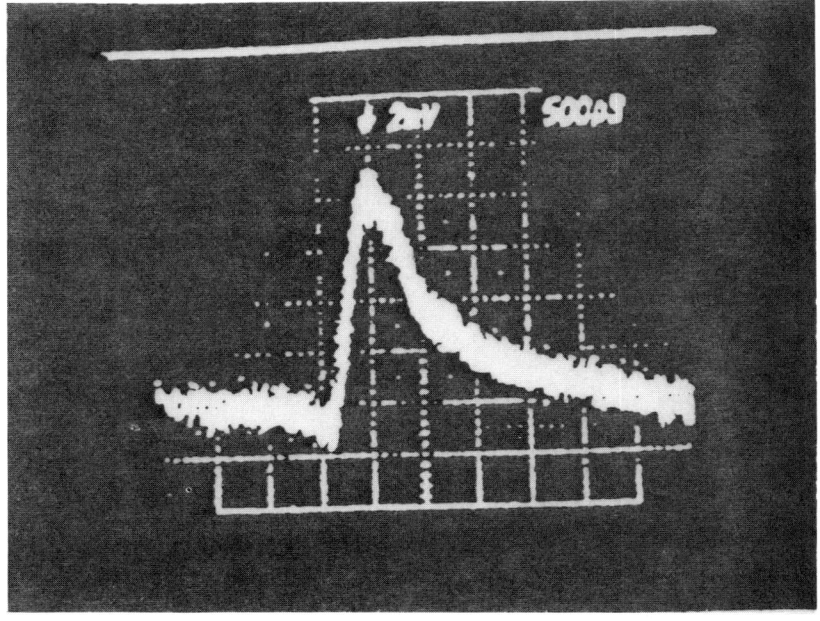

FIG. 73. Picosecond response of the photodiode.

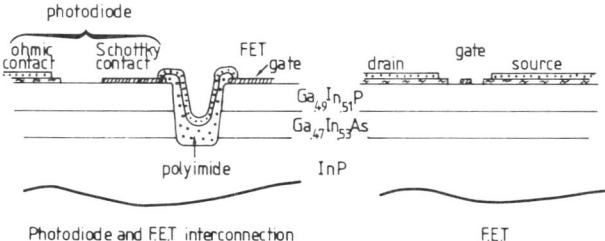

FIG. 74. Schematic cross section of the device (OEIC).

channel fabrication. The $Ga_{0.49}In_{0.51}P$ epilayer is undoped (residual n-type) and is 1000 Å thick. A photodiode, a FET and an OEIC have been realized on the same chip. The photodiode has a 40 μm × 40 μm photosensitive area. The Schottky contact has been realized by the deposition of a 200 Å thick Ti–Pt electrode. A responsivity close to 0.1 A/W has been measured for a 1.3 μm wavelength optical signal, and the picosecond response is given in Fig. 73.

The FET gate has a 3 μm length and 300 μm width. The saturation current is about 10 mA and the pinch-off voltage is close to -3 V. The average measured transconductance is 30 mS/mm.

The OEIC (schematic cross sections are shown in Fig. 74) associates the same photodiode and FET (600 μm gate width, in this case) with that

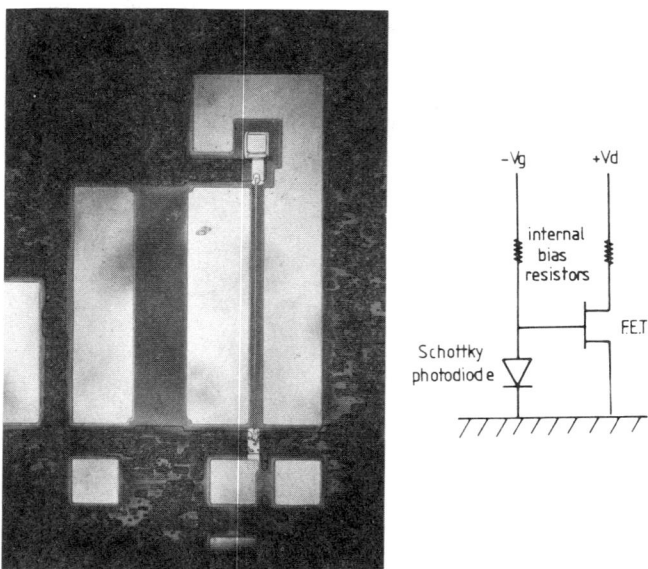

FIG. 75. SEM photograph and electrical equivalent circuit of the photoreceiver.

previously presented. A SEM microphotograph and equivalent electrical circuit are shown in Fig. 75. Static, dynamic, and noise properties of the IC have been evaluated. As an example, a responsivity of 50 A/W has been achieved at 1.3 μm for a 1.5 KΩ bias resistor.

XI. Conclusion

In this chapter, we have presented recent advances in LP-MOCVD growth of InP and strained layers based on this compound. The purest InP reported in the literature has been obtained with residual doping level: $N_d - N_A = 3.10^{13}$ cm^{-3}, and $T = 50$ K mobility: $\mu(50\text{ K})$: 200,000 cm^2 V^{-1} s^{-1}.

Growth and characterization of strained heterostructures InP/GaAs, InP/InAs, and InP/Si are also described. It is of great interest for integrated device applications.

Various examples have been presented including InP on GGG substrate; 1.3 μm BRS laser grown on GaAs substrate; planar monolithic integrated photoreceiver using $Ga_{0.47}In_{0.53}As$/GaAs heteroepitaxies; and $Ga_{0.47}In_{0.53}As$/GaAs photoconductor integrated with Ga FET.

Acknowledgments

I would like to thank D. Leguen and J. Antoine for their excellent technical assistance and A. Huber, J. Olivier, R. Bisaro, P. Alnot, B. Vinter, P. Maurel, F. Omnes, M. Defour, O. Acher, R. Blondeau, D. Decoster, G. Neu, B. Decremoux and C. Weishuch for their participation and helpful discussions.

References

Akiyama, M., Kawarada, Y., and Kaminishi, K. (1984). *J. Crystal Growth* **68**, 21.
Aksum, M. I., Morkoc, H., Lester, L. F., Duh, K. H. G., Smith, P. M., Chao, P. C., Longabone, M., and Erickson, L. P. (1986). *Appl. Phys. Lett.* **49**, 24.
Ambridge, T., and Faktor, M. M. (1974). *J. Appl. Electrochem.* **4**, 135–142.
Ambridge, T., and Faktor, M. M. (1975). *J. Appl. Electrochem.* **5**, 319–328.
Ambridge, T., Elliott, C. R., and Faktor, M. M. (1973). *J. Appl. Electrochem.* **3**, 1–15.
Anderson, D. (1986). Proceedings of the International Microwave Conference, Paris.
Antypas, G. A., and Moon, R. L. (1973). *J. Electrochem. Soc.* **120**, 1574.
Arthur, J. R. (1968). *J. Appl. Phys.* **39**, 4032.
Asuka, N., Kolodziejski, L. A., Gunshor, R. L., Datta, S., Bicknell, R. N., and Schetzina, J. F. (1985) *Appl. Phys. Lett.* **46**, 860.
Barnard, J., Ohno, H., Wood, C. E. C., and Eastman, L. F. (1980). *IEEE Electron. Device Lett.* **EDL9** (9), 174.
Bass, S. J. (1975). *J. Crystal Growth* **31**, 172.
Benz, K., Hagleko, H., and Bosch, R. (1982). *J. Phys.* **43**, C5.
Blood, P. (1985). *Semicond. Sci. Technol.* **1**, 7–27.

Bogatov, A. P., Dolginov, L. M., Druzhinina, L. V., Eliseev, P. G., Sverdlov, B. N., and Shevchenko, E. G. (1975). *Sov. J. Quantum Electron.* (Engl. Transl.) **4**, 1281.
Bourbin, Y., Enard, A., Blondeau, R., Razeghi, M., Papuchon, M., and Decremoux, B. (1988). *Appl. Phys. Lett.*
Chang, K. Y., Cho, A. Y., Wagner, W. R., and Bonner, W. A. (1981). *J. Appl. Phys.* **52**, 1015.
Chen, C. Y., Pang, Y. M., Alavi, K., Cho, A. Y., and Garbinski, P. A. (1984). *Appl. Phys. Lett.* **44**, 99.
Chen, C. Y., Cho, A. Y., Garlinski, P. A., (1985). IEEE Electron. Device Lett. **EDL6** (1), 20.
Cho, A. Y., and Ballamy, W. C. (1975). *J. Appl. Phys.* **46**, 783.
Christou, A., Wilkins, B. R., Tsang, W. T. (1985). *Electron. Lett.* **21**, 9.
Corlett, R. (1975). Indium phosphide CW transferred electron amplifiers. Inst. Phys. Conf. Series N°24, Chap. 2.
Crowley, J. D. (1980). High efficiency 90 GHz InP Gunn oscillators. *Electron. Lett.* **16**, 705–706.
Cullen, G. W. (1978). Heteroepitaxial Semiconductors for Electronic Devices, (G. W. Cullen and C. C. Wang, eds.) Springer, Berlin.
Dapkin, P., Manasevit, H., and Hess, K. (1981). *J. Crystal Growth* **55**, 10.
Davey, J. E., and Pankey, T. J. (1968). *J. Appl. Phys.* **39**, 1941.
Davies, G. J., Heckingbottom, R., Ohno, H., Wood, C. E. C., and Calawa, A. R. (1983). *Appl. Phys. Lett.* **37**, 290.
Delahaye, F., Dominguez, D., Alexandre, F., André, J. P., Hirtz, J. P., and Razeghi, M. (1986). *Metrologia.* **22**, 103.
Delacourt, D., Papuchon, M., Poisson, M. A., and Razeghi, M. (1987). *Electron. Lett.* **23**, 451.
Doi, A., Aoyagi, Y., and Namba, S. (1986). *Appl. Phys. Lett.* **48**, 1787.
Duchemin, J. P. (1977). *Rev. Res. Thomson-C.S.F.* **9**, 1.
Duchemin, J. P. (1977). *Rev. Res. Thomson-C.S.F.* **9**, 2.
Duchemin, J. P., Bonnot, M., Koelsch, F., and Huighe, D. (1979). *J. Electrochem. Soc.* **126**, 1134.
Escher, J. S., and Sankaran, R. (1976). *Appl. Phys. Lett.* **29**, 87.
Escher, J. S., Antypas, G. A., and Edgecumbe, J. (1976). *Appl. Phys. Lett.* **29**, 153.
Forrest, S. R. (1985). *J. Light Technol.* **LT3** (2), 347.
Fraas, L. M. (1981). *J. Appl. Phys.* **52**, 6939.
Gammel, J. C., Ohno, H., and Ballantyne, J. M. (1981). *IEEE J. Quantum Electron.* **17**, 269.
Gazaux, J., Etienne, P., Razeghi, M. (1986). *J. Appl. Phys.*
Giles, P., Davies, P., and Hardell, N., (1984). *J. Crystal Growth* **65**, 351.
Goodman, C. H. L., and Perra, M. V. (1986). *J. Appl. Phys.* **60**, R 65.
Guldner, Y., Vieren, J. P., Voos, M., Delahaye, F., Dominguez, D., Hirtz, J. P., and Razeghi, M. (1986). *Physical Review B.* **23** (6), 3990.
Guldner, Y., Vicren, J. P., Voisin, P., Voos, M., Razeghi, M. (1982). *Appl. Phys. Lett.* **40**, 877.
Gunther, K. G. (1958). Patent N° 2, 937–816.
Hamilton, R. J. (1976). InP Gunn-effect devices for millimeterwave amplifiers and oscillators. *IEEE Trans.* **MTT-24**, 775–780.
Hata, S., Ikeda, M., Amano, T., Motosugi, G., and Kurmuda, K. (1984). *Electron. Lett.* **20**, 947.
Hilsum, C. (1962). *Proc. IRE.* **50**, 185.
Hirch, J. J. (1976). *Appl. Phys. Lett.* **28**, 283.
Horikoshi, L. (1986). "The Physics and Fabrication of microstructures." Les Houches.
Huber, A. (1986). *Rev. Tech. Thomson-C.S.F.* **18**, 1.
Huber, A., Morillot, G., Bonnet, M., Derenda, P., Bessonneau, G. (1982). *Appl. Phys. Lett.* **41**, 638.
James, L. W., Antypass, G. A., Moon, R. L., Edgecumbe, J., and Bell, R. L. (1973). *Appl. Phys. Lett.* **22**, 270.
Kawamura, Y., Asahi, H., Ikeda, M., and Okamoto, H. (1981). *Appl. Phys.* **52**, 3445.

Kobayashi, N., Makimoto, T., and Horikoshi, Y. (1985). *Japan. J. Appl. Phys.* **24**, L962.
Kolbas, R. M., Abrokwah, J., Carney, J. K., Bradshaw, D. H., Elmre, H. R., and Briard, J. R. (1983). *Appl. Phys. Lett.* **43**, 821.
Lambert, M., Bonnerie, D., and Huet, D. (1983). Proc. Eur., Workshop MBE, 2nd, Brighton, Engl., Paper 40.
Lee, M. K., Wuu, D. S., Tung, H. H. (1987). *Appl. Phys. Lett.* **50**, 24.
Lepareux, M. (1980). *Rev. Tech. Thomson-C.S.F.* **12**, 225.
Lim, H., Sagres, G., Bastide, G., Gouskov, L. (1982). *J. Appl. Phys.* **53**, 3085.
Long (1968).
Ludowise, M. I. (1985). *J. Appl. Phys.* **58**, R31.
McFee, J. H., Miller, B. I., and Bachmann, K. J. (1977). *J. Electrochem. Soc.* **125**, 259.
Magee, C. (1979). *J. Electrochem. Soc.* **126**, 660.
Makimoto, T., Kobayashi, N., and Horikoshi, Y. (1986). *Japan J. Appl. Phys.* **25**, L513.
Manasenit, H. M., and Simpson, W. I. (1971). *J. Electrochem.* **118**, C291.
Matsumoto, T., and Usui, A. (1986). Abstracts of 1986, Fall Meetings of Japan Soc. Appl. Phys.
Miller, B. I., McFee, H. H., Martin, R. J., and Tien, P. K. (1978). *Appl. Phys. Lett.* **33**, 44.
Nagle, J., Hersee, S., Razeghi, M., Krakowski, M., Decremoux, B., and Weisbuch, C. (1986). *Surf. Sci.* **174**, 148.
Neave, J. H., and Joyce, B. A. (1978). *J. Crystal Growth* **44**, 387.
Neave, J. H., Joyce, B. A., Dobson, P. J., and Norton, N. (1983). *Appl. Phys.* A31.1.
Nelson, H. (1963). *R.C.A. Rev.* **19** (24), 603.
Nicholas, R. J., Brunel, L. C., Huant, S., Karrai, K., Portal, J. C., Brummell, M. A., Razeghi, M., Chang, K. Y., and Cho, A. Y. (1985). *Physical Review Letters* **55**, 8.883.
Nishi, S., Inomata, H., Akiyama, M., Kaminishi, K. (1985). *J. Appl. Phys.* **24**, 6.
Nishizawa, J., and Kurabayashi, T. (1986). *J. Crystallographic Soc. Japan* **28**, 133.
Noyanuma, M., and Takahashi, K. (1975). *Appl. Phys. Lett.* **27**, 342.
Olego, D., Chang, T. Y., Silberg, E., Caridi, E. A., and Pinczuk, A. (1982). *Appl. Phys. Lett.* **41**, 476.
Olsen, G., Zamerowski, J., Hawrylo, F. (1982). *J. Crystal Growth* **59**, 654.
Osbourn, G. C. (1985). *Mat. Res. Soc. Symp. Proc.* **37**, 220–237.
Panish, M. B. (1980).
Panish, M. B., and Hamm, R. A. (1986). *J. Crystal Growth* **78**, 445.
Pearsall, T. P., Miller, B. I., Capik, R. J., and Bachmann, K. J. (1976). *Appl. Phys. Lett.* **28**, 499.
Poisson, M. A., Brilinsky, C., Colomer, G., Osselin, D., Duchemin, J. P., Azan, F., Le Chevalier, D., and Lacombe, J. (1985). Inst. Phys. Conf. Ser. N°74, Chap. 9, p. 677.
Poulain, P., Razeghi, M., Hirtz, P., Kcymierski, K., and Decremoux, B. (1985). "9th International Conf. on Semiconduction lasers," 7–11 August 1984, Rio, Brazil.
Prandtl, L., and Tietgens, O. C. (1934). "Applied Hydro- and Aeromechanics," McGraw-Hill, Doner.
Raulin, J. Y., Vassilakis, E., Poisson, M. A., Razeghi, M., and Colomer, G. (1987). *Appl. Phys. Lett.* **50**, 535.
Razeghi, M., Blondeau, R., Hirtz, P., and Duchemin, J. P. (1982). *Electron. Lett.* **18**, 132.
Razeghi, M., Hirtz, P., Blondeau, R., and Duchemin, J. P. (1983a). *Electron. Lett.* **19**, 481.
Razeghi, M., Hersee, S., Blondeau, R., Hirtz, P., and Duchemin, J. P. (1983b) *Electron. Lett.* **19**, 336.
Razeghi, M., Hirtz, J. P., Ziemelis, U. D., Delalande, C., Etienne, B., and Voos, M. (1983c). *Appl. Phys. Lett.* **43**, 585.
Razeghi, M. (1983). *Rev. Tech. Thomson-C.S.F.* **15**, 1.
Razeghi, M. (1984a). *Rev. Thomson-C.S.F.* **16**, 1.
Razeghi, M. (1984b). MRT Report, private communication.

Razeghi, M., and Duchemin, J. P. (1984). *J. Crystal Growth* **70**, 145.
Razeghi, M., Blondeau, R., Bouley, J. C., Decremoux, B., and Duchemin, J. P. (1984a). "Proceedings of the 9th IEEE International Laser Conference," 1984.
Razeghi, M., Blondeau, R., and Duchemin, J. P. (1984b). Inst. Phys. Conf. Sci. N°74, Chap. 9, p. 679.
Razeghi, M. (1985) in "Lightwave technology for communication" W. T. Tsang, (ed.) Academic Press, New York.
Razeghi, M., Duchemin, J. P., and Portal, J. C. (1985a). *Appl. Phys. Lett.* **46**, 46.
Razeghi, M., Blondeau, R., Decremoux, B., and Duchemin, J. P. (1985b). *Appl. Phys. Lett.* **46**, 131.
Razeghi, M., Duchemin, J. P., Portal, J. P., Donovski, L., Remeni, G., Nicholas, R. J., and Briggs, A. (1986a). *Appl. Phys. Lett.* **48**, 712.
Razeghi, M., Maurel, Ph., Tardella, A., Dinevski, L., Gauthier, D., and Portal, J. C. (1986b). *J. Appl. Phys.* **60**, 2454.
Razeghi, M., Noyle, J., Maurel, Ph., Omnes, F., and Pocholle, J. P. (1986c). *Appl. Phys. Lett.* **49**, 1110.
Razeghi, M., Ramdani, J., Verriele, H., Decoster, D., Constant M., and Vanbremeersh, J. (1986d). *Appl. Phys. Lett.* **49**, 215.
Razeghi, M., Meunier, P. L., and Maurel, Ph. (1986e). *J. Appl. Phys.* **59**, 2261.
Razeghi, M., Maurel, Ph., Omnes, F., and Defour, M. (1987a). *Appl. Phys. Lett.*
Razeghi, M., Maurel, Ph., and Omnes, F. (1987b). To be published in Nato Advanced Research Workshop on "Properties of impurity states in semiconduction superlattices."
Razeghi, M., Tardella, A., Davies, X. A., Long, A. L., Kelly, M. J., Britton, E., Boothroyd, C., and Stobbes, W. A. (1987c). *Electronic Letters* **23**, 3.
Razeghi, M., Maurel, Ph., and Omnes, F. (1987d). *Appl. Phys. Lett.*
Razeghi, M., Maurel, Ph., Omnes, F., Defour, M., Tsui, D. C., Wei, H. P., Guldner, Y., and Vieren, J. P. (1987e). *Appl. Phys. Lett.*
Razeghi, M., Omnes, F., Maurel, Ph., Blondeau, R., and Krakowski, M. (1987f). SPIE Conference, Holland.
Razeghi, M., Blondeau R., Krakowski, M., Decremoux, B., Duchemin, J. P., Lozes, F., Martinet, M., and Bursoussan, (1987g). *Appl. Phys. Lett.* **50**, 230.
Razeghi, M., Defour, M., Omnes, F., and Maurel, Ph. (1987h). *Appl. Phys. Lett.*
Razeghi, M., Maurel, Ph., Omnes, F., and Defour, M. (1987i). *Appl. Phys. Lett.*
Razeghi, M., Maurel, Ph., Omnes, F., and Vassilakis-Thorngren, E. (1987j). "Optical properties of Narrow Gap Low Dimensional Structures" C. M. Sottomayor, (ed.) Plenum Publishing Corporation.
Razeghi, M., Ramdami, J., Legry, P., Vilcot, J. P., and Decoster, D. (1987k). "Proceedings of the 1987 GaAs and Related Compounds Conference," Creta.
Razeghi, M., Hosseini, A., Vilcot, J. P., and Decoster, D., (1987l). "Proceedings of the 1987 GaAs and Related Compounds Conference," Creta.
Ridley, B. K., and Watkins, T. B. (1961). *Proc. Phys. Soc., London* **78**, 293.
Rogers, D. C., Nicholas, R. J., Portal, J. C., and Razeghi, M. (1986). *Semicond. Sci. Technol.* **1**, 350.
Saxena, R., Cooper, C., Ludowise, M., Hilido, S., Borden, P. G. (1981). *J. Crystal Growth* **55**, 58.
Scott, M. D., Moore, A. H., Griffith, I., Griffith, R. J. M., Sussman, R. S., and Oxley, C. (1984). Inst. Phys. Conf. Ser. N°79, 475.
Shapiro, A. H. (1961). "Shape and Flow." Heinemann.
Shaw, D. (1968). *J. Electroch. Soc.* **115**, 405.
Sheldon, P., Jones, K. M., Hayes, R. E., Tsaur, B. Y., Fan, J. C. C. (1984). *Appl. Phys. Lett.* **45**, 3.
Soga, T., Hattori, S., Sakai, S., Takayasu, M., Umeno, M. (1985). *J. Appl. Phys.* **57**, 10.

Stormer, H., Dingle, R., Gossard, A., Gossard, W., Sturge, M. (1979). *Solid State Commun.* **29**, 705.
Suntola, T., Atron, J., Pakkala, A., and Linfors, S. (1980). SID 80, Digest 108.
Suntola, T. (1984). "Extended Abstract of the 16th 1984 International Conference on Solid State Devices and Materials, Tokyo, p. 647.
Sze, S. M. (1981). "Physics of Semiconductor Devices." John Wiley & Sons, New York.
Takahashi, K. (1984). "Proceedings of Research and Development Association for Future Electron Devices." Tokyo, Japan. N°2, pap. 1-2.
Teitjen, J. J., and Amick, J. A. (1966). *J. Electrochem. Soc.* **113**, 724.
Tsang, W. T. (1984). *Appl. Phys. Lett.* **45**, 1234.
Tsang, W. T. (1986). *Appl. Phys. Lett.* **48**, 511.
Tsang, W. T., Logan, R. A., and Ditzenberger, J. A. (1982). *Electron. Lett.* **18**, 123.
Van der Pauw, J. (1958). *Philips Res. Journ.*
Van der Ziel, J. P., Dupuis, R. D., Logan, R. A., Pinzone, C. J. (1987). *Appl. Phys. Lett.* **51**, 2.
Verriele, H., Maricot, S., Constant, M., Ramdani, J., and Decoster, D. (1985). *Electron. Lett.* **21**, 878.
Vilcot, J. P., Constant, M., Decoster, D., and Fauquembergue, R. (1985). *Physica B.* **129**, 488.
Vodjdani, N. (1982). Thèse de doctorat 3e Cycle Université Paris XI.
Wada, D., Mivra, S., Ito, M., Fujh, T., Sakurai, T., and Hiyamizu, S. (1983). *Appl. Phys. Lett.* **42**, 380.
Wang, W. I. (1985). *J. Vac. Sci. Technol.* **83**, 2.
Watanabe, H., and Usui, A. (1986). Inst. Phys. Conf. Ser. N°83, 1.
Watanabe, H. (1986). Springer Proceedings 13, 158, "The Physics and Fabrications of Microstructures and Microdevices," (N. Kelly and C. Weisbuch, eds.)
Wieder, H. H., Clawson, A. R., and McWilliams, G. E. (1977). *Appl. Phys. Lett.* **31**, 468.
Zur, A., and McGill. (1984). *J. Appl. Phys.* **55**, 378.

CHAPTER 7

Stoichiometric Defects in InP

T. A. Kennedy and P. J. Lin-Chung

ELECTRONICS SCIENCE AND TECHNOLOGY DIVISION
NAVAL RESEARCH LABORATORY
WASHINGTON, DISTRICT OF COLUMBIA

	LIST OF SYMBOLS	357
I.	INTRODUCTION	358
II.	THEORY OF STOICHIOMETRIC DEFECTS	359
	1. *Theory of Deep Levels*	359
	2. *Thermodynamics of Defects*	366
III.	EXPERIMENTS THAT REVEAL THE ATOMIC STRUCTURE OF STOICHIOMETRIC DEFECTS	368
	3. *Magnetic Resonance*	368
	4. *Positron Annihilation and EXAFS*	375
IV.	AREAS THAT MANIFEST STOICHIOMETRIC DEFECTS	376
	5. *Introduction*	376
	6. *Growth*	376
	7. *Irradiation and Implantation*	378
	8. *Annealing and Diffusion*	381
	9. *Dislocations and Surfaces*	382
V.	CONCLUSION	383
	ACKNOWLEDGMENTS	383
	REFERENCES	383

List of Symbols

A	hyperfine constant
A_1	singlet state
BZ	Brillouin zone
DLTS	deep level transient spectroscopy
ENDOR	electron nuclear double resonance
E_{nk}	energy of host band state
EPR	electron paramagnetic resonance
EXAFS	extended x-ray absorption fine structure
g	electron paramagnetic splitting factor
G	free energy of formation of defect
GF	Green's function
HF	hyperfine
$H_F{}^i$	formation enthalpy of the ith kind of defect
In_P	In at a P site

LEC	liquid encapsulation Czochralski
MBE	molecular beam epitaxy
MCD	magnetic circular dichroism
ODMR	optically detected magnetic resonance
OMCVD	organo-metallic chemical vapor deposition
P_i	P at an interstitial site
P_{In}	P at an In site
QB	quasiband
S_F^i	formation entropy of the ith kind of defect
$T_1(In)$	interstitial In surrounded by P
$T_1(P)$	interstitial P surrounded by P
T_2	triplet state
$T_2(In)$	interstitial In surrounded by In
$T_2(P)$	interstitial P surrounded by In
TB	tight binding
U^{ext}	external contribution to the defect potential
U^{scr}	screening contribution to the defect potential
U_H^{ps}	pure host crystal pseudopotential
U_D^{ps}	defective crystal pseudopotential
V_{In}	indium vacancy
$(V_{In})_2$	indium vacancy-pair defect
V_P	phosphorus vacancy
$\|X_{Rm}\rangle$	local orbital wave function
$\|\Phi\rangle$	defect wave function
$\|\phi_{nk}\rangle$	host crystal band state wave function
$\|\phi_{nk}^{QB}\rangle$	quasiband wave function
δH_M	migration enthalpy

I. Introduction

InP crystallizes in the zinc blende structure with each atom at the center of a tetrahedron of atoms of the opposite type. Real crystals are imperfect — they contain a variety of defects that affect the optical and electrical properties and thus limit the usefulness of InP for devices. Deviations from perfect stoichiometry can be accommodated by intrinsic point defects, which arise from an excess or deficiency of In and P, or impurities, which favor one sublattice or the other. The simple, intrinsic defects in InP are In and P vacancies, the antisite defects P on the In site (P_{In}) and In on the P site (In_P), and interstitial In or P. Defects can involve more than one atom or more than one lattice site. Such paired or cluster defects can be made up of any combination of the point defects. Line defects, dislocations, and defects at surfaces can also be distinguished. This family of defects is the subject of this review.

Stoichiometric defects can be studied in a variety of ways and can affect a host of material properties. This chapter attempts to provide a comprehensive review of the work that has been done in this area from 1979 to 1989. Because of the authors' expertise, there is an emphasis on theory and magnetic resonance results. No earlier review of this particular subject has been found, but there are a number of related reviews. The subject of defects in semiconductors has been covered in books by Lannoo and Bourgoin (1981) and Bourgoin and Lannoo (1983) and in books edited by Johnson et al., (1985) and by Pantelides (1986). Articles on the same subject are by Neumark and Kosai (1983), Milnes (1983), Masterov (1984) and Clerjaud (1985).

Our review is organized in four parts. Following this introduction, an overview of the theory of stoichiometric defects is presented in Part II. Theoretical approaches and results for InP defects are described. Experiments that reveal the atomic structure of simple defects, such as magnetic resonance and positron annihilation, are reviewed in Part III. Material aspects that have a strong interplay with stoichiometric defects are summarized in Part IV, and a conclusion is given in Part V.

II. Theory of Stoichiometric Defects

1. THEORY OF DEEP LEVELS

a. Various Approaches

Defects in semiconductors sometimes produce energy levels that lie inside the energy gap at a distance from the valence and conduction band edges of the pure crystal. Electrons in these states have wave functions highly localized in the vicinity of the defects, and they are attracted by the strong local potential of the "defect center." Traditionally, such defects are regarded as deep centers. Recently, however, it was found that a strong local potential may also produce fairly shallow energy levels. Isoelectronic impurity defects and rare-earth defects are such examples. Therefore, in a broader sense, deep centers are defined as centers that produce strong local potentials. In general, the effective mass theory is not valid in dealing with the deep centers. Stoichiometric defects (intrinsic defects) are a subgroup of deep centers, which are formed during crystal growth or particle or gamma ray irradiation. Under certain circumstances simple defects become mobile and are trapped by other defects to form stable complex defects.

Research on the electronic structure of such defects in semiconductors has made substantial progress over the past decade in advancing the one-electron theory of defect-induced states toward the same level of sophistication as the theory of electronic states in periodic structures. The various theoretical techniques capable of predicting defect-induced localized states can be grouped according to one of the following features: the choice for the defect potential, the choice of the defect wave functions, the approximation used to solve the Schroedinger equation, and the related question of the geometrical structure used to model the system (e.g., small cluster, infinite crystal). Each method has its own merits and is suitable to treat certain types of defects. In this section, we briefly review the various groups of techniques with particular emphasis on the advantages of each approach. Excellent reviews, focusing on basic questions and on detailed theoretical derivations of different methods can be found in several references (Pantelides, 1978; Jaros, 1980; Masterov, 1984; Lannoo and Bourgoin, 1981).

The formal treatment of defect states can be cast in three distinct steps. The first one is to choose the expression for the wave function $\|\Phi\rangle$ of the defect states. The second step is to introduce the representation for the short range defect potential U. The third step is to find the solution of the one-electron Schroedinger equation:

$$H\|\Phi\rangle = (H_o + U)\|\Phi\rangle = E\|\Phi\rangle, \qquad (1)$$

where H_o is the Hamiltonian for the unperturbed pure system whose electronic structure can be determined by solving the following eigenvalue equation using the standard techniques for extended states.

$$H_o\|\phi_{nk}\rangle = E_{nk}\|\phi_{nk}\rangle, \qquad (2)$$

Here E_{nk} and $\|\phi_{nk}\rangle$ represent the energy and wave function of the nth band state at point k of the Brillouin zone (BZ).

In the following, we shall discuss the essential features of several different approaches for each of the steps mentioned. We begin with the geometrical problem. In order to model defects in a perfect infinite crystal, two distinct approaches have thus far been pursued. The first is to consider a large molecule or cluster consisting of the defect surrounded by the host atoms (Messmer and Watkins, 1971; Hemstreet, 1977; Lin-Chung and Reinecke, 1983; Ferreira and De Siqueira, 1986). To reduce the boundary effect, either hydrogen atoms are used to saturate the bonds, or the size of the cluster is allowed to increase substantially. In such cases, periodicity no longer exists and a real-space approach is carried out. Alternatively, a repeated large cell configuration has been used to examine the defect states utilizing the resulting periodicity. The defects in different cells are assumed not to interact with one another. For the previous models, the electronic states of the defect are obtained to the same degree of accuracy as those for the surrounding host atoms in the same calculation. No use is made of the solutions of Eq. (2) for the perfect crystal. However, the similarity of these models to the real situation in the crystal depends upon the size of the clusters or cells, which are restricted by the computation time as well as by computer memory space. In the second approach, an infinite solid with only an isolated defect or defect complex is considered. The electronic structure of the unperturbed perfect crystal is assumed to be known from the solution of Eq. (2). Defect-induced features are then determined from Eq. (1) by perturbation theory taking U as the perturbation.

A number of expressions have been used to describe the defect wave function $\|\Phi\rangle$. One of these is inspired by the localized nature of deep levels. The wave function is expanded in terms of a set of local orbitals $\{\|X_{Rm}\rangle\}$ centered on each atomic site R in analogy to the tight-binding (TB) method used in band structure calculations

$$\|\Phi\rangle = \sum_{R}^{N} \sum_{m} a_{Rm} \|X_{Rm}\rangle. \qquad (3)$$

The summations over R and m are over the atomic sites in the system and over the orbitals associated with each site, respectively. The $\{\|X_{Rm}\rangle\}$ may be the atomic orbital of impurity or ligand atoms, or it may also be a set of orthogonalized local orbitals such as the orbitals used in the Slater-Koster model, which better describes the extended host states in a semiempirical approach.

Eq. (3) transforms Eq. (1) to a secular equation:

$$\sum_{R}^{N} \sum_{m} a_{Rm}[\langle X_{Rm}\|H_o + U\|X_{R'm'}\rangle - E < X_{Rm}\|X_{R'm'}\rangle] = 0. \qquad (4)$$

The size of the secular determinant reflects the size of the cluster through the number N. This is the so-called "local method."

In the "band method," the defect wave function $\|\Phi\rangle$ is expanded in terms of the wave functions of the band states $\|\Phi_{nk}\rangle$ of the host crystal

$$\|\Phi\rangle = \sum_{n}^{M} \sum_{k}^{BZ} A_{nk} \|\phi_{nk}\rangle. \qquad (5)$$

Based on Eq. (5), Eq. (1) becomes a secular equation of the following form

$$\sum_{n}^{M} \sum_{k}^{BZ} A_{nk}[(E_{nk} - E) \delta_{nn'}\delta_{kk'} + \langle \phi_{n'k'}\|U\|\phi_{nk}\rangle] = 0. \qquad (6)$$

The physical appeal of the band method stems from the fact that solving Eq. (6) permits the analysis of the evolution of the $\|\Phi\rangle$ from the host wave functions $\|\phi_{nk}\rangle$. For highly localized states such as those induced by a transition metal impurity, however, the convergence of the calculated result requires the number M in Eq. (5) to be quite large (Khowash et al., 1985).

This is because the band wave functions are chemically and physically unrelated to those of the impurity, and thus a finite number of band wave functions cannot describe well the characteristics of the $\|\Phi\rangle$. As a result, the complexity, which already exists during the summation process over the BZ, is increased. After this drawback in the band method was recognized, another modified version was proposed, namely, the "quasiband method" (QB), in which the $\|\Phi\rangle$ is expanded in a set of quasiband wave functions $\|\phi_{nk}^{QB}\rangle$ (Lindefelt and Zunger, 1982; Zunger, 1983). The $\|\phi_{nk}^{QB}\rangle$ merely diagonalize the H_o matrix with a limited basis set and are not the eigenstates of H_o. We then have

$$\|\Phi\rangle = \sum_n^M \sum_k A_{nk}^{QB} \|\phi_{nk}^{QB}\rangle, \tag{7}$$

where

$$\|\phi_{nk}^{QB}\rangle = \sum_s^{M_1} b_{nsk} \|\phi_{sk}\rangle + \sum_R^{M_2} \sum_m a_{nkRm} \exp(ikR) \|X_{Rm}\rangle. \tag{8}$$

Mixing a few band functions $\|\phi_{nk}\rangle$ with a few local functions $\|X_{Rm}\rangle$ to form the $\|\phi_{nk}^{QB}\rangle$ improves the convergence of the calculation when $\|X_{Rm}\rangle$ are related to the impurity. Indeed, the $\|\phi_{nk}^{QB}\rangle$ is physically closer to the wave function of the impurity than to that of the host atoms. To reach the same degree of convergence, the number M in Eq. (7) is significantly reduced when compared with that in Eq. (5).

In general, there are two different ways to choose U. The pseudopotential formalism was used for U in the earlier work of Pantelides and Sah (1974). They combined the pseudopotential with the effective mass idea to solve Eq. (1) for a simple point defect, which produces moderately deep levels. Later on the pseudopotential has also been used in the Green's function (GF) method. In this formalism, the defect potential U can be separated into two parts. The external contribution U^{ext} is represented by the pseudopotential formalism, and the screening contribution U^{Scr} may be calculated self-consistently using the density-function formalism (Kohn and Sham, 1965)

$$U = U^{ext} + U^{Scr} \tag{9a}$$

$$U^{ext} = U_D^{ps} - U_H^{ps} \tag{9b}$$

$$U^{Scr} = U_D^{ps} - U_H^{Scr}, \tag{9c}$$

where the subscripts D and H represent the defective and pure host crystals, respectively.

The second way to choose U is to use the tight-binding description of defects and the host crystal Hamiltonian. This approximation has become quite popular in dealing with the deep-level problem. It is an extension of a successful semi-empirical technique previously used in quantum chemistry. Subsequent modifications have been made to improve the convergence of the calculation by using an orthogonalized local basis set of functions to represent the one-electron wave functions. The essence of this approach is to assume that each electron satisfies the same Schroedinger equation given by Eq. (1). Instead of considering H_o and U separately, one takes the matrix elements of the sum $H_o + U = H$ as parameters, which are adjusted to reproduce some known properties, such as the bulk band structure, and then with the help of some interpolation schemes, one predicts other defect properties. This choice of potential is especially effective in cluster calculations on complex defects in covalent semiconductors (Lin-Chung, 1983).

The final step in the formal treatment of the defect problem is to find the defect-induced levels, wavefunctions, or changes in the density of states. Most of the methods for the final step fall into one of the two main categories: (1) solving the full Hamiltonian H in Eq. (1), and (2) treating only the perturbed potential U without asking the details of the solution of H. In the first category, Eq. (1) must be solved either by means of the effective mass approximation or by means of direct diagonalization of the secular determinant when the expansion of $\|\Phi\rangle$ in Eq. (4) or Eq. (5) contains a manageable number of terms as is the case in a finite cluster. Unless the size of the cluster is reasonably large, the energy gap obtained from this method usually tends to be too large and is unrealistic. However, solving Eq. (1) directly has the advantage of producing both the wave functions and the energy levels. The second category of approaches is the Green's function (GF) methods (Bernholc et al., 1980; Besson et al., 1986), which produce the density of states of the system once the eigenstates of the pure crystal are known. In the GF formalism the solution of Eq. (1) is given by the Lippmann-Schwinger equation

$$\|\Phi\rangle = \sum_n \sum_k (1 - G_o U)^{-1} \|\phi_{nk}\rangle = \sum_n \sum_k G G_o^{-1} \|\phi_{nk}\rangle, \qquad (10)$$

where $\|\phi_{nk}\rangle$ again represent the solutions of Eq. (2) for the pure crystal, G, G_o are the GF for the perturbed and unperturbed systems, respectively

$$G_o(E) = \lim_{\varepsilon \to 0} (E - H_o + i\varepsilon)^{-1}$$

$$= \sum_n \sum_k (E - E_{nk})^{-1} \|\phi_{nk}\rangle\langle\phi_{nk}\|. \qquad (11)$$

The density of states $N(E)$ of the system can be calculated conveniently as

$$N(E) = -\pi^{-1} Tr\, Im\, G(E) = -\pi^{-1} Tr\, Im\, G_o(1 - G_o U)^{-1}. \qquad (12)$$

The zeros of $(1 - G_o U)$ determine the localized bound state energy levels, and the inverse of $(1 - G_o U)$ yields information about resonances and antiresonances. One attractive feature of the GF approach is that once the most expensive part of the calculation, the setup of G_o, is done, many local defects may be studied by simply changing the short range perturbation matrices U; therefore, it becomes a valuable tool. For extended defects or for defects that induce extensive lattice relaxations, however, the GF approach has drawbacks.

Published defect calculations have adapted different approaches for each previously mentioned step. For simple point defects with negligible lattice relaxation, the GF approach with self-consistent pseudopotential or TB potential has worked out well. When significant lattice relaxation is present, it is crucial to develop a scheme that can determine the change of the Hamiltonian matrix elements as the interatomic distance changes (Li and Lin-Chung, 1985). For transition metal impurities, the major difficulty in the conventional GF method has also been overcome by adapting the quasiband wave functions. For defects with low symmetry or complicated form, cluster or supercell approaches are better than the GF method.

b. Results

In recent years the localized states of defects in III–V compound semiconductors have been investigated using several theoretical methods. Here, we summarize the results for neutral intrinsic defects in InP.

For sp^3-bonded point defects in InP, the defect levels have nondegenerate A_1 (s-like) or triply degenerate T_2 (p-like) symmetries. In an early calculation of the neutral P vacancy, V_P, employing the GF method and an empirical pseudopotential as a perturbation, it was found that only an A_1-type vacancy state exists in the lower half of the energy gap at $E(A_1) = E_v + 0.12$ eV. The T_2 state on the other hand resonates with the conduction band (Srivastava, 1979).

Subsequently, a large number of substitutional defects and nearest-neighbor pairs of defects in III–V crystals were studied in calculations based on a GF method in a nearest neighbor empirical TB model (Hjalmarson et al., 1980; Sankey and Dow, 1981). From this investigation, the physics of the states of a nearest-neighbor defect complex, which has reduced C_{3v} symmetry, was understood. Defects with C_{3v} symmetry have nondegenerate a_1 (σ-like) and doubly degenerate e (π-like) states. The e states of the pair have energies virtually identical to the point-defect T_2 energies, whereas the a_1 levels of the pair defect can differ markedly in energy from the A_1 and T_2 point-defect energies of their parent atomic defects. Thus, instead of removing an intrinsicly isolated defect, it is possible to use pairing as a method of driving a deep A_1 isolated defect level either shallow or out of the gap. The calculated results give energies for the A_1 levels of the isolated vacancies V_{In} and V_P at -1.3 and 1.6 eV, respectively, where the zero of energy is at the valence band edge (Hjalmarson et al., 1980). The T_2 levels of V_{In} and V_P are at 0 and 1.9 eV, and the divacancy (V_P-V_{In}) induced levels are predicted to be at 0 and 1.9 eV for e levels and -1.0 and 1.75 eV for the a_1 levels (Sankey and Dow, 1981). The $V_P - P_{In}$ pair produces a level at 1.25 eV, which may correspond to the 1.14 eV level detected in DLTS (Wieder, 1980). For comparison, the photoluminescence values corresponding to the ionization levels for V_P, V_{In}, and their complex are at 0.99, 1.21, and 1.08 eV, respectively, assuming the assignments are correct (Temkin et al., 1981).

The antisite defects in InP are also predicted in a similar theoretical approach. The A_1 state of P_{In} (anion on cation site) is found at 0.56 eV and the T_2 state of In_P (cation on anion site) at 0.14 eV (Buisson et al., 1982).

Instead of the simple TB model, an empirical TB linear combination of atomic orbitals basis set in conjunction with the Koster-Slater Green's function technique was used to find the point defects in six III–V semiconductors (Das Sarma and Madhukar, 1981a; 1981b). The results show that the $T_2(V_P)$ and $T_2(V_{In})$ in InP are resonant states in conduction and valence bands, respectively. The A_1 state of V_P is at 1.35 eV above the valence band edge.

A slightly different approach for these defects was carried out using equivalent orbitals and including interaction of atoms up to the third-nearest neighbors (Makhmudov et al., 1985). The vacancy levels are found at $T_2(V_P) = 0.78$, $A_1(V_P) = 0.39$ eV, and $T_2(V_{In}) = 0.06$ eV. The divacancy ($V_P - V_{In}$) levels are found at 0.73, 0.53, and 0.12 eV.

The various methods of calculation give significantly different values for a given defect level. The main interest of some of the previous calculations does not center on quantitatively accurate values of the defect-induced levels in any one compound but rather in finding the major chemical trends in the behavior of such states as the bond ionicity changes among the III–V compound semiconductors. Reliable quantitative results for individual system probably require substantially more sophisticated calculations.

Alternatively, a continued fraction method (Haydock et al., 1972) was used in determining levels induced by intrinsic defects in III–V semiconductors (Lin-Chung and Reinecke, 1983). No gap states were found for V_P and In_P in InP. The gap states for V_{In} and for P_{In} were found to be at 0.44 eV and 1.33 eV, respectively. Further improvements were carried out in constructing a new set of semi-empirical expressions for the two-center integrals, which enter the Slater-Koster parameters and have explicit dependences on the interatomic separations as well as on other atomic and host crystal characteristics (Li and Lin-Chung, 1985). Then recursion calculations for self-interstitials in InP become more reliable (Lin-Chung, 1989). There are two different tetrahedral interstitial sites in the zinc-blende structure. The T_1 site is surrounded by anion atoms, and the T_2 site is surrounded by cation atoms. The calculated levels for the isolated P or In interstitials situated at T_1 or T_2 sites and at some of the pair defect complexes are summarized in Table I.

The previous calculations are all for neutral defects. For neutral vacancy levels, the T_2 state contains one or three electrons. For the cation vacancy, the T_2 state is half filled and spin alignment could produce a many electron orbital singlet state as in GaP (Kennedy et al., 1983). For the anion vacancy, the T_2 state is occupied by only one electron, and, therefore, a Jahn-Teller distortion is expected to occur giving rise to states with lower symmetry and a splitting in energy. This distortion energy is of the order of a tenth of an eV range and is often offset by the Coulomb repulsive energy when the defect becomes negatively charged. The determination of charged deep levels requires a self-consistent calculation incorporating the Coulomb effects. Such a calculation was carried out for GaAs (Baraff and Schlüter, 1985). For InP, however, much progress in this area needs to be made.

2. THERMODYNAMICS OF DEFECTS

a. Formation of Isolated Defects and Pairs

In this section we examine briefly the formation of intrinsic defects at thermal equilibrium in InP. The energy required to move an atom from one site in a pure crystal to another site in order to form a defect is called the

TABLE I

CALCULATED DEEP LEVELS
FOR INTERSTITIAL DEFECTS
AND PAIR DEFECT COMPLEXES
IN InP. (Energy is measured in
eV from the top of valence
band.

Defect structure	E(eV)
$T_1(In)$	1.34
$T_2(In)$	1.52
$T_1(P) - V_P$	1.28 1.63
$T_1(P)$	1.55
$T_2(P)$	0.7
$(V_{In})_2$	0.36 0.71
$In_P P_{In}$	1.75
$In_P V_{In}$	0.16 0.40

formation enthalpy H_F of the defect. It is related to the bond dissociation energy and the lattice relaxation and distortion energies. At thermal equilibrium, n_i, the number of defects of the ith kind is related to N, the number of available sites for the defect in the system by the following relation

$$\frac{n_i}{N} = \exp\left[\frac{-H_F^i + TS_F^i}{kT}\right], \quad (13)$$

where T is the temperature, and H_F^i and S_F^i are the formation enthalpy and entropy of the ith kind of defect, respectively. The latter is associated with the disorder induced by the lattice vibrations in the vicinity of each isolated defect. The theoretical H_F^i for the vacancy and for the self-interstitial in Si, C, and Ge have been given from both classical and quantum mechanical approaches (Swalin, 1961; Seeger and Swanson, 1968; Benneman, 1965). For compound semiconductors, there are additional complications because of the existance of two types of vacancies and two types of antisite defects. Thus, such a calculation has not yet been carried out for InP.

Equation (13) is valid when there is no interaction between intrinsic defects. Under such a condition, at thermal equilibrium isolated defects or vacancy-interstitial pairs have a higher concentration than complex defects. When long-range interactions between defects is present, a defect that induces a lattice compression will tend to pair with a defect that induces lattice expansion in order to reduce the strain energy created through the interaction. When short-range interaction exists, the defects can form small stable aggregates such as divacancies. They can again be treated as independent entities in Eq. (13).

At thermal equilibrium a given defect may exhibit different charge states. For example, the vacancy defects in Si exist in V^+, V°, V^- (a negative U system) (Baraff et al., 1980). Their relative concentrations depend upon the chemical potential E_F, which is subject to the constraint of global charge neutrality and is rather complicated to determine. The relative concentrations have the following expression:

$$\frac{[V^-]}{[V^\circ]} \approx \exp[-\{G(V^-) - G(V^\circ) - E_F\}/kT], \quad (14)$$

where $G(V) = H_F(V) - TS_F(V)$ is the free energy of formation of the charged defect V.

b. Defect Migration and Diffusion

Defect migration in III-V compound semiconductors is permitted by one of several mechanisms. They include vacancy hopping to nearest neighbor site, divacancy or multivacancy hopping, interstitial-substitution hopping, and interstitial kick-out mechanism (Van Vechten and Wager, 1985a). Semiempirical treatments of vacancy migration have been available for several years. The Glyde–Flynn model gives an excellent fit to experimental values for elemental crystals (Glyde, 1967; Flynn, 1968). In that model, the enthalpy of migration for point defects is regarded as a pure potential energy. Later, a ballistic model was proposed, which assumed that the kinetic energy contribution is greater than the potential energy when the migrating atom is passing through the region around the saddle point of the potential along the migration path. This model is especially suitable in treating compounds containing cations that are much heavier than their anions (Van Vechten and Wager, 1985b). The migration enthalpy in this model is given by the following expression:

$$\delta H_M = \tfrac{1}{2} M (Fdv_D)^2, \quad (15)$$

where M is the mass of the mobile atom, d is the distance between the initial and final sites of each step of hopping. v_D is the empirical Debye frequency, and F is a structure-dependent fitting parameter, which is 0.9 for the faced-center cubic structure. For vacancy migration in InP, Eq. (15) gives $\delta H_M(V_P) = 1.2$ eV and $\delta H_M(V_{In}) = 0.3$ eV.

Similarly, interstitial migration also requires a strong kinetic energy contribution at high temperature. At low temperature, the rapid self-interstitial migration in Si has been explained as being due to the "Bourgoin-Corbett" athermal mechanism, which invokes rapid changes of the ionization states of the interstitials (Bourgoin and Corbett, 1972). Presumably the same mechanism prevails for the interstitial migration in InP.

III. Experiments That Reveal the Atomic Structure of Stoichiometric Defects

3. MAGNETIC RESONANCE

a. Introduction

Electron paramagnetic resonance (EPR) and related experiments have one major advantage and one major disadvantage in the study of defects in InP. The advantage derives from the fact that a perfect InP crystal has no unpaired electron spins and thus exhibits no EPR. The technique is responsive only to defects, in particular, defects in a paramagnetic charge state. These states split linearly with magnetic field and resonance can be performed between the levels. The disadvantage derives from the fact that the constituents of InP have nuclear spins. ^{31}P (100% abundant) has nuclear spin 1/2 and ^{113}In (4% abundant) and ^{115}In (95.8% abundant) have nuclear spin 9/2. Thus, like other group III-V compounds, the matrix is full of the magnetism produced by nuclear spins. Since the wave function of any defect is spread over at least a few neighbors, the EPR lines are broadened by this nuclear spin matrix. The result is that the resolution of the typical EPR spectrum is not very good compared to Si or a II-VI compound.

These ideas are conveniently expressed by introducing a spin Hamiltonian, which describes the magnetic interactions. For most of the cases to be considered here, the following simple Hamiltonian suffices:

$$\mathcal{H} = g\mu_B \vec{B}\cdot\vec{S} + A\,\vec{I}\cdot\vec{S}, \tag{16}$$

where the first term describes the splitting of a defect with spin S by the magnetic field B, and the second term describes the hyperfine interactions between the electronic spin S and the nuclear spins I. The hyperfine interactions function as a blessing as well as a curse. When there is a very strong interaction with the central atom of a defect, the splitting may overcome the weaker neighbor interactions and produce a spectrum with resolved hyperfine structure. From its multiplicity, one deduces the nuclear spin and often the chemical identity of the defect. In this case it is possible to discriminate intrinsic defects from the matrix of like atoms that reside on their proper lattice sites. This is the unique advantage of EPR to the study of intrinsic defects.

By considering the expected strength of the largest hyperfine interaction, the simple intrinsic defects of InP can be divided into two groups. The first group contains defects that would be expected to have resolved HF structure in EPR. The large HF structure denotes a Fermi contact interaction involving a central nucleus and an A_1 state in the forbidden gap. Two defects satisfy these requirements: the In interstitial (In_i) and the P antisite (P_{In}). All the other defects fall into the second group and would exhibit only a single unresolved EPR line. The P interstitial (P_i) and the In antisite (In_P) have T_2 states. The vacancies lack a central atom.

Two extensions of EPR with greatly enhanced power have been applied to the study of stoichiometric defects in InP. The first is optically detected magnetic resonance (ODMR) (Cavenett, 1981; Davies, 1985). In this technique, changes in an optical property, either absorption or photoluminescence, produced by magnetic resonance of the defect are detected. ODMR has greater sensitivity than EPR and often provides a measurement of an optical transition energy to go along with the structural (EPR) information. The second extension is electron nuclear double resonance (ENDOR) (Spaeth, 1986). In this technique, nuclear magnetic resonance is performed with EPR leading to greatly enhanced resolution of the hyperfine interactions. In the one example to be discussed, the ENDOR is optically detected.

b. Antisites

Due to their characteristic, fully resolved hyperfine splitting, the P_{In} antisites are readily detected and identified. A spectrum is shown in Fig. 1c. The identification proceeds as follows (Kennedy and Wilsey, 1984). The two broad lines are split by 98 mT. The multiplicity indicates that the hyperfine is due to a nucleus with spin 1/2. Because the intensity of the spectrum indicates a fairly high concentration of defects, only the common impurities (C, O, Si, B, N, and S) and constituents (In and P) need to be considered in assigning the spectrum. Since only ^{31}P is 100% spin 1/2, the spectrum is assigned to a defect with ^{31}P at its center. The only such defects are the P_{In} antisite and the P_i interstitial. As discussed above, however, only the antisite has an A_1 state in the gap, which would give rise to the large hyperfine splitting. Thus, the assignment to P_{In} is complete.

The observations by optically detected magnetic resonance have provided many more details about P-antisites. The first such experiments detected the changes in the absorption of circularly polarized light when the defects undergo magnetic resonance (Deiri et al., 1984, Kana-ah et al., 1985). These magnetic circular dichroism (MCD) experiments revealed excited states 1.3 eV above the ground state for P_{In} and proved to be 100 times more sensitive than the EPR. Antisites with slightly different magnetic resonance parameters were observed for irradiated Zn-doped and Sn-doped materials (See Table II). Later magnetic resonance was detected as a change in the

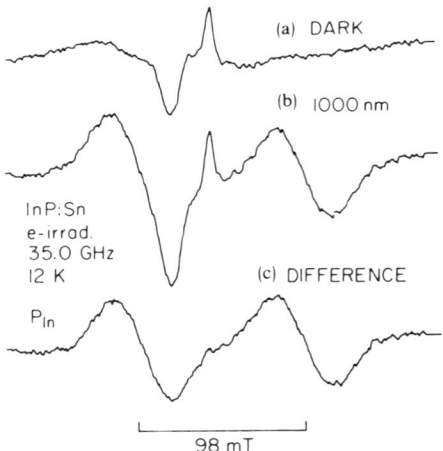

FIG. 1. EPR spectrum of P_{In}. Spectrum (a) is dominated by unwanted background signals. Photoexcitation in (b) populates the paramagnetic charge state. The difference spectrum (c) is due only to the antisite and is dominated by the ^{31}P hyperfine splitting into two lines. [From Kennedy and Wilsey (1984).]

TABLE II

P_{In} RESONANCE PARAMETERS

Dopant	Processing	g	A (cm^{-1})	T (K)	Method	Reference
Sn	2 MeV e-irrad.	1.992 ± 0.008	0.092 ± 0.005	12	EPR	Kennedy and Wilsey (1984)
Zn	2 MeV e-irrad.	2.012 ± 0.005	0.104 ± 0.002	2	PL	Cavenett et al., (1985)
Sn	2 MeV e-irrad.	1.991 ± 0.005	0.099 ± 0.002	2	MCD	Kana-ah et al., (1985)
Zn	As grown	1.992 ± 0.008	0.097 ± 0.005	1.6	PL	Kennedy and Wilsey (1987)
Zn	As grown & e-irrad.	2.000 ± 0.003	0.098 ± 0.002	1.7	MCD	Jeon et al., (1987)

photoluminescence intensity (Cavenett et al., 1985; Kennedy and Wilsey, 1987; and Deiri et al., 1988). These observations were made in electron-irradiated InP:Zn and also as-grown InP:Zn, indicating that the P_{In} antisite is a native defect in Zn-doped, LEC InP. The dependence of the ODMR on wavelength in as-grown samples showed that the emission peaks at 0.8 eV. The process is probably deep donor (P_{In}) to shallow acceptor pair recombination since the sample was Zn-doped. Thus, the P_{In}^+/P_{In}^{++} energy level is about 0.8 eV above the valence band.

The EPR and ODMR results for P_{In} are summarized in Table II. Note that the parameters for Zn-doped and Sn-doped samples differ. This suggests that there may be a difference in the structure of the nearest neighbors for the two cases. Recent optically detected ENDOR experiments have studied the hyperfine interactions of the nearest neighbors (Jeon et al., 1987), which are unresolved in EPR. These experiments show that in both as-grown and electron-irradiated Zn-doped InP the antisite is surrounded by four P neighbors. Perhaps the antisite in Sn-doped material is only surrounded by three P neighbors, as is the case for antisites in GaP (Kennedy and Wilsey, 1985). The second difference evident from the table is a slight temperature dependence to the central hyperfine constant. This effect is not understood at present. The P_{In} antisite is the only defect in InP that exhibits resolved hyperfine structure in EPR and in which ligand hyperfine structure was resolved using ENDOR. Thus, P_{In} is currently the best understood intrinsic defect in InP. However, information can be gained concerning defect structure when no resolved hyperfine is observed, as will be described in the following sections.

No reports were found of spectra attributed to the acceptor antisite, In_P. However, a report appeared of a spectrum that is tentatively attributed to the acceptor antisite paired with boron ($Ga_{As} - B_{Ga}$) in GaAs (Kaufmann et al., 1986). The spectrum is anisotropic with trigonal symmetry and a g-tensor, which is consistent with a d^9 or a p^5 one-hole configuration. The linewidth is broad, but does not exhibit a resolved hyperfine structure, which is consistent with the p-like (or T_2) symmetry of an acceptor antisite. The spectrum was observed in LEC grown GaAs.

c. Vacancies and Interstitials

There are a number of observations of single line spectra in InP that are usually attributed to vacancies. Arguments are given by the authors to support these assignments. One must keep in mind, however, that impurities with nuclei of zero spin and defect states with T_2 symmetry would exhibit similar spectra. Eventually ENDOR experiments should clarify the assignments of single-line spectra.

TABLE III

SINGLE LINE SPECTRA IN IRRADIATED ZN-DOPED SAMPLES.

g	Linewidth (mT)	T (K)	Method	Model	Reference
2.05 ± 0.04	180	4	EPR	V_P	von Bardeleben (1986)
2.00 ± 0.01	35	4	EPR		von Bardeleben (1986)
1.93 ± 0.01	75	2	MCD	V	Deiri et al., (1984)

A very broad single line has been observed in electron-irradiated InP:Zn by von Bardeleben (1986). Its resonance parameters are given in Table III. Arguments and spectral simulations lead to the assignment of the broad linewidth to In superhyperfine interactions. Thus, the spectrum is assigned to the phorphorus vacancy (V_P). The defect is introduced at a high rate by high energy electrons and is also unstable at room temperature. A second, less broad single line (See Table III) is reported in the same paper but no details are given.

A single line with a different linewidth has been observed by MCD-ODMR in electron-irradiated InP:Zn (Deiri et al., 1984). Its linewidth of 75 mT could also be attributed to a weak central hyperfine or a strong ligand hyperfine interaction. Its g-value is smaller than the other single lines (See Table III), which may suggest that it is donorlike. This line was attributed to vacancies.

There have been no spectra reported that may be attributed to self-interstitials in InP. Recently Ga self-interstitials have been reported in $Al_xGa_{1-x}As$ grown by molecular beam epitaxy (MBE) (Kennedy and Spencer, 1986). This observation suggests that for low substrate temperatures, the MBE material is Ga rich. Isolated Ga self-interstitials have been reported in bulk GaP grown with B_2O_3 (Lee, 1988) and isolated Zn interstitials in electron-irradiated ZnSe (Rong and Watkins, 1987).

d. Pairs that include an Intrinsic Defect

Point defects often exist in InP as pairs. In certain cases, the existence of a pair is evident from its magnetic resonance spectrum. ENDOR, with its much higher resolution, will certainly reveal further pairing. The following are results obtained to date in which the existence of an intrinsic defect paired with a second defect is deduced using magnetic resonance.

When InP:Fe is electron irradiated, changes occur in the magnetic resonance of the Fe, which indicate pairing (Kennedy and Wilsey, 1981). The spectrum of Fe^{3+} on the cubic In site diminishes with irradiation and is replaced with a spectrum of trigonal symmetry. The new spectrum still has a high spin, characteristic of transition metals, but both the value of the spin and the symmetry are new. *In situ* irradiation of the sample with infrared light causes the symmetry to change again but does not restore the original Fe spectrum. Thus, the cubic Fe spectrum is not disappearing due to a Fermi level change caused by the electron irradiation. The conclusion is that a radiation-induced defect, probably an interstitial, is mobile at room temperature and pairs with the Fe. Brailovskii *et al.*, (1982) have studied the isochronal annealing of the Fe-unknown defect pair, which takes place from 400 to 450 K following first-order kinetics with an activation energy of 1.0 ± 0.1 eV.

Annealing of undoped, LEC InP produces high resistivity material in some cases that exhibits an ODMR signal due to a P-antisite paired with an unknown defect (Kennedy *et al.*, 1986). The observations were made in two crystals that increased in resistivity following an anneal at 900–940 °C under 6 atm of phosphorus for about three weeks. The ODMR signals (See Fig. 2) have triplet ($S = 1$) character with an angular dependence, which indicates a center of monoclinic (C_s) symmetry. The spectrum also contains an isotropic hyperfine splitting due to a nucleus of spin 1/2. The strength of this coupling is 0.041 ± 0.002 cm^{-1}. This is close to half the value for the $S = 1/2$ P_{In} centers (See Table II), which is the ratio expected when the $S = 1$ excited state is observed. From the hyperfine splitting, the spectrum is attributed to a P_{In}

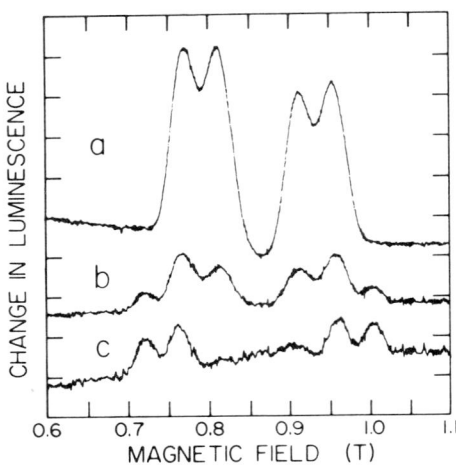

FIG. 2. ODMR spectra of the $P_{In} - X_{In}$ pair for B parallel to [110]. Spectrum (a) is in phase and spectrum (b) in quadrature with the chopped microwaves. Both show the fine structure splitting into low and high field groups indicating that $S = 1$. Spectrum (c), a linear combination of (a) and (b), and (a) itself show that there is a smaller hyperfine splitting from ^{31}P. (Reprinted with permission from Trans Tech Publications, Kennedy, T. A., Wilsey, N. D., Klein, P. B., and Henry, R. L. (1986). Triplet spin ODMR from phosphorus antisites in undoped InP. *In* "Defects in Semiconductors" (H. J. von Bardeleben, ed.). 271–276.)

antisite. From the symmetry, pairing with a next-nearest neighbor defect is inferred. Thus the structure is $P_{In} - X_{In}$ The fine structure parameters of the spectrum (D = 0.110 ± 0.003 cm^{-1} and E = 0.038 ± 0.001 cm^{-1}) should provide further information on the defect when a more complete analysis is performed.

4. POSITRON ANNIHILATION AND EXAFS

Positron annihilation can be used to detect vacancies in metals and semiconductors. Positrons are trapped in regions of reduced atomic density such as vacancies, vacancy clusters and complexes, and dislocations, since the Coulombic repulsion between the positron and the positive atom cores is reduced in these regions. The net charge of the trap should be neutral or negative. The trapping is detected by an increased positron lifetime or by changes in the angular correlations of the two photons emitted. By its nature, this technique is specific to vacancy-related defects, which are difficult to identify by other methods.

Positron annihilation of as-grown crystals revealed vacancies in heavily Zn-doped samples (Dlubek and Brummer, 1985). Undoped, Sn-, S- and Fe-doped material exhibits a positron lifetime of 242 to 252 ps. Zn-doped crystals with concentration greater than 4.5×10^{18} cm^{-3} have a second lifetime of 320 to 330 ps, which is attributed to Zn-vacancy complexes. Correlation with transport results indicate that the number of complexes ($1-2 \times 10^{18}$ cm^{-3}) is equal to the number of inactive Zn atoms. Annealing studies show that these complexes are stable up to 400 °C.

Positron lifetime and angular correlation measurements have been performed with Hall effect measurements on electron-irradiated InP (Brudnyi et al., 1982ab). The lifetime measured in as-grown material is 246 ps. Irradiations to high fluences (up to 1×10^{19} cm^{-2}) produced changes in the angular correlation curves, which are attributed to vacancies. Correlations between the angular curves and Hall effect in isochronal anneals suggest that the negatively charged acceptor is V_{In} with an energy level of $E_c - 0.33$ eV.

Extended x-ray-absorption fine structure (EXAFS) shows great promise in elucidating the structure of defects, especially impurities, in semiconductors. S in GaAs has recently been studied through the S K-alpha fluorescence yield (Sette et al., 1986). Heavily doped (2×10^{19} cm^{-3} or more) samples were studied; although the detection limit is considerably lower (5×10^{17} cm^{-3}). The technique provides measurements of the impurity-neighbor bond lengths. In the GaAs:S case changes from normal bond lengths led to a model in which approximately half of the sulfur was paired with an As vacancy ($S_{As}-V_{As}$). This heavy doping result is similar to the InP:Zn case discussed previously. EXAFS studies in InP can be expected in the future.

IV. Areas That Manifest Stoichiometric Defects

5. INTRODUCTION

The remainder of this review is organized around different aspects of growth and processing of InP. Many of the manifestations of stoichiometric defects are less specific than a magnetic resonance spectrum that contains hyperfine interaction. Thus, the attribution of a signature to a stoichiometric defect depends heavily on the circumstances of growth or processing, which enhance the experimental signature. Comparisons with recent theory should help to strengthen the assignments of these effects to specific defect models.

6. GROWTH

InP is currently grown using a variety of techniques, as described in other chapters in this volume. Due to the high vapor pressure of P at the melting point, the liquid encapsulation Czochralski (LEC) method is generally used to grow bulk crystals following an InP synthesis step (Tomzig, 1990; Farges, 1990; Inada and Fukuda, 1990). Other methods, such as vertical gradient freeze growth (Monberg, 1990), are also being explored. InP and its alloys are grown epitaxially on InP by liquid phase, molecular beam, hydride (Stillman and McCollum, 1990) and organo-metallic chemical vapor deposition (OMCVD) (Razeghi, 1990) techniques.

The stoichiometry of InP under LEC growth conditions has been studied by precise measurements of the lattice constant and density of single crystals. In-rich melts were found to produce nonstoichimetric crystals with microinclusions of In and high phosphorus vacancy concentrations (Morozov et al., 1983). Hall measurements showed that the electrically active defects are at much lower concentration and, thus, the phosphorus vacancies may be electrically inactive. Further analysis of the point defects formed in LEC growth was presented with the prediction that indium precipitates form upon cooling of single crystals of InP grown from stoichiometric melts (Morozov et al., 1984).

Many of the intrinsic defects in InP have deep levels in the energy gap and can be expected to have radiative transitions producing deep luminescence. Many such spectra have been reported and assigned to intrinsic defects on the basis of correlations with growth conditions or annealing treatments. These results are summarized in Table IV. An emission band around 1.10 eV is often observed and has been assigned to a V_{In}-V_P pair in polycrystalline InP (Temkin et al., 1981) and to a V_P-Fe_{In} pair in epitaxial and bulk crystalline material (Yu, 1980). Studies including ion implantation suggest that this line, or one at the same energy, can be attributed either to V_P or a V_P-impurity pair (Georgobiani et al., 1983 a; b).

TABLE IV

Proposed Structure of Deep-level Emission Bands in InP.

Emission band (eV)	Defect structure Model	Reference
1.21	V_{In}-impurity	Temkin and Dutt (1983)
	P_i or P_i-impurity	Temkin and Dutt (1983)
		Georgobiani et al., (1983)
1.15	V_{In}-impurity	Temkin and Dutt (1983)
	Mn^{2+}	Eaves et al., (1982)
1.08/1.10	$V_P - Fe$	Yu (1980)
	$V_{In} - V_P$	Temkin et al., (1982)
	V_P or V_P-impurity	Georgobiani et al., (1983)
0.99	V_P-impurity	Temkin and Dutt (1983)
0.8	P_{In}	Kennedy and Wilsey (1987)

By studying the photoluminescence along a melt-grown, polycrystalline ingot, Temkin et al., (1981) also observed lines at 0.99 and 1.21 eV (See Fig. 3). Gradient freezing of the InP is expected to induce a gradient in the vacancy concentration when a vapor pressure of P is present. Anneals were performed to confirm assignments of the 0.99 eV band to a donorlike P vacancy and the 1.21 eV line to an acceptorlike In vacancy or P interstitial. Later analysis suggested these defects could be paired with impurities (Temkin and Dutt, 1983).

Some uncertainty remains whether some of these spectra might be due to isolated transition metal impurities. Eaves et al., (1982) assign a line at 1.15 eV to Mn^{2+}, and it is also quite possible that some of these defects are intrinsic defect-transition metal pairs. Temkin and Dutt (1983) assign the 1.15 eV line to an In vacancy-native acceptor pair. Yan et al., (1983) also discuss the assignment of 1.15 eV photoluminescence in the light of ODMR experiments on Mn-diffused InP.

Optically detected magnetic resonance directly links deep luminescence with atomic structure information. Zn-doped LEC crystals exhibit the P_{In} ODMR (See Section 3b) on 0.8 eV emission (Kennedy and Wilsey, 1987). Native group V antisites are also observed in LEC GaP and GaAs. It is interesting to speculate on whether they reveal a deviation in stoichiometry toward the P-rich side (Morozov et al., 1984) or occur due to a self-compensation process.

InP grown epitaxially by MBE or OMCVD exhibits a 1.36 eV band due to a bound exciton (Wakefield et al., 1984). Since either growth method is expected to produce epilayers depleted of phosphorus, these authors assign the bound exciton to a complex incorporating a phosphorus vacancy. This same emission was also observed in heat-treated n-type LEC InP (Duncan et

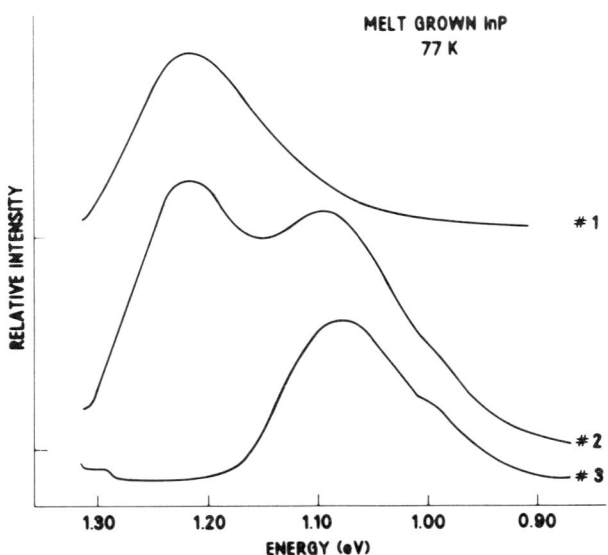

FIG. 3. Deep photoluminescence bands in InP. The numbers refer to different positions in the ingot. The shoulder at 0.99 eV was assigned to a P vacancy and the peak at 1.21 eV to an In vacancy or P interstitial. [From Temkin et al., (1981).]

al., 1984) and polycrystalline InP prepared from an In-rich solution in a Bridgman reactor (Barthruff, 1979). The deep levels associated with intrinsic defects can also be detected electrically by deep-level transient spectroscopy (DLTS). A DLTS signal with activation energy of 0.42 eV has been correlated with the 1.1 eV deep luminescence band in LEC material (Yamazoe et al., 1981). Heat treatment provided the correlation and led to the assignment to complexes involving P vacancies. The same DLTS signature is observed in InP grown by synthesis, solute diffusion (SSD) but only after heat treatment (Chung et al., 1983). In Zn-doped, LEC-grown material, a hole trap was observed with an activation energy of 0.52 eV above the valence band (Li et al., 1984). The trap anneals at 170°C and is attributed to either a phosphorus vacancy or a phosphorous interstitial.

7. Irradiation and Implantation

Energetic particles and gamma rays incident on InP displace atoms from their lattice sites producing vacancies and interstitials. Depending on the energy and mass of the incident particles, either individual atoms or groups of atoms are displaced in each collision. The irradiation leaves the crystal in a state far from thermal equilibrium, which can be restored by thermal or other annealing. The resulting defects in the crystal at any stage depend on the diffusion characteristics of the vacancies and interstitials. Often the dopants and residual impurities in the material also play a role.

Various experiments have shown that there is a marked difference in the defect introduction rate between electron-irradiated n- and p-type InP, with a larger rate for p-type. Positron annihilation experiments do reveal indium vacancies in heavily electron-irradiated n-type InP (Brudnyi et al., 1982a; b). DLTS studies have revealed that the dominant damage in p-type material consists of three hole traps labelled H2–H4 (Sibille and Bourgoin, 1982). See Fig. 4. A great deal of information on the nature of the defects has been gained by using the DLTS signatures in conjunction with irradiations and anneals under different conditions. By lowering the temperature, one can study more closely the primary nature of the damage; that is, before diffusion leads to defect reactions. Furthermore, by lowering the energy of the electrons one can selectively displace only P atoms because they are much lighter than In atoms. Sibille and Suski (1985) have performed these experiments that reveal that the dominant damage involves V_P and P_i and that one of these defects can migrate long distances below room temperature. Annealing to room temperature shows that the dominant defects become pairs of the mobile P-lattice defect with the acceptor impurity (Sibille et al., 1986). By annealing to 200°C, the crystal is returned to near its original state (Sibille and Bourgoin, 1982).

These DLTS results can be readily reconciled with some of the magnetic resonance results on electron- and neutron-irradiated InP. Phosphorus vacancies have been assigned to EPR spectra in electron-irradiated (von Bardeleben, 1986) and neutron-irradiated (Goltzene et al., 1987) materials. Irradiation of Fe-doped material seems to be similar to irradiation of p-type.

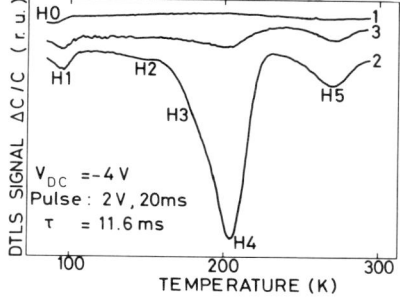

FIG. 4. Deep level transient spectra of p-type InP. Spectrum (1) is before irradiation, spectrum (2) following irradiation with 10^{16} electrons cm^{-2}, and spectrum (3) after a 470 K anneal. H2 – H4 are hole traps formed by the pairing of P-sublattice defects with acceptors. [From Sibille and Bourgoin (1982).]

Room temperature n- or e-irradiation produces an Fe-radiation-induced defect pair (Kennedy and Wilsey, 1981), where the radiation-induced member could be either an In_i (Kennedy and Wilsey, 1981; Goltzene et al., 1987) or a divacancy (Brailovskii et al., 1982). The EPR experiments require that the defect paired with Fe have $S = 1$, which could be satisfied by either In or P interstitials. Annealing of this complex takes place from 130 to 180°C (Brailovskii et al., 1982).

All evidence suggests that the introduction rate for P_{In} antisites is small. In the original work, large electron fluences were required before the antisites were observed (Kennedy and Wilsey, 1984). Subsequent work has not yet resolved whether the antisites are radiation induced or not, while anitisites have been observed in as-grown p-type material (Kennedy and Wilsey, 1987; Deiri et al., 1987).

Radiation-induced defects in InP exhibit metastability in their atomic structure, which has been revealed in two experiments. The Fe-defect pair observed in EPR at 4.2 K is sensitive to light (Kennedy and Wilsey, 1981). Following near-bandgap excitation, the trigonal EPR spectrum is completely quenched and replaced by a spectrum of lower symmetry. Thermal cycling (annealing) to room temperature restores the trigonal spectrum. On the atomic scale, this could easily be due to motion of the interstitial, which is paired with the substitutional Fe. In the n-irradiated InP:Fe, the trigonal spectrum is observed along with a second Fe-pair spectrum, which is analyzed revealing the symmetry of an Fe-next-nearest neighbor interstitial pair (Goltzene et al., 1987). Metastability in the DLTS signature has been observed for a defect called the M center in e-irradiated InP (Levinson et al., 1983a). Here the spectrum depends on the voltage bias conditions and is attributed to a charge-state controlled configurational metastability. Atomic models are difficult to infer from DLTS results, but the structure may be a shallow-donor-intrinsic defect complex (Levinson et al., 1983b) or involve the P_{In} antisite with phosphorus and indium vacancies (Wager and Van Vechten, 1985).

Because of the heavy mass of the incident particles, ion implantation results in displacements of more than one atom. Annealing is necessary to enable the implanted species to adopt lattice sites. Interesting stoichiometric changes occur because of the large mass difference between In and P. Calculations based on a Boltzmann transport equation showed that implantations with heavy ions result in an In-rich shallow region and a P-rich deeper region (Christel and Gibbons, 1981). Experiments subsequently showed, however, that sputtering was also important, and the different In and P sputtering rates led to a P-rich surface region (Haberland et al., 1984). These effects are especially important when a group IV element is implanted that can occupy either the In- or P-sublattice.

8. ANNEALING AND DIFFUSION

Following growth, InP is subjected to a variety of thermal treatments in order to form ohmic contacts, grow epitaxial layers, activate ion implants, or simply alter the state of the material. During these treatments, the defects in the material are changed by the diffusion and possibly reaction of the impurities and intrinsic defects present. In addition to the temperature, the ambient is important to the types and distribution of defects following the treatment (Kröger, 1964; Flynn, 1970). The most important factor relating to this review is the potential loss of P in thermal anneals. This occurs at 425 °C for a dynamic vacuum ambient, 575 °C for a flowing gas, and 600 °C for a sealed ampoule (Wong and Bube, 1984). Laser processing can also deplete the surface region of phosphorus (Moison and Bensoussan, 1982). Two simple types of diffusion can be distinguished. Interstitial atoms diffuse readily. Substitutional atoms may be hindered by the preference of an atom for either the In site or P site. More complicated diffusion processes may be important (Van Vechten and Wager, 1985 a; b).

Fragments of annealing results were presented in the sections on growth (6) and irradiation and implantation (7). Two new topics emphasizing annealing and diffusion are summarized in the following paragraphs.

Zn exhibits anomalies in diffusions and heat treatments. Early work suggested that some Zn was complexed in InP, and the substitutional Zn diffused by a substitutional-interstitial mechanism (Tuck and Zahari, 1977). This diffusion requires mobile Zn interstitials reacting with In vacancies. Recent work has shown that the Zn diffusion profile depends on the substrate doping concentration (Serreze and Marek, 1986), and double diffusion fronts are observed (Kazmierski and de Cremoux, 1985). The results can be explained by including the possibility of different charge states for the interstitial, as in the following dissociation reaction

$$Zn_{In}^{-1} \Leftrightarrow V_{In}^{\circ} + Zn_i^{+m} + (m+1)e^-. \tag{17}$$

The diffusion rate for the interstitial depends on the charge state, which in turn depends on the doping of the starting material. More work needs to be done to obtain a firm model and to understand the role of Zn in high resistivity and semi-insulating InP.

Heat treatments have been applied to bulk InP crystals in order to improve its qualities as a semi-insulating substrate material. The effect of out-diffusion of impurities into epitaxial layers during growth can be lessened if the substrates are preannealed and etched to remove the contaminated surface layer (Brown et al., 1984). A new photoluminescence line at 1.391 eV is observed after the annealing. Improvement in InP:Fe substrates after annealing with phosphosilicate glass (PSG) encapsulation was attributed to a reduction in the number of phosphorus vacancies (Kamijoh et al., 1984). PSG anneals were found to suppress an emission line at 1.393 eV, which was attributed to a phosphorus-vacancy-related defect (Oberstar and Streetman, 1982). Annealing of undoped InP in a P overpressure was found to produce semi-insulating (3.6×10^5 ohm-cm) material in some cases (Klein et al., 1986). This annealing produced bound exciton luminescence bands at 1.3975 and 1.408 eV, which correlate with ODMR of a P-antisite complex observed in the same samples (Kennedy et al., 1986). More work is also necessary in this interesting and important area.

9. Dislocations and Surfaces

Stoichiometric defects coalesce, possibly with impurities, to form dislocations and occur at and near surfaces. Thermal stresses during growth and cooling of bulk InP lead to dislocations. Their formation may be associated with departures from stoichiometry (Brown et al., 1981). Dislocations in bulk material adversely affect device yield since they are replicated in the active device layer during its epitaxial growth. Loss of phosphorus from the surface region of InP is also important to the performance of Schottky barriers and metal-insulator-semiconductor (MIS) devices formed on InP.

The formation of dislocation loops in InP by coalescence of point defects has been shown graphically by transmission-electron microscope (TEM) studies (Reynaud and Legros-de Mauduit, 1986). Electron irradiations at 2 MeV were carried out in a high-voltage electron microscope (HVEM) to very high fluences (19^{19} cm^{-2}) with the sample between 200 and 350 °C. Subsequent TEM studies revealed interstitial loops that were formed by diffusion of intrinsic interstitials. The irradiation produces only single vacancies and interstitials. Both the high fluence and high temperature are required to allow the formation of dislocations, which can be seen in the TEM experiments. Some photoluminescence has also been performed on samples irradiated to even higher fluences (10^{21} cm^{-1}) in a HVEM (Frandon et al., 1986). An emission at 1.392 eV is attributed to a radiation-induced defect.

Efforts have been made to understand the Fermi level pinning at the surface of InP. This effect has been attributed to P-vacancies (Daw and Smith, 1980) and to P-antisites (Dow and Allen, 1982). Greater Schottky barrier heights were obtained by incorporating a thin oxide layer (Wada et al., 1982). These authors attributed the surface state density to P-vacancies at the oxide-InP interface. Annealing the starting InP material in an As overpressure reduced the surface state density in metal-insulator-InP structures (Blanchet et al., 1985). As-atoms are believed to compensate the P-vacancies near the surface more strongly than P-atoms. The density of states at the surface has also been reduced by deposition of a P-overlayer

(Schachter et al., 1985). Again this improvement is attributed to a reduction in the number of P-vacancies at the P-InP interface.

V. Conclusion

The current understanding of stoichiometric defects in InP has benefited in many cases from earlier work done on elemental and other III–V semiconductors. The time is ripe for such advances as calculations of energies for charged defects and further resonance study (ENDOR) of broad EPR spectra. The further development of this field will depend heavily on the amount of effort expended to produce technologically useful InP materials and devices.

In some cases, the defects found in InP were unique or revealed properties quite new to defects in semiconductors. The low defect introduction rate for electron irradiation is unique to n-type InP among wide-bandgap III–Vs and may be technologically important. The discovery of the metastable defects in InP stimulated the search for such effect in other semiconductors. The challenge remains to confirm the atomic structure of these centers.

We hope that this review will lead to fruitful crossfertilization between different subfields. It is interesting to compare the diffusion work on InP:Zn with the electron irradiation studies of the same material. The annealing experiments suggest that Zn goes from substitutional to interstitial quite readily. Irradiation studies suggest that pairs are formed with Zn acceptors (substitutional Zn) by diffusion of either V_P or P_i. Perhaps Zn becomes interstitial under the highly excited conditions of electron irradiation and diffuses to form the pairs. Further work in this and other areas is eagerly anticipated.

Acknowledgments

We thank P. B. Klein, T. L. Reinecke, and N. D. Wilsey for carefully reading the manuscript. Work by the authors was supported in part by the Office of Naval Research.

References

Baraff, G. A., and Schlüter, M. (1985). Electronic structure, total energies and abundances of the elementary point defects in GaAs. *Phys. Rev. Lett.* **55**, 1327–1330.

Baraff, G. A., Kane, E. O., and Schluter, M. (1980). Simple parameterized model for Jahn-Teller systems: Vacancy in *p*-type silicon. *Phys. Rev. B* **21**, 3563–3570.

Barthruff, D., Benz, K. W., and Antypas, G. A. (1979). Photolumininescence characterization of solution and LEC grown InP. *J. Electron. Mater.* **8**, 485–491.

Benneman, K. H. (1965). New method for treating lattice point defects in covalent crystals. *Phys. Rev.* **A137**, 1497–1514.

Bernholc, J., Lipari, N. O., and Pantelides, S. T. (1980). Scattering-theoretic method for defects in semiconductors. II. Self-consistent formulation and application to the vacancy in silicon. *Phys. Rev. B* **21**, 3545–3562.

Besson, M., DeLeo, G. G., and Fowler, W. B. (1986). Tight-binding Green's-function approach to off-center defects: Nitrogen and oxygen in silicon. *Phys. Rev. B* **33**, 8188–8195.

Blanchet, R., Vitroovitch, P., Chave, J., and Santinelli, C. (1985). Reduction of fast interface states and suppression of drift phenomena in arsenic-stablized metal-insulator-InP structures. *Appl. Phys. Lett.* **46**, 761–763.

Bourgoin, J. C., and Corbett, J. W. (1972). A new mechanism for interstitial migration. *Phys. Lett.* **38A**, 135–137.

Bourgoin, J., and Lannoo, M. (1983). *"Point Defects in Semiconductors II.* Springer-Verlag, Berlin.

Brailovskii, E. Yu., Megela, I. G., Pambuchchyan, N. H., and Teslenko, V. V. (1982). EPR study of electron-irradiated InP:Fe. *Phys. Stat. Sol. A* **72**, K109–111.

Brown, A. S., Palmateer, S. C., Wicks, G. W., Eastman, L. G., Calawa, A. R., and Hitzman, C. (1984). The heat treatment of Fe-doped InP substrates for the growth of higher purity $Ga_{0.47}In_{0.53}As$ by MBE. In *"Semi-Insulating III–V Materials"* (D. C. Look, and J. S. Blakemore, eds.). 36–40. Shiva Publishing, Cheshire, England.

Brown, G. T., Cockayne, B., and MacEwan, W. R. (1981). The identification of faulted prismatic dislocation loops in single crystals of undoped InP. *J. Mater. Sci.* **16**, 2867–2876.

Brudnyi, V. N., Vorob'ev, S. A., and Tsoi, A. A. (1982a). Positron annihilation in electron-irradiated n-type InP. *Sov. Phys. Semicond.* **16**, 121–123.

Brudnyi, V. N., Kuznetsov, V. D., Vorobiev, S. A., and Tsoi, A. A. (1982b). Positron annihilation and Hall effect in electron-irradiated n-InP crystals. *Appl. Phys. A* **29**, 219–223.

Buisson, J. P., Allen, R. E., and Dow, J. D. (1982). Antisite defects in $In_{1-y}Ga_yAs_{1-x}P_x$. *Solid State Comm.* **43**, 833–836.

Cavenett, B. C. (1981). Optically detected magnetic resonance (ODMR) investigations of recombination processes in semiconductors. *Adv. Phys.* **30**, 475–538.

Cavenett, B. C., Kana-ah, A., Deiri, M., Kennedy, T. A., and Wilsey, N. D. (1985). On the prospect of as-grown semi-insulating InP: ODMR of the P_{In} antisite. *J. Phys. C* **18**, L473–476.

Christel L. A., and Gibbons, J. F. (1981). Stoichiometric disturbances in ion implanted semiconductors. *J. Appl. Phys.* **52**, 5050–5055.

Chung, C. H., Noh, S. K., Park, S. C., and Kim, C. K. (1983). Deep levels in InP grown by SSD method. *Inst. Phys. Conf. Ser.* **65**, 641–646.

Clerjaud, B. (1985). Transition-metal impurities in III–V compounds. *J. Phys. C* **18**, 3615–3661.

Das Sarma, S., and Madhukar, A. (1981a). Study of the ideal-vacancy induced netural deep levels in III–V compound semiconductors and their ternary alloys. *Phys. Rev. B* **24**, 2051–2068.

Das Sarma, S., and Madhukar, A. (1981b). Cation and anion ideal vacancy induced gap levels in some III–V compound semiconductors. *Solid State Commun.* **38**, 183–186.

Davies, J. J. (1985). ODMR studies of recombination emission in II–VI compounds. *J. Crystal Growth* **72**, 317–325.

Daw, M. S., and Smith, D. L. (1980). Surface vacancies in InP and GaAlAs. *Appl. Phys. Lett.* **36**, 690–692.

Deiri, M., Kana-ah, A., Cavenett, B. C., Kennedy, T. A., and Wilsey, N. D. (1984). Optical detection of the P_{In} antisite resonances in InP. *J. Phys. C* **17**, L793–797.

Deiri, M., Kana-ah, A., Cavenett, B. C., Kennedy, T. A., and Wilsey, N. D. (1988). Optically detected magnetic resonance investigation of antisite and vacancy centres in InP. *Semicond. Sci. Technol.* **3**, 706–714.

Dlubek, G., and Brummer, O. (1985). Vacancy-Zn complexes in InP studied by positrons. *Appl. Phys. Lett.* **46**, 1136–1138.
Dow, J. D., and Allen, R. E. (1982). Surface defects and Fermi-level pinning in InP. *J. Vac. Sci. Technol.* **20**, 659–661.
Duncan, K. R., Eaves, L., Ramdane, A., Roys, W. B., Skolnick, M. S., and Dean, P. J. (1984). An investigation of the 1.36 eV photoluminescence spectrum of heat-treated InP using Zeeman spectroscopy and strain effects. *J. Phys. C.* **17**, 1233–1245.
Eaves, L., Smith, A. W., Skolnick, M. S., and Cockayne, B. (1982). An investigation of the deep level photoluminescence spectra of InP(Mn), InP(Fe) and of undoped InP. *J. Appl. Phys.* **53**, 4955–4963.
Farges, J. P. (1990). Growth of dislocation free crystals. *In "Semiconductors and Semimetals"* (R. K. Willardson and A. C. Beer, eds.). This volume. Academic Press, Orlando.
Ferreira, L. G., and De Siqueira, M. L. (1986). New technique in the calculation of defects in solids by molecular methods: Pure and Cu-doped ZnS. *Phys. Rev. B* **34**, 5315–5319.
Flynn, C. P. (1968). Atomic migration in monatomic crystals. *Phys. Rev.* **171**, 682–698.
Flynn, C. P. (1970). *"Defects and Diffusion."* Clarendon Press, Oxford.
Frandon, J., Fabre, F., Bacquet, G., Bandet, J., and Reynaud, F. (1986). Luminescence of heavily electron irradiated InP. *J. Appl. Phys.* **59**, 1627–1632.
Georgobiani, A. N., Mikulyonok, A. V., Stoyanova, I. G., and Tiginyanu, I. M. (1983a). Nonequilibrium carrier radiative recombination in indium phosphide single crystals. *Phys. Stat. Sol. A* **80**, 109–118.
Georgobiani, A. N., Mikulenok, A. V., Panasyuk, E. I., Radautsan, S. I., and Tiginyanu, I. M. (1983b). Deep centers in undoped and iron-doped indium phosphide single crystals. *Sov. Phys. Semicond.* **17**, 370–373.
Glyde, H. R. (1967). Relation of vacancy formation and migration energies to the Debye temperature in solids. *J. Phys. Chem. Solids* **28**, 2061–2065.
Goltzene, A., Meyer, B., and Schwab, C. (1987). Fast neutron induced defects in undoped and iron doped indium phosphide. *J. Appl. Phys.*, **62**, 4406–4412.
Haberland, D., Harde, P., Nelowski, H., and Schlaak, W. (1984). Stoichiometric disturbance in InP measured during ion implantation process. *Mat. Res. Sco. Symp. Proc.* **27**, 371–375.
Haydock, R., Heine, V., and Kelley, M. J. (1972). Electronic structure based on the local atomic environment for tight-binding bands. *J. Phys. C* **5**, 2845–2858.
Hemstreet, L. A. (1977). Electronic states of simple transition metal impurities in silicon. *Phys. Rev. B* **15**, 834–839.
Hjalmarson, H. P., Vogl, P., Wolford, D. J., and Dow, J. D. (1980). Theory of substitutional deep traps in covalent semiconductors. *Phys. Rev. Lett.* **44**, 810–813.
Inada, T., and Fukuda, T. (1990). Direct synthesis and growth of InP by the liquid phosphorus encapsulated Czochralski method. *In " Semiconductors and Semimetals,"* (R. K. Willardson, and A. C. Beer, eds.). This volume. Academic Press, Orlando.
Jaros, M. (1980). Deep levels in semiconductors. *Adv. in Phys.* **29**, 409–525.
Jeon, D. Y., Gislason, H. P., Donegan, J. F., and Watkins, G. D. (1987). Determination of the P_{In} antisite structure in InP by optically detected electron-nuclear double resonance. *Phys. Rev. B* **36**, 1324–1327.
Johnson, N. M., Bishop, S. G., and Watkins, G. W. (1985). *"Microscopic Identification of Electronic Defects in Semiconductors.* Materials Research Society Symposia Proceedings Vol. 46, Materials Research Society, Pittsburgh.
Kamijoh, T., Takano, H., and Sakuta, M. (1984). Heat treatment of semi-insulating InP:Fe with phophosilicate glass encapsulation. *J. Appl. Phys.* **55**, 3756–3759.
Kana-ah, A., Deiri, M., Cavenett, B. C., Wilsey, N. D., and Kennedy, T. A. (1985). Anti-site centres in e-irradiated InP:Zn. *J. Phys. C* **18**, L619–623.

Kaufmann, U., Baeumler, M., Windscheif, J., and Wilkening, W. (1986). New omnipresent electron paramagnetic resonance signal in as-grown semi-insulating liquid encapsulation Czochralski GaAs. *Appl. Phys. Lett.* **49** 1254–1256.

Kazmiersi, K., and de Cremoux, B. (1985). Double zinc diffusion fronts in InP: Correlation with models of varying charge transfer during interstitial-substitutional interchange. *Jpn. J. Appl. Phys. Part 1* **24**, 239–242.

Kennedy, T. A., and Spencer, M. G. (1986). Identification of the Ga interstitial in $Al_xGa_{1-x}As$ by optically detected magnetic resonance. *Phys. Rev. Lett.* **57**, 2690–2693.

Kennedy, T. A., and Wilsey, N. D. (1981). EPR of defects in electron-irradiated InP:Fe. *Inst. Phys. Conf. Ser.* **59**, 257–262.

Kennedy, T. A., and Wilsey, N. D. (1984). Electron paramagnetic resonance identification of the phosphorus antisite in electron-irradiated InP. *Appl. Phys. Lett.* **44**, 1089–1091.

Kennedy, T. A., and Wilsey, N. D. (1985). Antisite production by electron irradiation of InP and GaP. *J. Electro. Mater.* **14a**, 929–935.

Kennedy, T. A., and Wilsey, N. D. (1987). Applications of electron paramagnetic resonance and optically detected magnetic resonance to InP materials. *J. Crystal Growth* **83**, 198–201.

Kennedy, T. A., Wilsey, N. D., Krebs, J. J., and Strauss, G. H. (1983). Electronic spin of the Ga vacancy in GaP. *Phys. Rev. Lett.* **50**, 1281–1284.

Kennedy, T. A., Wilsey, N. D., Klein, P. B., and Henry, R. L. (1986). Triplet spin ODMR from phosphorus antisites in undoped InP. In "*Defects in Semiconductors*" (H. J. von Bardeleben, ed.). 271–276. Trans Tech Publications Ltd., Switzerland.

Khowash, P. K., Khan, D. C., and Singh, V. A. (1985). Theoretical study of the electronic states of substitutional transition-metal impurities of InP. *J. Phys. C* **18**, 6177–6184.

Klein, P. B., Henry, R. L., Kennedy, T. A., and Wilsey, N. D. (1986). Semi-insulating behavior in undoped LEC InP after annealing in phosphorus. In "*Defects in Semiconductors*" (H. J. von Bardeleben, ed.) 1259–1264. Tran Tech Publications Ltd., Switzerland.

Kohn, W., and Sham, L. J. (1965). Self-consistent equations including exchange and correlation effects. *Phys. Rev. A* **140**, 1133–1138.

Kröger, F. A. (1964). "*The Chemistry of Imperfect Crystals.*" North Holland, Amsterdam.

Lannoo, M., and Bourgoin, J. (1981). "*Point Defects in Semiconductors I.*" Springer-Verlag, Berlin.

Lee, K. M. (1988). Observation of the Ga self-interstitial defect in GaP. In "*Defects in Electronic Materials*" (M. Stavola, S. J. Pearton and G. Davies, eds.) 449–445. Materials Research Society, Pittsburgh.

Levinson, M., Benton, J. L., and Kimerling, L. C. (1983a). Electronically controlled metastable defect reaction in InP. *Phys. Rev. B* **27**, 6216–6220.

Levinson, M., Stavola, M., Benton, J. L., and Kimerling, L. C. (1983b). Metastable M center in InP: Defect-charge-state-controlled structural relaxation. *Phys. Rev. B* **28**, 5848–5855.

Li, S. S., Wang, W. L., and Shaban, E. H. (1984). Characterization of the grown-in defects in Zn-doped InP. *Solid State Commun.* **51**, 15–18.

Li, Y., and Lin-Chung, P. J. (1985). Universal LCAO parameters for III–V compound semiconductors. *J. Phys. Chem. Solids* **46**, 241–247.

Lin-Chung, P. J. (1983). Complex defects in GaAs and GaP. *Mat. Res. Soc. Sym. Pro.* **14**, 267–270.

Lin-Chung, P. J. (1989). Electronic structures of native defect complexes in GaP and InP, *Defect and Diffusion Forum* **62/63**, 161–166.

Lin-Chung, P. J., and Reinecke, T. L. (1983). Theoretical study of native defects in III–V semiconductors. *Phys. Rev. B* **27**, 1101–1113.

Lindefelt, U., and Zunger, A. (1982). Quasiband crystal-field method for calculating the electronic structure of localized defects in solids. *Phys. Rev. B* **26**, 846–895.

Masterov, V. F. (1984). Deep centers in semiconductors. *Sov. Phys. Semicond.* **18**, 1–13.
Makhmudov, A. Sh., Adilov, M. K., Khakimov, Z. M., and Levin, A. A. (1985). Localized states of defects in III–V crystals calculated by the Green function method. *Sov. Phys. Semicond.* **19**, 1201–1203.
Messmer, R. P., and Watkins, G. D. (1971). An LCAO–MO treatment of the vacancy in diamond. *In* "*Radiation Effects in Semiconductors*" (J. W. Corbett and G. D. Watkins, eds.) 23–28 Gordon and Breach, New York.
Milnes, A. G. (1983). Impurity and defects levels (experimental) in gallium arsenide. *Advances in Electronics and Electron Physics* **61**, 63–160.
Moison, J. M., and Bensoussan, M. (1982). Laser-induced order-disorder transition of the (100) InP surface. *J. Vac. Sci. Technol.* **21**, 315–318.
Monberg, E. M. (1990). Vertical gradient freeze crystal growth. *In* "*Semiconductors and Semimetals*," (R. K. Willardson and A. C. Beer, eds.). This volume. Academic Press, Orlando.
Morozov, A. N., Bublik, V. T., Osvenskii, V. B., Berkova, A. V., Mikryukova, E. V., Nashel'skii, A. Ya., Yakobson, S. V., and Popov, A. D. (1983). Nature and concentration of intrinsic point defects in undoped single crystals of InP. I. Influence of melt composition. *Sov. Phys. Crystallogr.* **28**, 458–461.
Morozov, A. N., Bublik, V. T., and Grigor'eva, T. P. (1984). Nature and concentration of intrinsic point defects in undoped single crystals of InP. II. Region of homogeneity of indium phosphide. *Sov. Phys. Crystallogr.* **29**, 447–450.
Neumark, G. F., and Kosai, K. (1983). Deep levels in wide band-gap III–V semiconductors. *In* "*Semiconductors and Semimetals. Vol. 19*" (R. K. Willardson, and A. C. Beer, eds.). 1–74. Academic Press, Orlando.
Oberstar, J. D., and Streetman, B. G. (1982). Annealing encapsulants for InP II: Photoluminescence studies. *Thin Solid Films* **94**, 161–170.
Pantelides, S. T. (1978). The electronic structure of impurities and other point defects in semiconductors. *Rev. Mod. Phys.* **50**, 797–858.
Pantelides, S. T., ed. (1986). "*Deep Centers in Semiconductors.*" Gordon and Breach, New York.
Pantelides, S. T., and Sah, C. T. (1974). Theory of localized states in semiconductors II. The pseudo-impurity theory application to shallow and deep donors in silicon. *Phys. Rev. B* **10**, 638–658.
Razeghi, M. (1990). Low pressure MOCVD growth. *In* "*Semiconductors and Semimetals*" (R. K. Willardson and A. C. Beer, eds.). This volume. Academic Press, Orlando.
Reynaud, F., and Legros-de Mauduit, B. (1986). Studies of the dislocation loops produced in III–V semiconducting compounds of B3 structure by irradiation in a high voltage electron microscope. *Radiat. Eff.* **88** 1–16.
Rong, F., and Watkins, G. D. (1987). Optically detected magnetic-resonance observation of the isolated zinc interstitial in irradiated ZnSe. *Phys. Rev. Lett.* **58**, 1486–1489.
Sankey, O. F., and Dow, J. D. (1981). Deep levels produced by pairs of impurities in InP. *J. Appl. Phys.* **52**, 5139–5142.
Schachter, R., Olego, D. J., Baumann, J. A., Bunz, L. A., Raccah, P. M., and Spicer, W. E. (1985). Interfacial properties of InP and phosphorus deposited at low temperture. *App. Phys. Lett.* **47**, 272–274.
Seeger, A., and Swanson, M. L. (1968). Vacancies and diffusion mechanisms in diamond structure semiconductors. "*Lattice Defects in Semiconductors*" (R. R. Hasiguti, ed.) 93–130. University of Tokyo Press, Tokyo and The University of Pennsylvania State University Press, University Park.
Serreze, H. B., and Marek, H. S. (1986). Zn diffusion in InP: Effect of substrate dopant concentration. *Appl. Phys. Lett.* **49**, 210–211.

Serreze, H. B., and Marek, H. S. (1986). Zn diffusion in InP: Effect of substrate dopant concentration. *Appl. Phys. Lett.* **49**, 210–211.

Sette, F., Pearton, S. J., Poate, J. M., and Rowe, J. E. (1986). Local structure of S impurities in GaAs. *Phys. Rev. Lett.* **56**, 2637–2640.

Sibille, A., and Bourgoin, J. C. (1982). Electron irradiation induced deep levels in p-InP. *Appl. Phys. Lett.* **41**, 956–958.

Sibille, A., and Suski, J. (1985). Defect reactions on the phosphorus sublattice in low-temperature electron-irradiated InP. *Phys. Rev. B* **31**, 5551–5553.

Sibille, A., Suski, J., and Gilleron, M. (1986). A model of deep centers formation and reactions in electron irradiated InP. *J. Appl. Phys.* **60**, 595–601.

Spaeth, J.-M. (1986). Application of optically detected magnetic resonance to the characterization of point defects in semiconductors. *Materials Science Forum* **10-12**, 505–514.

Srivastava, G. P. (1979). Electronic structure of a neutral phosphorus vacancy in GaP and InP. *Phys. Stat. Sol.* **b93**, 761–765.

Stillman, G. E., and McCollum, M. J. (1990). High purity InP growth by hydride vapor phase epitaxy. In "*Semiconductors and Semimetals*" (R. K. Willardson and A. C. Beer, eds.). This volume. Academic Press, Orlando.

Swalin, R. A. (1961). Theoretical calculations of the enthalpies and entropies of diffusion and vacancy formation in semiconductors. *J. Phys. Chem. Solids* **18**, 290–296.

Temkin, H., and Dutt, B. V. (1983). Deep radiative transitions in InP. *Mat. Res. Soc. Symp. Proc.* **14**, 253–265.

Temkin, H., Dutt, B. V., and Bonner, W. A. (1981). Photoluminescence study of native defects in InP. *Appl. Phys. Lett.* **38**, 431–433.

Temkin, H., Dutt, B. V., Bonner, W. A., and Keramidas, V. G. (1982). Deep radiative levels in InP. *J. Appl. Phys.* **53**, 7526–7533.

Tomzig, E. (1990). High pressure LEC growth of InP. In "*Semiconductors and Semimetals*" (R. K. Willardson and A. C. Beer, eds.). This volume. Academic Press, Orlando.

Tuck, B., and Zahari, M. D. (1977). Isoconcentration diffusion of zinc in indium phosphide at 800°C. *Inst. Phys. Conf. Ser.* **33a**, 177–184.

Van Vechten, J. A., and Wager, J. F. (1985a). Consequences of anion vacancy nearest-neighbor hopping in III–V compound semiconductors: Drift in InP metal-insulator-semiconductor field effect transistors. *J. Appl. Phys.* **57**, 1956–1960.

Van Vechten, J. A., and Wager, J. F. (1985b). Asymmetry of anion and cation vacancy migration enthalpies in III–V compound semiconductors: Role of the kinetic energy. *Phys. Rev. B* **32**, 5259–5264.

von Bardeleben, H. J. (1986). Identification of the phosphor vacancy defect in electron irradiated p-type InP. *Solid State Commun.* **57**, 137–139.

Wada, O., Majerfeld, A., and Robson, P. N. (1982). InP Schottky contacts with increased barrier height. *Solid-State Electron.* **25**, 381–387.

Wakefield, B., Eaves, L., Prior, K. A., Nelson, A. W., and Davies, G. J. (1984). The 1.36 eV radiative transition in InP: its dependence on growth conditions in MBE and MOCVD material. *J. Phys. D* **17**, L133–136.

Wager, J. F., and Van Vechten, J. A. (1985). Atomic model for the M center in InP. *Phys. Rev. B* **32**, 5251–5258.

Wieder, H. H. (1980). Surfaces and dielectric-semiconductor interfaces of binary and quaternary alloy III–V compounds. In "*Insulating Films on Semiconductors*" (G. G. Roberts and M. J. Morant, eds.). 234–250. Institute of Physics Conference Series No, 50, Bristol and London.

Wong, C-C. D., and Bube, R. H. (1984). Bulk and surface effects of heat treatment of p-type InP crystals. *J. Appl. Phys.* **55**, 3804–3812.

Yamazoe, Y., Sasai, Y., Nishino, T., and Hamakaway, Y. (1981). Deep impurity levels in InP LEC crystals. *Jpn. J. Appl. Phys.* **20**, 347–354.

Yan D., Cavenett, B. C., and Skolnick, M. S. (1983). ODMR and EPR in InP:Mn. *J. Phys. C* **16**, L647–653.

Yu, P. W. (1980). A model for the -1.10 eV emission band in InP. *Solid State Commun.* **34**, 183–186.

Zunger, A. (1983). Electronic structure of transition-atom impurities in semiconductors: Substitutional 3d impurities in silicon. *Phys. Rev. B* **27**, 1191–1227.

Index

A

Absorptiometry, analysis of sulfur in phosphorus, 140
Analysis
  atomic absorption (AA), 81, 140, 214
  auger electron spectroscopy, 157, 158, 164, 293, 294, 320–322
  glow discharge mass spectroscopy (GDMS), 212–214
  inductively coupled radio frequency plasma atomic emission spectroscopy (ICP), 81, 214
  x-ray diffractometry, 84, 87, 89
Annealing, 220, 381–382
Antisite, 365, 370–372, 380, 383
Applications, 176, 226
Arsine, 265
Atomic layer epitaxy (ALE), 261, 263

B

Boron oxide, 15, 16, 111–114, 146, 184, 196, 198
Bow, wafer, 152, 155
Brillouin zone, 249–257

C

Carrier concentration, 55–57, 81, 85, 103, 104, 105, 108, 110
Cathodoluminescence, 143–146
Chemical properties, TEG, TMI, TEI, TME, DEZ, 265–269
Cleaning wafer, 155–159, 234, 236
Cleavage plane, 223
Compensation ratio, 110
Contact angle, 156
Critical resolved shear stress, 9, 11, 25, 26, 32, 115, 120, 180
Crystal growth, 72, 73, 75, 78, 82–90
  electrodynamic gradient freeze (EDG), 12, 13
  horizontal bridgman (HB), 1, 3, 11, 21, 72, 78

liquid encapsulated Czochralski (LEC), 5, 14–16, 24, 32, 33, 72, 75, 89, 90, 181, 182, 184, 186, 188, 190, 195–200
  automatic growth control, 190
  diameter control, 190, 191
  necking, 23, 25, 30, 31
  orientation, 30
  rotation rate, 19
LP-CZ (liquid phosphorus encapsulated Czochralski), 73, 82–90
  cooling rate, 84–86
  dendrites, 86, 87, 89
  etch pit density, 85, *see also* Etch pit density
  liquid phosphorus, 82, 83, 86, 87, *see also* Phosphorus
  single crystal, 84, 89
  temperature gradient, 83
vertical Bridgman (VB or VGF), 11, 12, 14, 32, 181, 182
Cutting, 231
C-V profiling, 287–289

D

Damaged layers, 159–163, 235
Deep acceptors, 64–69, 130–136, 293, 370, 379–381
Deep level transient spectroscopy
  hydride VPE InP, 59
  MOCVD, 291, 293
  spectra, 64, 65
Defects
  experiments, 368–383
  growth-related, 376–378
  intrinsic, 358–359
  metastable, 380
  microscopic, 146–148
  pairs, 171–175
  stoichiometric, 357–389
  theory, *see* Theory
Diffusion, 367–368, 381–382

391

392　INDEX

Diffusion coefficient
    As, 29
    Fe, 20
    Ga, 26, 29
    P, 109, 111
Direct synthesis, 24, 97, 99
Dislocation-free InP, 3, 4, 6, 7, 16, 26, 29–31, 128–131
Dislocation
    density, 10, 11, 14, 29–31, 115, 128–132, 382–383
    generation, 6, 7, 8
    pinning effect, 199
    propagation, 6, 8
    velocity, 9
Distribution coefficients
    As, 29
    Ga, 26, 29
    Fe, 20, 135, 193
Dopant incorporation, 140–141, 301–307

E

Edge rounding, 231
Effective mass, 252, 264
Electron mobility, 81, 85, 110, 276, 277, 289–292, 295
    anomalously high, 53
    hydride VPE, 55–58
Energy bandgap, 249, 264, 341
Etch pit density
    automatic counting system, 206
    of S doped InP, 30, 31, 127–131, 161
    of Sn doped InP, 124, 126
    of Zn doped InP, 132
Etchant, 10, 147, 154, 205, 231, 235
Evaluation
    electrical, 81, 85, 214
    off-line, 205
    on-line, 204, 205
    optical, 217
    practical usage, 218
    quasi-device, 220
    real device, 220
EXAFS, 375

F

Fermi energy, 8, 265, 285
FET, 341
Free carrier absorption, 25, 28, 218

G

GaAs, 259
GaAsP, 259
Grinding, 228
Gas source molecular beam epitaxy (GSMBE), 260
Gunn diodes, 307, 313

H

Hall effect measurements, 54–58, 205–208
Heat transfer coefficient, 4, 5, 17
Heterojunction, 313–317
High purity InP
    purity enhancement, 48–53
    summary by growth method, 39, 40
Hydride vapor phase epitaxy
    growth procedure, 47, 48
    reactions
        growth, 41
        leading to Si incorporation, 49
    substrate preparation, 44
    system design, 41–47

I

Implantation, 378–380
Impurities
    codoping, 203
    electrically active
        Fe, 29, 31, 130, 133, 134, 136, 177, 216–219
        S, 25, 26, 28, 30, 125–131, 161, 177, 192, 208, 212, 218
        Sn, 123–125, 177, 178, 192, 207, 212, 218
        Te, 24
        Zn, 26, 129–132, 177, 192, 208, 218
    isoelectronic
        Ga, 26, 29
        Sb, 26, 27
    multiple
        Ga + Sb, Ga + As, Ga + As + Sb, 27, 29, 202
Impurity hardening, 24–30, 120–123, 192, 199–203, 208
Inclusion
    indium, 78, 85
    phosphorus, 78, 85
Infrared light scattering tomography, 198, 218
InGaAs, 246, 259, 339–351

InGaAsP, 247, 341
InP/InGaAsP, 247, 340
Interstitial, 365-366, 373, 381-382
Ionized impurity scattering, 289-291
Iron-doped InP, 130, 136
Irradiation, 378-380

L

Lapping, 151, 152, 233
Lasers, 339-341
Lattice constant, 29, 264
Lattice-matched layers, 341, 349
Lattice matching, 253
Liquid-phase epitaxy, 254, 261
Low-energy electron diffraction, 164
Low-pressure MOCVD, 248, 264-276

M

Magnetic resonance, 368-375
    ENDOR, 369
    EPR, 368-369
    hyperfine interactions, 369
    ODMR, 369, 377
Magneto-photoluminescence
    hydride VPE, InP, 59
        spectra, 62, 63
MESFET, 341
Metallo-organic atomic layer epitaxy (MO-ALE), 261
Metallo-organic chemical beam deposition (MOCBD), 260, 262
Metallo-organic chemical vapor deposition (MOCVD), 257
Migration enhanced epitaxy (MEE), 263, 264
Multiple heater furnaces, 17

N

Native defects, 7, 8

O

Orientation flat, 228
Orientation, 230
    effects, 274
    laser, 231, 232
    x-ray, 231, 232

P

Parallelism, 155, 236
Phase diagram, 89, 90, 179

Phosphorus (P), 72-80, 82, 83, 86-88
    allotropes, 73
    liquid, 75, 77-80, 82, 86, 87
        liquid flow, 75, 77-79, 82
    layer, 79, 80, 82, 86, 87
    phase transition, 73-75, 83
    red phosphorus, 72-74, 76, 80, 82
    solubility of indium, 88
    sublimation, 73, 76
    white phosphorus, 73, 74
Photoluminescence, 377, 381
Photothermal ionization spectroscopy, 56, 59, 61, 140-142, 276, 298-300, 317, 326, 327
    hydride VPE spectra
        GaAs, 52
        InP, 53, 60
Physical properties, InP, GaAs, Si, 178, 179, 223-225
Polishing, 231
Positron annihilation, 375
Purity
    indium, 139
    phosphorus, 140
    InP polycrystal, 140
    InP single crystal, 140

Q

Quality control, 218
Quaternary compositional plane, 253

R

Refractive index, 264
Resistivity distribution, microscopic, 215
Rocking curve, 316

S

Schottky barrier, 286-288, 341, 349
Secondary-ion mass spectroscopy (SIMS), 276, 278, 285, 305
Segregation, 88, 193, 198, 202
Semi-insulating InP, 130-136, 178, 218, 219
Shockley diagram
    Fe-doped InP, 134
    Ti-doped InP, 134
Silicon contamination, in InP polycrystal, 102-105
Single crystal
    growth, 111-138
    yield, 12, 27, 32

Sn-doped InP
  substrate, 178
  InP wafer, 208
Solid liquid interface, InP crystal growth, 113
Spark source mass spectroscopy analysis
  Indium, 139
  InP polycrystal, 140
  InP single crystal, 140
Stacking fault energy, 112, 180
Stoichiometry
  InP melt, 7, 8, 21, 119
Substrate
  GaAs, 248, 314, 320–322, 327–339, 341, 342
  GGG, 247, 323–326
  InP, 44–47, 148–166, 222–238, 246, 247, 314, 339
  Si, 247, 327–339
Substrate processing, 148–159, 222, 225, 227
Sulfur-doped InP, 25, 26, 125–129, 131
Surfaces, 161–163, 235, 382, 383
Surface waviness of InP wafers, 163, 165–166
Synthesis
  autoclave-synthesis, 21, 72, 97–111, 181, 184
  direct synthesis, 21–23, 72–82
  indium inclusion, 21, 78, *see also* Inclusion
  liquid phosphorus, 75, 77–80, 82, *see also* Phosphorus
  phosphorus inclusion, 78, 85, *see also* Inclusion
  phosphorus injection, 21, 73, 181
  solute diffusion (SSD), 106–111, 181–183
  stoichiometry, 75, 76, 80, 82, 87

T

Temperature gradients, 184, 195, 197
Theory
  defects, 359–368
  methods, 359–364
  results, 364–366
  thermodynamics, 366–368
Thermal
  convection, 18, 19
  gradients, 6, 7, 11–17
  models, 4–6
  shields, 16
Thermal stability of InP, 135
Thermal stress, 6–8, 11
Tin-doped InP, 123–125
Transmittance, 218
Twinning prevention, 111–115, 180, 191, 198

V

Vacancy, 364–367, 372–373, 375, 381, 382

X

X-ray
  double crystal rocking curve, 160, 211, 235
  image, 77, 83, 86
  lattice parameter, 84
  observation system, 75
  orientation, 231, 232
  photoemission spectroscopy, 157
  topography, 10, 11, 27, 161–163, 211
  transmission image, 75, 77, 83, 86

Z

Zinc blend structure, 9, 223, 224

Contents of Previous Volumes

Volume 1 Physics of III–V Compounds
C. Hilsum, Some Key Features of III–V Compounds
Franco Bassani, Methods of Band Calculations Applicable to III–V Compounds
E.O. Kane, The $k \cdot p$ Method
V.L. Bonch-Bruevich, Effect of Heavy Doping on the Semiconductor Band Structure
Donald Long, Energy Band Structures of Mixed Crystals of III–V Compounds
Laura M. Roth and Petros N. Argyres, Magnetic Quantum Effects
S.M. Puri and T.H. Geballe, Thermomagnetic Effects in the Quantum Region
W.M. Becker, Band Characteristics near Principal Minima from Magnetoresistance
E.H. Putley, Freeze-Out Effects, Hot Electron Effects, and Submillimeter Photoconductivity in InSb
H. Weiss, Magnetoresistance
Betsy Ancker-Johnson, Plasmas in Semiconductors and Semimetals

Volume 2 Physics of III–V Compounds
M.G. Holland, Thermal Conductivity
S.I. Novkova, Thermal Expansion
U. Piesbergen, Heat Capacity and Debye Temperatures
G. Giesecke, Lattice Constants
J.R. Drabble, Elastic Properties
A.U. Mac Rae and G.W. Gobeli, Low Energy Electron Diffraction Studies
Robert Lee Mieher, Nuclear Magnetic Resonance
Bernard Goldstein, Electron Paramagnetic Resonance
T.S. Moss, Photoconduction in III–V Compounds
E. Antončik and J. Tauc, Quantum Efficiency of the Internal Photoelectric Effect in InSb
G.W. Gobeli and F.G. Allen, Photoelectric Threshold and Work Function
P.S. Pershan, Nonlinear Optics in III–V Compounds
M. Gershenzon, Radiative Recombination in the III–V Compounds
Frank Stern, Stimulated Emission in Semiconductors

Volume 3 Optical of Properties III–V Compounds
Marvin Hass, Lattice Reflection
William G. Spitzer, Multiphonon Lattice Absorption
D.L. Stierwalt and R.F. Potter, Emittance Studies
H.R. Philipp and H. Ehrenreich, Ultraviolet Optical Properties
Manuel Cardona, Optical Absorption above the Fundamental Edge
Earnest J. Johnson, Absorption near the Fundamental Edge
John O. Dimmock, Introduction to the Theory of Exciton States in Semiconductors
B. Lax and J.G. Mavroides, Interband Magnetooptical Effects

CONTENTS OF PREVIOUS VOLUMES

H.Y. Fan, Effects of Free Carries on Optical Properties
Edward D. Palik and George B. Wright, Free-Carrier Magnetooptical Effects
Richard H. Bube, Photoelectronic Analysis
B.O. Seraphin and H.E. Bennett, Optical Constants

Volume 4 Physics of III–V Compounds

N.A. Goryunova, A.S. Borschevskii, and D.N. Tretiakov, Hardness
N.N. Sirota, Heats of Formation and Temperatures and Heats of Fusion of Compounds $A^{III}B^{V}$
Don L. Kendall, Diffusion
A.G. Chynoweth, Charge Multiplication Phenomena
Robert W. Keyes, The Effects of Hydrostatic Pressure on the Properties of III–V Semiconductors
L.W. Aukerman, Radiation Effects
N.A. Goryunova, F.P. Kesamanly, and D.N. Nasledov, Phenomena in Solid Solutions
R.T. Bate, Electrical Properties of Nonuniform Crystals

Volume 5 Infrared Detectors

Henry Levinstein, Characterization of Infrared Detectors
Paul W. Kruse, Indium Antimonide Photoconductive and Photoelectromagnetic Detectors
M.B. Prince, Narrowband Self-Filtering Detectors
Ivars Melngailis and T.C. Harman, Single-Crystal Lead-Tin Chalcogenides
Donald Long and Joseph L. Schmit, Mercury-Cadmium Telluride and Closely Related Alloys
E.H. Putley, The Pyroelectric Detector
Norman B. Stevens, Radiation Thermopiles
R.J. Keyes and T.M. Quist, Low Level Coherent and Incoherent Detection in the Infrared
M.C. Teich, Coherent Detection in the Infrared
F.R. Arams, E.W. Sard, B.J. Peyton, and F.P. Pace, Infrared Heterodyne Detection with Gigahertz IF Response
H.S. Sommers, Jr., Macrowave-Based Photoconductive Detector
Robert Sehr and Rainer Zuleeg, Imaging and Display

Volume 6 Injection Phenomena

Murray A. Lampert and Ronald B. Schilling, Current Injection in Solids: The Regional Approximation Method
Richard Williams, Injection by Internal Photoemission
Allen M. Barnett, Current Filament Formation
R. Baron and J.W. Mayer, Double Injection in Semiconductors
W. Ruppel, The Photoconductor-Metal Contact

Volume 7 Application and Devices
PART A

John A. Copeland and Stephen Knight, Applications Utilizing Bulk Negative Resistance
F.A. Padovani, The Voltage-Current Characteristics of Metal-Semiconductor Contacts
P.L. Hower, W.W. Hooper, B.R. Cairns, R.D. Fairman, and D.A. Tremere, The GaAs Field-Effect Transistor
Marvin H. White, MOS Transistors

CONTENTS OF PREVIOUS VOLUMES

G.R. Antell, Gallium Arsenide Transistors
T.L. Tansley, Heterojunction Properties

PART B

T. Misawa, IMPATT Diodes
H.C. Okean, Tunnel Diodes
Robert B. Campbell and Hung-Chi Chang, Silicon Carbide Junction Devices
R.E. Enstrom, H. Kressel, and L. Krassner, High-Temperature Power Rectifiers of $GaAs_{1-x}P_x$

Volume 8 Transport and Optical Phenomena

Richard J. Stirn, Band Structure and Galvanomagnetic Effects in III–V Compounds with Indirect Band Gaps
Roland W. Ure, Jr., Thermoelectric Effects in III–V Compounds
Herbert Piller, Faraday Rotation
H. Barry Bebb and E.W. Williams, Photoluminescence 1: Theory
E.W. Williams and H. Barry Bebb, Photoluminescence II: Gallium Arsenide

Volume 9 Modulation Techniques

B.O. Seraphin, Electroreflectance
R.L. Aggarwal, Modulated Interband Magnetooptics
Daniel F. Blossey and Paul Handler, Electroabsorption
Bruno Batz, Thermal and Wavelength Modulation Spectroscopy
Ivar Balslev, Piezooptical Effects
D.E. Aspnes and N. Bottka, Electric-Field Effects on the Dielectric Function of Semiconductors and Insulators

Volume 10 Transport Phenomena

R.L. Rode, Low-Field Electron Transport
J.D. Wiley, Mobility of Holes in III–V Compounds
C.M. Wolfe and G.E. Stillman, Apparent Mobility Enhancement in Inhomogeneous Crystals
Robert L. Peterson, The Magnetophonon Effect

Volume 11 Solar Cells

Harold J. Hovel, Introduction; Carrier Collection, Spectral Response, and Photocurrent; Solar Cell Electrical Characteristics; Efficiency; Thickness; Other Solar Cell Devices; Radiation Effects; Temperature and Intensity; Solar Cell Technology

Volume 12 Infrared Detectors (II)

W.L. Eiseman, J.D. Merriam, and R.F. Potter, Operational Characteristics of Infrared Photodetectors
Peter R. Bratt, Impurity Germanium and Silicon Infrared Detectors
E.H. Putley, InSb Submillimeter Photoconductive Detectors
G.E. Stillman, C.M. Wolfe, and J.O. Dimmock, Far-Infrared Photoconductivity in High Purity GaAs
G.E. Stillman and C.M. Wolfe, Avalanche Photodiodes

CONTENTS OF PREVIOUS VOLUMES

P.L. Richards, The Josephson Junction as a Detector of Microwave and Far-Infrared Radiation
E.H. Putley, The Pyroelectric Detector–An Update

Volume 13 Cadmium Telluride

Kenneth Zanio, Materials Preparation; Physics; Defects; Applications

Volume 14 Lasers, Junctions, Transport

N. Holonyak, Jr. and M.H. Lee, Photopumped III–V Semiconductor Lasers
Henry Kressel and Jerome K. Butler, Heterojunction Laser Diodes
A. Van der Ziel, Space-Charge-Limited Solid-State Diodes
Peter J. Price, Monte Carlo Calculation of Electron Transport in Solids

Volume 15 Contacts, Junctions, Emitters

B.L. Sharma, Ohmic Contacts to III–V Compound Semiconductors
Allen Nussbaum, The Theory of Semiconducting Junctions
John S. Escher, NEA Semiconductor Photoemitters

Volume 16 Defects, (HgCd)Se, (HgCd)Te

Henry Kressel, The Effect of Crystal Defects on Optoelectronic Devices
C.R. Whitsett, J.G. Broerman, and C.J. Summers, Crystal Growth and Properties of $Hg_{1-x}Cd_xSe$ Alloys
M.H. Weiler, Magnetooptical Properties of $Hg_{t-x}Cd_xTe$ Alloys
Paul W. Kruse and John G. Ready, Nonlinear Optical Effects in $Hg_{t-x}Cd_xTe$

Volume 17 CW Processing of Silicon and Other Semiconductors

James F. Gibbons, Beam Processing of Silicon
Arto Lietoila, Richard B. Gold, James F. Gibbons, and Lee A. Christel, Temperature Distributions and Solid Phase Reaction Rates Produced by Scanning CW Beams
Arto Lietoila and James F. Gibbons, Applications of CW Beam Processing to Ion Implanted Crystalline Silicon
N.M. Johnson, Electronic Defects in CW Transient Thermal Processed Silicon
K.F. Lee, T.J. Stultz, and James F. Gibbons, Beam Recrystallized Polycrystalline Silicon: Properties, Applications, and Techniques
T. Shibata, A. Wakita, T.W. Sigmon, and James F. Gibbons, Metal-Silicon Reactions and Silicide
Yves I. Nissim and James F. Gibbons, CW Beam Processing of Gallium Arsenide

Volume 18 Mercury Cadmium Telluride

Paul W. Kruse, The Emergence of $(Hg_{t-x}Cd_x)Te$ as a Modern Infrared Sensitive Material
H.E. Hirsch, S.C. Liang, and A.G. White, Preparation of High-Purity Cadmium, Mercury, and Tellurium
W.F.H. Micklethwaite, The Crystal Growth of Cadmium Mercury Telluride
Paul E. Petersen, Auger Recombination in Mercury Cadmium Telluride
R.M. Broudy and V.J. Mazurczyck, (HgCd)Te Photoconductive Detectors
M.B. Reine, A.K. Sood, and T.J. Tredwell, Photovoltaic Infrared Detectors
M.A. Kinch, Metal-Insulator-Semiconductor Infrared Detectors

CONTENTS OF PREVIOUS VOLUMES

Volume 19 Deep Levels, GaAs, Alloys, Photochemistry

G.F. Neumark and K. Kosai, Deep Levels in Wide Band-Gap III–V Semiconductors
David C. Look, The Electrical and Photoelectronic Properties of Semi-Insulating GaAs
R.F. Brebrick, Ching-Hua Su, and Pok-Kai Liao, Associated Solution Model for Ga-In-Sb and Hg-Cd-Te
Yu. Ya. Gurevich and Yu. V. Pleskov, Photoelectrochemistry of Semiconductors

Volume 20 Semi-Insulating GaAs

R.N. Thomas, H.M. Hobgood, G.W. Eldridge, D.L. Barrett, T.T. Braggins, L.B. Ta, and S.K. Wang, High-Purity LEC Growth and Direct Implantation of GaAs for Monolithic Microwave Circuits
C.A. Stolte, Ion Implantation and Materials for GaAs Integrated Circuits
C.G. Kirkpatrick, R.T. Chen, D.E. Holmes, P.M. Asbeck, K.R. Elliott, R.D. Fairman, and J.R. Oliver, LEC GaAs for Integrated Circuit Applications
J.S. Blakemore and S. Rahimi, Models for Mid-Gap Centers in Gallium Arsenide

Volume 21 Hydrogenated Amorphous Silicon
Part A

Jacques I. Pankove Introduction
Masataka Hirose, Glow Discharge; Chemical Vapor Deposition
Yoshiyuki Uchida, dc Glow Discharge
T.D. Moustakas, Sputtering
Isao Yamada, Ionized-Cluster Beam Deposition
Bruce A. Scott, Homogeneous Chemical Vapor Deposition
Frank J. Kampas, Chemical Reactions in Plasma Deposition
Paul A. Longeway, Plasma Kinetics
Herbert A. Weakliem, Diagnostics of Silane Glow Discharges Using Probes and Mass Spectroscopy
Lester Guttman, Relation between the Atomic and the Electronic Structures
A. Chenevas-Paule, Experiment Determination of Structure
S. Minomura, Pressure Effects on the Local Atomic Structure
David Adler, Defects and Density of Localized States

Part B

Jacques I. Pankove, Introduction
G.D. Cody, The Optical Absorption Edge of a-Si:H
Nabil M. Amer and Warren B. Jackson, Optical Properties of Defect States in a-Si:H
P.J. Zanzucchi, The Vibrational Spectra of a-Si:H
Yoshihiro Hamakawa, Electroreflectance and Electroabsorption
Jeffrey S. Lannin, Raman Scattering of Amorphous Si, Ge, and Their Alloys
R.A. Street, Luminescence in a-Si:H
Richard S. Crandall, Photoconductivity
J. Tauc, Time-Resolved Spectroscopy of Electronic Relaxation Processes
P.E. Vanier, IR-Induced Quenching and Enhancement of Photoconductivity and Photoluminescence
H. Schade, Irradiation-Induced Metastable Effects
L. Ley, Photoelectron Emission Studies

CONTENTS OF PREVIOUS VOLUMES

Part C

Jacques I. Pankove, Introduction
J. David Cohen, Density of States from Junction Measurements in Hydrogenated Amorphous Silicon
P.C. Taylor, Magnetic Resonance Measurements in a-Si:H
K. Morigaki, Optically Detected Magnetic Resonance
J. Dresner, Carrier Mobility in a-Si:H
T. Tiedje, Information about Band-Tail States from Time-of-Flight Experiments
Arnold R. Moore, Diffusion Length in Undoped a-Si:H
W. Beyer and J. Overhof, Doping Effects in a-Si:H
H. Fritzche, Electronic Properties of Surfaces in a-Si:H
C.R. Wronski, The Staebler-Wronski Effect
R.J. Nemanich, Schottky Barriers on a-Si:H
B. Abeles and T. Tiedje, Amorphous Semiconductor Superlattices

Part D

Jacques I. Pankove, Introduction
D.E. Carlson, Solar Cells
G.A. Swartz, Closed-Form Solution of I-V Characteristic for a-Si:H Solar Cells
Isamu Shimizu, Electrophotography
Sachio Ishioka, Image Pickup Tubes
P.G. LeComber and W.E. Spear, The Development of the a-Si:H Field-Effect Transitor and Its Possible Applications
D.G. Ast, a-Si:H FET-Addressed LCD Panel
S. Kaneko, Solid-State Image Sensor
Masakiyo Matsumura, Charge-Coupled Devices
M.A. Bosch, Optical Recording
A.D'Amico and G. Fortunato, Ambient Sensors
Hiroshi Kukimoto, Amorphous Light-Emitting Devices
Robert J. Phelan, Jr., Fast Detectors and Modulators
Jacques I. Pankove, Hybrid Structures
P.G. LeComber, A.E. Owen, W.E. Spear, J. Hajto, and W.K. Choi, Electronic Switching in Amorphous Silicon Junction Devices

Volume 22 Lightwave Communications Technology
Part A

Kazuo Nakajima, The Liquid-Phase Epitaxial Growth of InGaAsP
W.T. Tsang, Molecular Beam Epitaxy for III-V Compound Semiconductors
G.B. Stringfellow, Organometallic Vapor-Phase Epitaxial Growth of III–V Semiconductors
G. Beuchet, Halide and Chloride Transport Vapor-Phase Deposition of InGaAsP and GaAs
Manijeh Razeghi, Low-Pressure Metallo-Organic Chemical Vapor Deposition of $Ga_xIn_{1-x}As_yP_{1-y}$ Alloys
P.M. Petroff, Defects in III–V Compound Semiconductors

Part B

J.P. van der Ziel, Mode Locking of Semiconductor Lasers
Kam Y. Lau and Amnon Yariv, High-Frequency Current Modulation of Semiconductor Injection Lasers
Charles H. Henry, Spectral Properties of Semiconductor Lasers

CONTENTS OF PREVIOUS VOLUMES

Yasuharu Suematsu, Katsumi Kishino, Shigehisa Arai, and Fumio Koyama, Dynamic Single-Mode Semiconductor Lasers with a Distributed Reflector
W.T. Tsang, The Cleaved-Coupled-Cavity (C^3) Laser

Part C

R.J. Nelson and N.K. Dutta, Review of InGaAsP/InP Laser Structures and Comparison of Their Performance
N. Chinone and M. Nakamura, Mode-Stabilized Semiconductor Lasers for 0.7–0.8- and 1.1–1.6-μm Regions
Yoshiji Horikoshi, Semiconductor Lasers with Wavelengths Exceeding 2 μm
B.A. Dean and M. Dixon, The Functional Reliability of Semiconductor Lasers as Optical Transmitters
R.H. Saul, T.P. Lee, and C.A. Burus, Light-Emitting Device Design
C.L. Zipfel, Light-Emitting Diode Reliability
Tien Pei Lee and Tingye Li, LED-Based Multimode Lightwave Systems
Kinichiro Ogawa, Semiconductor Noise-Mode Partition Noise

Part D

Federico Capasso, The Physics of Avalanche Photodiodes
T.P. Pearsall and M.A. Pollack, Compound Semiconductor Photodiodes
Takao Kaneda, Silicon and Germanium Avalanche Photodiodes
S.R. Forrest, Sensitivity of Avalanche Photodetector Receivers for High-Bit-Rate Long-Wavelength Optical Communication Systems
J.C. Campbell, Phototransistors for Lightwave Communications

Part E

Shyh Wang, Principles and Characteristics of Integratable Active and Passive Optical Devices
Shlomo Margalit and Amnon Yariv, Integrated Electronic and Photonic Devices
Takaaki Mukai, Yoshihisa Yamamoto, and Tatsuya Kimura, Optical Amplification by Semiconductor Lasers

Volume 23 Pulsed Laser Processing of Semiconductors

R.F. Wood, C.W. White, and R.T. Young, Laser Processing of Semiconductors: An Overview
C.W. White, Segregation, Solute Trapping, and Supersaturated Alloys
G.E. Jellison, Jr., Optical and Electrical Properties of Pulsed Laser-Annealed Silicon
R.F. Wood and G.E. Jellison, Jr., Melting Model of Pulsed Laser Processing
R.F. Wood and F.W. Young, Jr., Nonequilibrium Solidification Following Pulsed Laser Melting
D.H. Lowndes and G.E. Jellison, Jr., Time-Resolved Measurements During Pulsed Laser Irradiation of Silicon
D.M. Zehner, Surface Studies of Pulsed Laser Irradiated Semiconductors
D.H. Lowndes, Pulsed Beam Processing of Gallium Arsenide
R.B. James, Pulsed CO_2 Laser Annealing of Semiconductors
R.T. Young and R.F. Wood, Applications of Pulsed Laser Processing

Volume 24 Applications of Multiquantum Wells, Selective Doping, and Superlattices

C. Weisbuch, Fundamental Properties of III–V Semiconductor Two-Dimensional Quantized Structures: The Basis for Optical and Electronic Device Applications

CONTENTS OF PREVIOUS VOLUMES

H. Morkoc and H. Unlu, Factors Affecting the Performance of (Al, Ga)As/GaAs and (Al, Ga)As/InGaAs Modulation-Doped Field-Effect Transistors: Microwave and Digital Applications
N.T. Linh, Two-Dimensional Electron Gas FETs: Microwave Applications
M. Abe et al., Ultra-High-Speed HEMT Integrated Circuits
D.S. Chemla, D.A.B. Miller, and P.W. Smith, Nonlinear Optical Properties of Multiple Quantum Well Structures for Optical Signal Processing
F. Capasso, Graded-Gap and Superlattice Devices by Band-gap Engineering
W.T. Tsang, Quantum Confinement Heterostructure Semiconductor Lasers
G.C. Osbourn et al., Principles and Applications of Semiconductor Strained-Layer Superlattices

Volume 25 Diluted Magnetic Semiconductors

W. Giriat and J.K. Furdyna, Crystal Structure, Composition, and Materials Preparation of Diluted Magnetic Semiconductors
W.M. Becker, Band Structure and Optical Properties of Wide-Gap $A_{1-x}^{II}Mn_xB^{VI}$ Alloys at Zero Magnetic Field
Saul Oseroff and Pieter H. Keesom, Magnetic Properties: Macroscopic Studies
T. Giebultowicz and T.M. Holden, Neutron Scattering Studies of the Magnetic Structure and Dynamics of Diluted Magnetic Semiconductors
J. Kossut, Band Structure and Quantum Transport Phenomena in Narrow-Gap Diluted Magnetic Semiconductors
C. Riqaux, Magnetooptics in Narrow Gap Diluted Magnetic Semiconductors
J.A. Gaj, Magnetooptical Properties of Large-Gap Diluted Magnetic Semiconductors
J. Mycielski, Shallow Acceptors in Diluted Magnetic Semiconductors: Splitting, Boil-off, Giant Negative Magnetoresistance
A.K. Ramdas and S. Rodriquez, Raman Scattering in Diluted Magnetic Semiconductors
P.A. Wolff, Theory of Bound Magnetic Polarons in Semimagnetic Semiconductors

Volume 26 III-V Compound Semiconductors and Semiconductor Properties of Superionic Materials

Zou Yuanxi, III-V Compounds
H.V. Winston, A.T. Hunter, H. Kimura, and R.E. Lee, InAs-Alloyed GaAs Substrates for Direct Implantation
P.K. Bhattacharya and S. Dhar, Deep Levels in III-V Compound Semiconductors Grown by MBE
Yu. Ya. Gurevich and A.K. Ivanov-Shits, Semiconductor Properties of Superionic Materials

Volume 27 High Conducting Quasi-One-Dimensional Organic Crystals

E.M. Conwell, Introduction to Highly Conducting Quasi-One-Dimensional Organic Crystals
I.A. Howard, A Reference Guide to the Conducting Quasi-One-Dimensional Organic Molecular Crystals
J.P. Pouget, Structural Instabilities
E.M. Conwell, Transport Properties
C.S. Jacobsen, Optical Properties
J.C. Scott, Magnetic Properties
L. Zuppiroli, Irradiation Effects: Perfect Crystals and Real Crystals

CONTENTS OF PREVIOUS VOLUMES

Volume 28 Measurement of High-Speed Signals in Solid State Devices

J. Frey and D. Ioannou, Materials and Devices for High-Speed and Optoelectronic Applications
H. Schumacher and E. Strid, Electronic Wafer Probing Techniques
D. H. Auston, Picosecond Photoconductivity: High-Speed Measurements of Devices and Materials
J. A. Valdmanis, Electro-Optic Measurement Techniques for Picosecond Materials, Devices, and Integrated Circuits
J. M. Wiesenfeld and R. K. Jain, Direct Optical Probing of Integrated Circuits and High-Speed Devices
G. Plows, Electron-Beam Probing
A. M. Weiner and R. B. Marcus, Photoemissive Probing

Volume 29 Very High Speed Integrated Circuits: Gallium Arsenide LSI

M. Kuzuhara and T. Nozaki, Active Layer Formation by Ion Implantation
H. Hashimoto, Focused Ion Beam Implantation Technology
T. Nozaki and A. Higashisaka, Device Fabrication Process Technology
M. Ino and T. Takada, GaAs LSI Circuit Design
M. Hirayama, M. Ohmori, and K. Yamasaki, GaAs LSI Fabrication and Performance

Volume 30 Very High Speed Integrated Circuits: Heterostructure

H. Watanabe, T. Mizutani, and A. Usui, Fundamentals of Epitaxial Growth and Atomic Layer Epitaxy
S. Hiyamizu, Characteristics of Two-Dimensional Electron Gas in III-V Compound Heterostructures Grown by MBE
T. Nakanisi, Metalorganic Vapor Phase Epitaxy for High-Quality Active Layers
T. Mimura, High Electron Mobility Transistor and LSI Applications
T. Sugeta and T. Ishibashi, Hetero-Bipolar Transistor and Its LSI Application
H. Matsueda, T. Tanaka, and M. Nakamura, Optoelectronic Integrated Circuits